Tourism Governance

De Gruyter Studies in Tourism

Series editor
Jillian M. Rickly

Volume 9

Tourism Governance

A Critical Discourse on a Global Industry

Edited by
Amir Gohar

DE GRUYTER

ISBN 978-3-11-135312-8
e-ISBN (PDF) 978-3-11-063814-1
e-ISBN (EPUB) 978-3-11-063406-8
ISSN 2570-1657
e-ISSN 2570-1665

Library of Congress Control Number: 2022930261

Bibliographic information published by the Deutsche Nationalbibliothek
The Deutsche Nationalbibliothek lists this publication in the Deutsche Nationalbibliografie; detailed bibliographic data are available on the Internet at http://dnb.dnb.de.

This volume is text- and page-identical with the hardback published in 2022.
Typesetting: Integra Software Services Pvt. Ltd.
Printing and binding: CPI books GmbH, Leck

www.degruyter.com

Foreword

These reports provide a unique comparative perspective on agencies, bureaus, and offices charged with tourism governance world-wide. It is important to note that they are not press releases or mission statements provided by the governing organizations. Each report was written by independent researchers and scholars who are knowledgeable about tourism governance in the country under review. Amir Gohar, our editor, conceived of the project, commissioned the reports, and shepherded us through the various drafts and revisions.

This volume covers most of the variation in global tourism governance. The only part of the story that is not told here is tourism governance in countries with centrally planned economies – Cuba, Vietnam, North Korea, People's Republic of China. With the exception of sub-Saharan Africa, every region of the world is represented. There are chapters on countries like Italy and France with deep historical engagement with tourism, and other chapters on countries that are just beginning to attract tourists, like Oman. Some of the entries, e.g., Lebanon and Colombia, report on governmental efforts to overcome recent internal conflict that tarnished their reputations as tourist destinations. Others, like Turkey and Thailand, describe governmental efforts to increase tourism even though their recent success suggest they may have already reached their peak.

Reading these reports, we soon discover that governmental action, no matter how well intended, does not always lead to increasing and/or improved tourism – two goals that do not always align. There are almost as many negative examples here as positive ones. This volume will prove to be indispensable to government officers tasked with managing tourism development. It will provide them with alternative responses to problems similar to the ones they face. It will be helpful to local and foreign entrepreneurs who want to learn about the governance landscape before undertaking any tourist business dealings in these countries. It will also be valuable to anyone interested in comparing governing strategies in general in our fraught, globalizing world.

There are a few basic matters that every system of tourism governance must deal with. These include who may and who may not cross the border as a tourist? How long do they get to stay as a tourist? Must they pay a fee for the right to visit? Which, if any, of their bank cards will work in your country? In his very helpful last chapter in this volume, Nelson Graburn outlines these baseline issues that every country must deal with as well as a number of related matters that require multinational agreements – airline and cruise ship safety protocols, for example.

There are some commonalities across the patterns described in these chapters. As a general rule, the governing bodies engage in the following: They assume responsibility for promoting national identity abroad and constructing and disseminating a positive image of tourism in their countries. They set quality standards for guide training and grading of hotels and restaurants. They establish preservation requirements for natural and cultural heritage. They designate sub-national places,

https://doi.org/10.1515/9783110638141-202

regions, and sites as having touristic value and establish rules for their protection and presentation. They set goals for future tourism development at the national level and propose policy to reach those goals. They commission studies and/or support tourism research in their universities. Not every system of tourism governance does all of these things, but most of them do most of these things.

Beyond these commonalities, tourism governance is heterodox. The chapters in this volume make it clear that anything human beings have ever done in appreciable numbers is now the basis for a "type" of tourism. These "types" include medical tourism, extreme tourism, sex tourism, nature tourism, conflict tourism, pro-poor tourism, MICE tourism (Meetings, Incentives, Conferences, Exhibitions), WOOF tourism (World-Wide Working On Organic Farms), etcetera. This list can be as long as anyone would wish to make it. Some countries have pristine, warm beaches and there is notorious competition between them for tourists seeking sun, sea, and sand. But specific combinations of attractions are unique to each of these countries or shared with only one or two others. Turkey and Tunisia, for example, feature low priced plastic surgery that attracts "medical tourists." But there is no mention of plastic surgery tourism as an important component of the tourist economy in any of the other reports. Poland uniquely provides its medical tourists with dental work that is less costly than other EU nations.

In Colombia armed guards offer "Favela Tours" of neighborhoods controlled by drug cartels. The federal government worries that these might project an unwanted image of the country. Egypt highlights its UNESCO designated "A-list" attractions, as does every country that has one. But World Heritage Sites are, by definition, completely different from one another. In Finland, it is the Northern Lights, and down-list but worthy of mention, its famous "Wife Carrying World Championship" and "Swamp Soccer." Almost every country makes claims for the attractiveness of its local, distinctive cuisine. France double underscores its cuisine as among its top attractions. India features its tigers, and Italy its art. From 1950 to 1970, Lebanon was proud of its reputation as "Paris of the Middle East." After its civil war, multiple invasions by Israel, and an explosion that destroyed much of downtown Beirut, Lebanon has scrambled to build thousands of new hotel rooms. Now it is trying to come up with some reason for tourists to visit. Mexico is trying to shift emphasis from warm beach tourism to cultural tourism featuring distinctive rural villages, customs, festivals, and folk art. Oman has just begun to recognize the potential touristic value of its natural beauty, mountain and beach areas, and open and welcoming Omani culture, and old towns. Poland has a wide range of attractions, but it continues to emphasize the old Soviet equation of tourism with youth festivals, exercise, physical education, health, and sports. Portugal enjoys its reputation as "the best destination country in the world," but worries that as annual tourist visits outnumber the Portuguese population by 2:1, the sheer numbers of tourists may be destroying the country. Thailand has had enormous success as a global destination by emphasizing the happiness, warmth, and hospitality of its people and the generosity of its Royal Family. Turkey is seeking to enhance its attractiveness beyond its UNESCO sites (the Blue Mosque Hagia Sophia, Topkapi Palace) by adding bird watching, hiking,and scuba diving to its offerings. The direct contribution of foreign and domestic tourism to the GDP of the United States is US$581 billion, topping the second place People's Republic of China (US$403 billion) by almost one-third. But the United States uniquely has no federal agency, bureau, or office of tourism anywhere in its federal system.

Out of this diversity, these reports broadly recognize two general types: (1) "mass tourists," who favor sun, sea, and sand, who come in pre-paid packages, and/or spend their time in resorts that provide for all of their vacation needs, wants, and desires; and (2) "cultural/natural tourists" who search out cultural attractions (museums, distinctive local traditions, archeology, architecture, etc.) and/or scenic landscapes and opportunities for expressive self-testing in nature (scuba diving, rock climbing, skiing).

"Mass tourism" is favored by private sector developers because the flow of income from tourism can be contained and controlled. A resort can be built to include gift shops, bars, restaurants, golf courses, horseback riding, surfing, scuba diving, parasailing, and other services and amenities to the point that no one need leave the premises for the duration of their stay. This allows the corporate owner(s) to collect every tourist dollar that is spent.

"Nature/culture tourism" is diffuse and potentially more intimately connected to local peoples and economies. A family that hires a car and moves through the countryside following their own itinerary from nature preserve, to ethnic crafts market, to museums and archeological sites, might actually spend more per person per day than someone in an all-inclusive resort. But they will purchase meals, buy petrol, stay in bed and breakfasts, and get snacks and souvenirs, from dozens of different local entrepreneurs and providers, not from a single resort owning corporation or consortium.

The role of tourism governance in "mass tourism" development usually involves initiatives to attract and facilitate capital investment in building resorts, hotels, golf courses, amusement parks, and so on. Some of these are complex public/private projects. Others involve licensing agreements, tax breaks, granting exemptions from building codes, easing tourist industry labor laws, waiving environmental protections, and providing public funding for infrastructure upgrades – new transportation, broadband, electrical grid, sewer and water systems, to increase carrying capacity for private resort development.

The role of tourism governance in "cultural/natural tourism" usually involves setting aside nature preserves, protecting distinctive plant and animal species, restoring historic architecture, shrines, and technologies, and subsidizing remote communities so they can fit themselves into the global economy as exemplary of traditional and colorful ways of life, worthy of tourist attention.

These reports make it clear that in order to succeed, different types of tourism require different human skills in the host communities, different environmental and natural contexts, and different organizational support frameworks. There is no overlap between the support required and special equipment needed to service skiing, face-lifts, working on organic farms, sightseeing, enjoying a drink on a beach, among others. It is important that tourist governing organizations not make the mistake of assuming that types of tourism equate to types of tourists. The tourists themselves are no more specialized than humanity in general. Some may arrive with specific preferences and expectations, but they can be lured into other experiences that they didn't necessarily anticipate. World-wide, tourists are found gathered around every recreational activity, every kind of natural formation and every knot in the social fabric. They multi-task and switch easily between attractions visiting a museum in the morning and relaxing on the beach in the afternoon. Having a well-established "tourism brand" should not blind the host country to its own potentially attractive undeveloped features and qualities.

A profound lesson to be learned from these reports, taken together, is that tourism cannot be contained in organizational charts of government bureaucracy. Every chart reveals organizational porosity with aspects of tourism governance overflowing into finance, transportation, security, health, education and welfare, natural resources, housing, security, and so on. This suggests that the next stage of tourism governance will be for host countries to develop ethical guidelines for tourism development that can be operationalized across all branches of government. Will tourism policy lead to greater income equality for the citizen hosts? Will it improve labor conditions for tourism support workers? Will it promote greater mutual respect in the interactions between tourists and locals? Will it protect the integrity of natural areas and cultural minorities, especially those that are designated as attractions? Are planned new tourist developments as accessible and interesting to domestic tourists as to foreign visitors? Does the policy encourage tourists and locals to respect history and heritage? Does it understand and accept limits in the human carrying capacity of regional infrastructure and natural environments?

One country that appears to have overcome or at least minimized the inherent problem of bureaucratic fragmentation is Thailand. While the Thai organizational chart is as complicated as the others, they have come up with a transcendent criterion for tourism development across all bureaus, ministries, and offices: "Thainess." Governance does not begin with a listing of its standard tourism assets, its various separate attractions. Instead, Thailand prioritizes the overall happiness of Thai people as what is most crucially important for tourism. It leads with the fact that it is the least unequal economy, a "land of smiles." It is religiously open and tolerant, has a wonderful cuisine, not just in its five-star restaurants but everyday, and is unfailingly polite and gracious toward strangers. In Thailand, the question asked about every tourism development initiative is, "how does it leverage these components of Thainess?" Its beach may be no better than the beach in a competing country, but the other country lacks Thainess.

Tourism is perhaps the least fraught of a series of global problems that cannot be contained conceptually or politically. International migration and flows of refugees, pandemics, global warming, economic modernization, and tourism, are precisely the kinds of problems that government bureaucracy was not designed to solve. They are not to subject to expert control by hierarchically organized specialists in a single department or agency. They have become imbricated in every detail of life on this planet and should be the work prioritized in every specialized department and initiative of government. We may someday look back upon successful tourism governance as the beta test version of our capacity to respond to the other global challenges facing nations today.

Dean MacCannell

Preface

As a global industry, and throughout time, most tourism occurs in physical environments that got shaped and re-shaped by the dynamic and forces that respond to such an industry across the globe. Hotels, resorts, cities, and larger destinations are would never look the way they are if tourism wasn't part of their economy. This book project was conceived in 2016 in one of the Berkeley's Tourism Studies Working Group (TSWG), in University of California at Berkeley, California. I have been observing both tourists and tourism critics occupied with the final tourism product or tourism development while paying very little attention to the energies, forces, and dynamics that contribute to shaping this tourism development that they have been dissatisfied with. While this ambitious project started in Berkeley, it continued to expand and evolve to include a variety of countries with very different tourism governance profile.

Today's tourism is viewed, studied, and researched under a wide range of disciplines such as social sciences, hospitality, business, anthropology, marketing, parks, geography, and environment. Because of its interdisciplinary nature, it involves thinking across boundaries (local, regional, and international). The diversity of backgrounds of the contributors is a testimony that this book engages with such interdisciplinarity and diversity of regions. The book compares and contrasts the governance system in Colombia, Egypt, Finland, France, India, Italy, Lebanon, Mexico, Oman, Poland, Portugal, Thailand, Tunisia, Turkey, and United States. It provides systematic critical comparisons between different governance bodies and stakeholders across these nation-states where each country is a stand-alone chapter. In addition, there are two extra chapters that covers non-Governmental tourism, such as AirB&B, and international governance, such as the role of UNWTO, UNESCO, AgaKhan Foundation and other international players, that either shape the decision about tourism destinations or influence the strategies of central governments.

The book explores the relationship between governance systems and the physical forms of the built landscape and tourism infrastructure. Of the total 20 chapters, 16 chapters focus on individual countries, each describing the country's tourism policies and the resulting tourism infrastructure. This represents the first time such a systematic perspective on governance and tourism has been undertaken, and it revealed some surprising patterns. In countries where tourism authorities are represented in the central/federal government such as Turkey, France, and Spain, tourism development has an embodied conflict with other authorities such as environment, antiquities, fishery, among others. This has led to more hotels and resorts that created significant ecological and cultural impacts. On the other hand, tourism is shaped by antagonism when there are no tourism authorities in central or federal government such as the case of the United States where most tourism development (hotels, resorts, leisure facilities, etc . . .) are private-sector led.

While the book started with a list of countries that is not identical to the one it ended with, there were substantial contributions in the TSWG seminars from those

https://doi.org/10.1515/9783110638141-203

who departed the project at earlier for other reasons. The research journey from all contributors have benefited from the engagement of intellects from TSWG. The authors vary in their academic level, angle in which they look at tourism, and their scholarly backgrounds. This book project was not going to see the light without the collective effort by all contributors in addition to the TSWG intellectual community and the assistance of Josie Miller, a dedicated graduate researcher at Berkeley and Reem El Desouky, a highly motivated graduate researcher in Cairo, Egypt.

Contents

Acronyms

AAMA	American Mexican Automobile Association
APT	Aziende di Promozione Turistica – Agencies for the Promotion of Tourism
AWHSF	Association of World Heritage Sites in Finland
BCD	Beirut Central District
CDR	Council of Development and Reconstrunction
CIPS (in English)	Integrally Planned Tourist Centers
CIPS (in Spanish)	Los Centros Integralmente Planeados (para el Turismo)
CIT	Interministerial Council for Tourism (Conseil interministériel du tourisme)
CRZ	Costal Regulation Zone
CSR	Corporate Social Responsibility
DGUP	General Directorate of Urban Planning
EAFRD	European Agricultural Fund for Rural Development
EEAA	Egyptian Environmental Affairs Agency
EIA	Environmental Impact Assessment
EMFF	European Maritime and Fisheries Fund
ENIT	Ente Nazionale Turismo Italiano – Italian government tourist board
EU	European Union
EFDR	European Reginal Development Fund
ESF	European Social Fund
EPR	European Recovery Plan
EPT	Ente Promozione Turistica – Tourism Promotion Board
ETIS	European Tourism Indicator System Toolkit for Sustainable Destinations
FEE	Foundation for Environmental Education
GDP	Gross Domestic Product
GEF	Global Environment Facility
GOPP	General Organization of Physical Planning
GWSP	Gulmarg Winter Sports Project
HCUP	Higher Council for Urban Planning
HEPCA	Hurghada Environmental Protection and Conservation Assiciation
HTL	High Tide Line
IAT	Informazioni Accoglienza Turistica – Tourist Welcome and Information Office
IATA	International Air Transportation Association
ICAO	International Civil Aviation Organization
ICCROM	International Center for the Study of the Presevation and Restoration of Cultural Property
IDAL	Investment Development Authority in Lebanon
IFS	Indian Forest Services
IISM	Indian Institute of Skiing and Mountaineering
IITTM	Indian Institute for Tourism and Travel Management
IPCC	Intergovernmental Panel on Climate Change
IPS	Indian Police Services
ISI	Import Substitution Industrialization
ISO	International Organization for Standardization
ITDC	India Tourism Development Corporation
IUNC	International Union for the Conservation of Nature
MEAE	Ministry of Economic Affairs and Employment

https://doi.org/10.1515/9783110638141-205

MiBAC	Ministero per i Beni e le Attività Culturali – Ministry of Cultural Heritage and Activities
MTA	Mexican Tourism Association
NCHMCT	National Council for Hotel Management and Catering Technology
NCS	Nature Conservation Sector
NDZ	No Development Zone
NGT	National Green Tribunal
NIWS	National Institute of Water Sports
NGO	Non-Governmental Organizations
ODIT	Observation, développement et ingénierie touristiques France; Tourism Observation, Development, and Engineering France
OEA	Order of Engineers and Architects
OECD	Organization for Economic Co-operation and Development
ONT	Office National du Tourisme
PLU	Plans locaux d'urbanisme
PRI	Partido Revolucionario Institucional
SETE	Société d'Exploitation de la tour Eiffel
SMAL/AFTA	Association of Finish Travel Agents
STCAS	Tourism, Commerce, and Artisan Service
STL	Sistemi Turistici Locali (Local Tourist Systems)
TDA	Tourism Development Authority
TTCI	Travel and Tourism Competitive Index
TUGEV	Tourism Development Foundation
TUREB	Tourist Guides Association
TUROB	Turkish Hotel Association
TÜRSAB	Turkish Travel Agencies Association
TUTAV	Turkish Promotion Foundation
TYD	Association of Tourism Investors
UNDP	United Nations Development Program
UNEP	United Nations Environmental Program
UNESCO	United Nations Educational, Scientific, and Cultural Organization
UNWTO	United Nations World Tourism Organization
USAID	United States Agency for International Development
USTTA	United States Travel and Tourism Adminstration
VF	Visit Finland
WHSs	World Heritage Sites
WTO	World Tourism Organization

1 Introduction

1.1 The Concept of Governance

The origins of current notion of governance are long-standing since ancient past yet had the chance to evolve throughout time. The foundartions of principal forms of governance, as stated by Finer (1997), can be traced as far back as 3200 BC along the Nile Valley and Southern Mesopotamia. Finer affirms that the earliest forms of governance that can be traced relates back to the ancient states of Sumer, Egypt, Persia, and Assyria, as well as the classical states of Greece and Rome, the Byzanitine and Caliphate Empires of the near East, The Han, Tang and Ming states of China, Tokugawa Japan, and the "modern" states of Europe and North America.

With regard to its contemporary definition, social scientists and public administration scholars make clear distinctions between government and governance, and they explore the relationship between the two concepts. For instance, Fasenfest (2010) argues that government is the office, authority, or function of governing; governing is having control or rule over oneself; and governance is the activity of governing. In a more recent definition is by UNESCO (2016), governance refers to structures and processes that are designed to ensure accountability, transparency, responsiveness, rule of law, stability, equity and inclusiveness, empowerment, and broad-based participation.

The concept of governance is an overarching one that encompasses government. Weil (2015) asserts that government is only one arm of modern society and that it derives its legitimacy and powers from its taxes, spending, laws, and regulations. The other two arms of modern society are the business/for-profit sector, which derives its power from creating jobs and paying taxes, and the non-profit sector, which serves the public interest without profit.

Because these three facets of civil society were never explicitly trifurcated, the distinctions between them are oftentimes porous. Such blurred lines apply to overall governance and sector-specific governance (such as the governance of tourism), and governance must be considered in order to fully understand tourism development.

Spirou (2011) asserts that the introduced tourism infrastructure coupled with intense marking efforts and promotional practices exercised by municipal government, corporations, and business leaders has lured thousands if not millions of visitors towards destinations. This volume aims to understand the evolving nature of tourism governance and compare how tourism is shaped in different countries as a result of its governance system. Each chapter is about one country, describing its recent contemporarily tourism policies and how it is shaped. Each follows a consistent structure (with some flexibility) to allow for each chapter to address similar issues, concerns, recommendations across wide range of countries from all over the world. The chapters are presented in alphabetical order by country name.

https://doi.org/10.1515/9783110638141-001

1.2 Tourism Governance Across Countries

In this volume, different authors offer a critical overview on tourism governance across varying scales (central, regional, and local outreach) and also across different governing bodies (government, non profit, and private sector). The authors also explore how tourism governance is led in their respected chapters. A quick reference guide can be found in the Appendix at the end of the book, while a list of common terms and their acronyms is presented.

In Chapter 2 (Colombia), Patrick Naef presents how Colombia managed to completely reverse its global image with strong political will and consistent work of the government and how the international community such as UNWTO has supported this notion by only declaring Colombia as a visitable destination.

Bertram Gordon in Chapter 5 (France) uses the Eiffel Tower to explain how France's government collaborates on different levels (central government of France and local government of Paris) in supporting Paris as tourism detonation. He also uses Euro Disney as an example to the coordination between government agencies and private sector initiatives. Euro Disney is the most visited site in France, and it would not have achieved this success without the backing of the government.

Amir Gohar in Chapter 3 (Egypt) and Priyanka Ghosh in Chapter 6 (India) explain the dynamics between central government as a regulator and local state government as an executive overseeing the hotel industry. The two countries also share the intersection of tourism authorities' efforts with the environmental authorities and forests/park management. Although the overall composition of government in this area is similar, India seems to have higher influence from the environmental authorities on shaping its tourism activities than Egypt. This appears in looking at the extent in which they control tourism development in coastal areas which shapes tourism by antagonism.

Monica Pascoli in Chapter 7 (Italy) explains that in Italy tourism in central government is combined with heritage and culture which, for a country like Italy, makes it the largest economic sector. One can navigate tourism governance in Italy through multiple levels: from the EU, central government, regional government, and subregional. The central government main role is to orchestrate the scene and negotiate roles with other stakeholders such as private sector and community organizations. Italy seems to be unique in understanding the importance of tourism on the local level. And how local government is more aware of what is happening in their specific regions. However, this decentralized approach has revealed some challenges that local tourism is facing due to the absence of central government role such as lack of infrastructure, lack of strong country level brand, lack of quality standards to monitor local operations and absent of statistical surveys that can inform future tourism decisions.

In Chapter 8 (Lebanon) Kamil Hamaty reveals the complexity of tourism governance in Lebanon. The lack of funding and power of the ministry of tourism in the central government have resulted into multiple tourism projects that are popping up under different authorities. These popped-up projects can be funded by Arab investors, real estate developers, and international funding from external organizations. Therefore, multiples of projects are labelled under leisure, hotels, restaurants, and other activities of that nature. This has led to the involvement of other non-tourism sectors such as investment development authority in Lebanon (IDAL), general directorate of urban planning (DGUP), and council of development and reconstruction (CDR). This has left the ministry of tourism with the role of promotion and branding. The weak role of ministry of tourism, rather absent one, has also encouraged the non-profit sector to play stronger role with some independent initiatives that seem to be very successful, such as Shouf Biosphere Reserve.

As for Chapter 9 (Mexico), Matilde Córdoba Azcárate highlights the shifting yet dominant role of the state in development and planning while consistently working side by side with the business community and private initiatives when it comes to tourism development. Tourism governance in Mexico is not unilinear, rather multidirectional with the influence of alliances between the government, private sector, transnational, and philanthropic actors. The continuous presence of a governmental body solely responsible for tourism is threatened due to the drop in resources for the Ministry of Tourism, and hence the agreement on the route of discontinuing tourism boards and governmental agencies responsible for designing, operating, and coordinating the tourism strategies across the nation.

One governance direction that can be detected from the upcoming chapters is the notion of merging a tourism governmental body under different ministries that regulate other aspects of society, just as in the case of Oman and Poland. Amna Al-Ruhelli and Rashid Al-Hinai in Chapter 10 (Oman) shed a light on the relatively young and recently developed governmental ministry of tourism in Oman. Oman recently decided to merge the ministry of tourism along with the ministry of heritage into one governmental body as to make use of Oman's diverse cultural and natural lifestyles as main tourism attraction points across the nation. Oman, as a nation, recently took the decision to shift its economic standpoint from mainly depending on oil and its by-products to a more diverse plan, which includes initiatives as preparing tourism masterplans, privatizing the management of natural sites, investing more into adventure activities, and other initiatives such as Oman Tourism Strategy 2040 and Tanfeedh National Program that aim mainly to diversify the touristic activities within the country to contribute towards the national economy. Oman is shifting its focus towards promoting the nation as a sustainable and eco-friendly touristic attraction.

As for Chapter 11 (Poland), Magdalena Banaszkiewicz and Sabina Owsianowska highlight how tourism was listed under the governance of different ministries such as Ministry of Transport and Marine Economy, Ministry of Economy and Labor, Ministry of Sport, and finally Ministry of Economic Development, Labor, and Technology. The authors also showcase how the Polish state is mainly concerned with the promotion of tourism by depending on regional and local tourist organizations as well as foreign branched of Polish tourist organizations by promoting brands of cities instead than promoting a brand of the country as a whole.

Joana Almeida and Pedro J. Pinto in Chapter 12 (Portugal) explain how tourism in Portugal is regulated under five different regional tourism bodies, where each body is dedicated for a specific territorial scope and holds administrative and financial autonomy over it, and report back to the national tourism authority under the Ministry of Economy. Moreover, tourism development and marketing emerged as a joint effort between the regional governance bodies and non-profit or private associations within each region. The authors also argue that despite the abundance in tourism resources, Portugal only recently started being labelled as a successful touristic destination as a result of the cooperation between the central government and municipalities for the development of attractive tourism products and investing in the upgrading of city enters and their infrastructure.

In Chapter 14 (Thailand), Louise Mozingo and Wilasinee Darnthamrongkul discuss Thai tourism governance and how it is highly centralized under the dominance of Department of Tourism, responsible for development of touristic sites, and Tourism Authority of Thailand, responsible for marketing and promoting Thailand as a touristic destination. The authors also note that even though Thailand has abundant geographical and other touristic magnets, the governmental approach towards tourism does not engage in a sustainability dialogue when it comes to how to preserve these touristic magnets, which can be a red flag when it comes to sustaining the touristic flow in Thailand specially when tourism is responsible for a chunk percentage of the country's GDP.

Ines Mestaoui and Amira Benali in Chapter 15 (Tunisia) discuss how Tunisia has tourism under the control of the central government, through the establishment of the Ministry of Tourism which seeks the implementation of governmental policies, and with the help of other multiple organizations working under the supervision of the ministry itself. The authors argue that even though there are governmental bodies responsible for tourism, the sector is lagging behind due to the unsuccessful use of Tunisian territories into promoting different events and forms of tourism; currently, Tunisia lacks entertainment schemes and events that significantly help stir tourists' interest into local traditions and customs, or what is referred to as Tunisian Paratourism.

Another example discussed in this collection is Chapter 16 (Turkey), where İsmail Kervankıran, Gülsel Çiftci, & Azade Özlem Çalıkalkin highlight how Turkey's current tourism governance is a joint-effort between the governmental Ministry of Culture and Tourism body and organizations or agencies working closely with it as well as the private business sector. Legislations, promotion, and control over tourism is traced as interaction between governmental and nongovernmental actors (private and social). The authors also discuss that regardless of Turkey's abundant touristic magnets, the stakeholders working around tourism have to evaluate and adjust their governance approaches to ensure the sustainabiliy of these magnets against rising new technologies.

The last nation-specific Chapter 17 of this book addresses tourism in the United States of America. Dean MacCannell discusses how the US tourism follows a fundamentally decentralized form of governance when it comes to tourism; the US does not house a federal department or national policy makers dedicated to tourism. Instead, tourism governance is achieved thorugh coordinated state or local offices and often promoted and managed through private sectors.

In addition to individual country chapters, this book discusses other forms of tourism that are not country-specific and do not necessarily follow the mold of tourism governance. One of these forms as discussed in Chapter 18 (Tourism Without Governance) is the notion of World Wide Opportunities on Organic Farms, or WWOOF. In the 21st century, WWOOF is a platform used to connect farm owners and prospect "helpers" or people who volunteer their help in exchange of being provided accommodation by farm owners. Such relationship allows volunteers to travel around the world without being bound to monetary ristrictions or the negative effects of mass tourism. In addition, WWOOF is listed as a type of volunteer tourism or voluntourism. The chapter unfolds into discussing how WWOOF is present in various countries such as Thailand, Japan, New Zealand, and USA.

No doubt that international governace play a significant role in shaping national and local tourism poslices. In Chapter 19, Nelson Graborn elaborates on how international organisations such as UNWTO, UNESCO, WB, USAID, UNEP, GEF, and many others contribute to infuncing tourism polcieis on both the national level, such as offering economic support to nation states that gets spent on the tourism sectorsa. As well as development of plans for specific sites on local level such as developing world heritage sites for tourism, tourism development within national parks and environmental sensitive areas, or sites with significant cultural attractions.

The coming chapters will take the reader to a journey through different tourism governance systems, where the reader will witness many similarities and differences between different nations. This volume is an attempt to compare/contrast the tourism governing bodies and provide insights on how tourism is shaped, measures, represented and managed by different stakeholders. The volume will meander through different tourism landscapes across a number of countries.

Patrick Naef

2 Colombia

2.1 Colombia as a Post-Conflict Destination?

Colombia, a country of about 50 million people situated in the north of South America, has been off the tourist map for decades due to its high level of violence. While cities such as Medellin and Cali were considered as among the most dangerous in the world until the end of the 20th century, the whole country has been plagued by a long-lasting armed conflict involving guerrillas, paramilitaries, narcos, and regular government forces. The security situation significantly improved during the last two decades, and in 2016, the Revolutionary Armed Forces of Colombia (FARC) and the government signed peace agreements promising to put these years of war behind them. However, the violence is still far from being over and since the peace process began, hundreds of social leaders, former guerillas and civilians have lost their lives. In view of this, many questions the label of "post-conflict" officially attributed to the country, and prefer to refer to "post-agreements" (González & Álvarez 2018; Guilland & Naef 2019). Yet, the general improvement in security, especially in large urban centers and specific areas like the Caribbean coast and the coffee region, resulted in an important increase of foreign visitors, as well as local tourists eager to rediscover their country. This rapid growth of the tourism sector is not happening without tensions. The public authorities are struggling to adapt tourism governance to a fast-changing setting. Indeed, in Colombia's "post-conflict" context, tourism is not always a priority: the formulation and application of laws and regulations regarding tour guidance, hospitality, transport, or natural conservation are still weak. Nonetheless, Colombia's tourism and promotional bodies have been very active in reshaping the image of a country presented for more than half of a century as an epicenter of violence and drug trafficking.

In 2017, the Colombian regional newspaper El Pais highlighted the fact that half of the map of Colombia presented in the Lonely Planet travel book was a grey area without any references (Linde, 2017). Working to color this map and thus participate in enhancing the attractiveness of a country plagued by decades of war implies the necessity of agency from stakeholders involved in the tourism sector. This chapter aims to explore the role and the limits of tourism governance in the construction and the promotion of Colombia's "post-conflict" image.

While actors involved in tourism and place-branding actively participate in building destination image, other dynamics contribute to shaping it. Representations from international media and popular culture significantly influence visitors' geographic imaginaries (Debarbieux, 2016; Gravari-Barbas & Graburn, 2012; Salazar, 2012; Salazar & Graburn, 2016). These imaginaries themselves may impact a country's image. Actors involved in the tourism sector may adapt their offer to globally marketed representations. An international movie portraying a specific city can for instance influence the way

https://doi.org/10.1515/9783110638141-002

tourism stakeholders will brand it. This can be even more the case when actors detached from official tourism bodies take part in the construction of such an image. Hence, the many private entrepreneurs increasingly participating in the growing tourism sector of Colombia play an important role in the creation of a new destination image. The many dimensions influencing the construction of a country's image and identity, as well as the diversity of actors involved, imply a high level of complexity for stakeholders associated with tourism governance and place branding.

Hence, the objective here is to explore some strategies developed by tourism stakeholders implicated in the construction of Colombia's new, post-conflict image. The questions underlying this analysis are the following: Who are the main actors involved in such a process? What are the obstacles related to the promotion of a country aiming to turn its back on a violent past? What are some of the main representations associated with Colombia nowadays and how can tourism governance influence them? This chapter first explores the role of tourism governance in a context of postwar place branding. It will briefly look back at the rebirth of the Colombian tourism sector, concentrating on stakeholders associated with the promotion of the country's new image. It will then focus on the city of Medellin as a case study, to demonstrate that diverse and competing representations are produced in the touristscape of a place that, just a decade ago, was still considered as one of the most violent in the world.

2.2 Tourism Governance and the Post-War Image

Tourism governance is defined by the World Tourism Organization (WTO) as a system and process to define strategies and implement them to achieve competitiveness and sustainable development in the tourism destination (UNWTO, 2011). As Scott and Marzano demonstrate, governance is a more encompassing phenomenon than government: "It embraces governmental institutions, but also subsumes informal, nongovernmental mechanisms whereby those persons and organizations within its purview move ahead, satisfy their needs, and fulfill their want" (2015, p. 3). Moreover, considering the tourism sector as an open and fragmented industry, Borges et al. (2014, as cited in Scott & Marzano) state that tourism governance is therefore a complex and multidimensional issue in which the state and other actors depend on each other. Various stakeholders thus interact in this process, implying initiatives from private and public sectors, as well as interventions from different levels: local, national, and international. An aggregation of diverging interests can lead to conflicting views of what should be expressed and represented in the touristscape of a country. This can be even more the case in a post-conflict context, where dissonant representations of the past are progressively incorporated into the tourism sector. Therefore, the promotion of tourism and the reconstruction of a country's image after an armed conflict represent a challenge for stakeholders often driven by diverging interests. The (re)construction of a post-conflict destination image has already generated a significant amount of work in tourism studies (Guilland 2012; Guilland & Naef 2019; Ndlovu & Chigora 2019; Shirley et al. 2018; Vitic & Ringer 2008). However, analysis looking specifically at the role of tourism governance is still rare; this chapter proposes to explore some of the governance challenges related to the construction of Colombia's new destination image. As Scott and Marzano (2015) put it, the development of a country brand represents an important issue for central governments. "It can provide an 'umbrella' under which the sub-national brands may function. Clearly the governance of tourism at the sub-national level must involve consideration of the potential to work under a national umbrella brand, to be effective" (Scott & Marzano, 2015, p. 14). However, as the Colombian example will demonstrate, the variety of actors involved in this process can lead to the diffusion of antagonistic images, and therefore clash with this umbrella brand. After a war, images of peace are in conflict with representations related to a violent and dissonant heritage. In this competing arena, tourism plays a central role in shaping Colombia's post-war image, through the various images and narratives it produces and diffuses.

2.3 Colombia's Touristscape

Internal violence in the country has seriously affected the tourism sector in recent decades, especially from the 1980s to the 2000s. Colombia saw the arrival of 3 million non-resident tourists in 2018, about three times more than a decade ago and twice as many as 2009, a point when the country was considered to be "back on the world tourism map," by the UNWTO (2009). In response to a decrease in homicides and kidnappings since 2002, Colombia's tourism trade has been growing every year and the sector is considered as a key area of development in a country struggling to put decades of war behind it. As stated in a WTO report on the reconstruction of the Colombian tourism sector, "Colombia is a country that has managed to come back from the edge of the abyss. With its very survival threatened – in a way that is without parallel in the world – by the combined effects of drug-trafficking, guerrillas, and terrorism for many years" (UNWTO, 2009, p. 3).

Several programs were developed by the authorities to secure the main tourism routes, such as "Vive Colombia. Viaja por ella" (Live Colombia. Travel all over it.) and "Rutas seguras" (Safe Routes). While this context encouraged an increasing number of foreign visitors eager to discover a country which was absent from the tourism map for decades, it also brought freer movement for the Colombians who started to rediscover their own country. International bodies praised the rebirth of tourism as a factor for "social cohesion and a reaffirmation of the self-confidence of Colombian society" (UNWTO, 2009, p. 10). The founder of the conservative "Democratic Center" political party and former president Álvaro Uribe Vélez, who ran the country from 2002 to 2010 with an iron hand, certainly played a role in securing the roads and strengthening the domestic tourism sector with the so-called "tourist caravans" (caravanas turísticas), a program included in his Democratic Security and Defense Policy. In accordance with the program, on specific days military forces would provide reinforced protection for roads reaching major holiday attractions (Hudson, 2010).

The government also worked hard to bring back international tourists, mainly through its executive branch ProColombia, a government agency in charge of promoting international tourism and foreign investment. The strategy appeared to be fruitful: since 2014, Colombia has been regularly recognized as a new hotspot for international tourists by travel guides and international media. Other government bodies (Figure 2.1) related to the tourism sector are the Ministry of Trade, Industry and Tourism and the Vice-Ministry of Tourism, who both support the management of tourism in the regions, as well as the competitiveness and the sustainability of the sector (OECD, 2018). However, in 1996 some regions gained more autonomy with the establishment of the General Tourism Law, which stated that regional and local authorities are responsible for developing tourism activities in their territories. Since 2013 and the Decree 1837, the Vice-Ministry of Tourism works in collaboration with other government entities. The Superior Tourism Council acts as a coordinating body and brings together relevant ministries in the field of tourism. The Vice-Ministry of Tourism is also responsible for the management, collection, and implementation of tourism resources, through the National Tourism Fund (FONTUR). The main sources of funding include fiscal resources, obligatory contributions from tourism service providers, tourism taxes from international visitors entering Colombia by air (US$15 per visitor), resources from the management of tourism properties by the State, income from the exploitation of tourism-related brands owned by the Ministry and penalties imposed on tourism service providers for legal infractions (OECD, 2018).

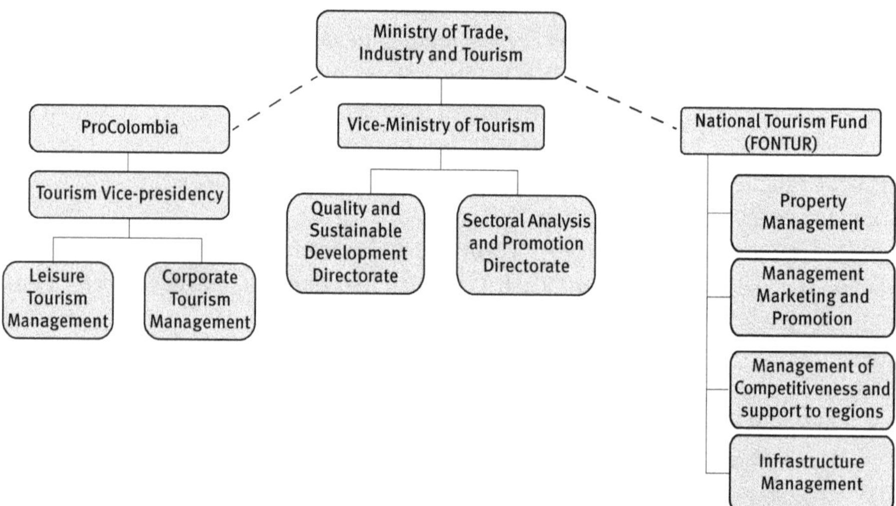

Figure 2.1: Organizational chart of tourism bodies. (Source: Organization for Economic Co-operation and Development, adapted from Ministry of Trade, Industry and Tourism, 2018).

The Colombian National Development Plan 2014–18 places tourism as a priority sector in the country's development strategy, through the establishment of the Tourism Plan 2014–18: Tourism for Peace Building. As implied in the title of the document, tourism is seen as a vector to consolidate peace. The plan is that by 2018 the sector will generate 300,000 new jobs and US$6 billion in foreign currency (OECD, 2018). Three tourism sub-sectors are identified: responsible and sustainable tourism, tourism culture, and peace tourism. The promotion of so-called "post-conflict destinations" is part of the priority action of the Tourism Plan. Tourism development in areas of former conflict is seen as a way to "rebuild the social fabric and the culture of the territories, as well as to develop value chains and improve the quality of life of host communities through responsible and sustainable practices" (OECD, 2018, p. 317).

2.4 Narcos in the Realm of Magical Realism

Colombian institutions are working hard on the construction of the country's post-war image. While ProColombia is the central public body in charge of this task, international bodies are also active here. The WTO has been significantly promoting the country and praising its national campaigns, which present "the true reality of the country," as it states in its report "Colombia, back on the map of world tourism" (UNWTO, 2009, p. 3).

The objectives of Colombia's communication strategy are not to redefine the profile of the country, or its identity or its ideology. Rather to highlight all of these identifying elements and framing them in the current reality of Colombia in order to transmit a new image, free of stereotypes and prejudices derived from a traumatic historical stage, which it has largely left behind. The WTO report highlights the fact that Colombia was already well-known by the worldwide public and that communication efforts were not needed to make the country known, but to correct a distorted image. In 2004, a national marketing strategy was put in place to create a brand image for the country. As Guilland (2012, p. 4) has demonstrated through its campaign "Colombia is passion," one of its main goals was to bring Colombians together behind a common cause: "'passion' was chosen as the theme to characterize the population. [. . .] it was all about getting the Colombian people to claim for themselves this image of the country. By creating a feeling of common identification, it was easier to convert the Colombians into real promotional actors." Guilland also shows how a few years later Colombian representatives took the notion of risk, previously associated with the country, and managed to convert it into an opportunity, with the launch of a new campaign: "Colombia, the only risk is wanting to stay."

In 2013, ProColombia inaugurated a new promotional campaign based on "magical realism." The main idea was to change the perception of potential visitors and "to show the charms of Colombian tourist destinations, as well as progress in terms of security and stability" (ProColombia, 2013, n.p.). Magical realism is a category encompassing literature, painting, and cinema and is closely associated with Latin American culture (Flores, 1955; Slemon, 1988). It can be summarized as the presence of irrational and magical elements in a rational and realistic environment. It has been closely linked to the Colombian novelist and Nobel laureate Gabriel García Marquéz, and in particular to his worldwide best-seller One Hundred Years of Solitude (1967). According to ProColombia former president, Maria Claudia Lacouture, things that might be normal for Colombians are "a magical image, a revealing moment, an unforgettable experience for a tourist" (ProColombia, 2013, n.p.). This dynamic has also been reinforced by the worldwide success of the movie Magia Salvaje (Wild Magic) portraying the wilderness of the country by associating it with notions of enchantment and strangeness. Although the movie was produced by Mike Slee, a British director, it is often used in the promotion of Colombia for tourists (Naef, 2018).

ProColombia's repertoire of images related to magical realism features elements like mysterious jungles (Figure 2.2), colorful seas, exotic animals, and enigmatic indigenous traditions. However, for some actors, magical realism can resonate differently. In Colombia, narcotraffickers were sometimes referred to as "magicians" due to their ability to accumulate huge fortunes in a very short time (Naef, 2018). The international Netflix series *Narcos*, often associates the notion of magical realism with narcos. "Colombia is where it began. And anyone who spent real time here knows why. It is a place where the bizarre shakes hands with the inexplicable on a daily basis." The worldwide success of the show, especially in the United States, which represents one of the main foreign tourist markets for Colombia, contributed to shaping the international reputation of the country. Here, opposing views of magical realism clash. ProColombia's images of an enchanting and peaceful land compete with the Netflix portrayal of a violent country ruled by criminals (Guilland & Naef, 2019). The description of the third city of Colombia by the main character of the *Narcos* is illustrative: "Cali is the Soviet Union with sun."

At the end of 2017, in parallel with its 25th anniversary, ProColombia launched a new campaign: "Colombia: Land of Sabrosura." After five market studies, this local word without a direct English translation was chosen to express emotions such as positive attitude, beauty and flavor, elements in tune with a context of peace. While tourism can without a doubt serve as a tribune for presenting the positive transformation of a country previously at war (Causevic & Lynch 2011; Rivera 2008), competing representations are diffused, as in the field of popular culture with the example of an international media like Netflix.

Figure 2.2: ProColombia Magical realism tourism campaign.

2.5 Tourism and the Branding of the New Medellin

The example of Medellin, the second city of Colombia, contains thought provoking elements related to tourism rebranding. After decades of being considered a capital for crime and drug trafficking, it is now promoted by international organizations and the media as a resilient and innovative city. Achievements in architecture, security, social development and mobility, like urban cable-cars and outdoor electric stairs (Figure 2.3), are regularly used to illustrate the transformation of Medellin. This discourse is widely shared at the municipality level by the official bodies in charge of city marketing and tourism, such as the Medellin Convention and Visitor Bureau and the Sub-Secretary for Tourism.

The transformation of the city is also an important resource for the local tourism sector, who promotes "tourism of transformation" as one of its most important products (Naef, 2016). However, the city's narco-related past is also central stage in the offer of local guides, who sell "pablo tours" or "narco-tours" to foreign visitors (Naef, 2018a). Here, again, representations of the city compete and clash. Official tourism institutions and private entrepreneurs depict disparate and often conflicting facets of Medellin. The municipality uses tourism as a window to promote its transformation into a resilient, innovative, and peaceful city, while some private tour guides cash in

Figure 2.3: Tourists visiting the outdoor electric stairs located at the heart of the commune 13. (Source: Photo by the author, 2017).

on the past violence by offering narco-tours as a tourism attraction. Governmental tourism agencies do not provide any support or encouragement for this practice. Tourism officials, as well as town representatives, clearly express discomfort, but never undertook any preventive measures until recently. In 2018, however, the mayor of Medellin, Federico Gutiérrez, took action against two emblematic sites of these narco-tours: The Mónaco building and the informal museum run by Roberto Escobar, the brother of the famous drug lord Pablo Escobar.

After a long period of controversial discussions, the municipality decided to tear down the Mónaco building, one of the former homes of Pablo Escobar and a key stop for most of the narco-tours. The Mónaco building was first adorned with placards in Spanish and English to explain to visitors the many atrocious crimes that Pablo Escobar ordered. The building was then destroyed on February 22nd, 2019, with great fanfare in front of a large media audience. Even the president Iván Duque made a quick trip from the capital to celebrate what he described as the triumph of the culture of legality. The 23,000 cubic meters of remains were disposed of by the city in a secret location. As stated in the daily El Tiempo: "The issue associated with the rubble is much more complex. They have been taken to an unknown and secret location under a confidentiality clause. [. . .] There is a kind of symbolism and they want to prevent people from entering the site to search for remains." (Gutiérrez, 2019). The site of the former Mónaco building is now in the process of being transformed into a memorial park for the victims of the past decades' violence. In the frame of the project "Medellin embraces its history", the "Inflexión Memorial Park" aims to change the symbolic value of this place, "by putting an end to the heroic perception of drug traffickers and instead, values and acknowledges those affected by the car bombs, targeted killings and other attacks committed between the eighties and the end of the 20th century" (Official website, 2019). As the New York Times stated when the decision to destroy the Mónaco building was made public, "The conflicting response to the building – municipal embarrassment or photo opportunity – is also a prime example of how Medellin still struggles over the Escobar narrative" (Casey, 2018, n.p.). In the same article, one collaborator of the Casa Memoria Museum added, "Pablo Escobar has become the pop icon of this story. The city saw no urgency to tell this part of history. It wasn't a priority for the government until there was a problem, until suddenly you had narco-tours led by Popeye[1]" (Casey, 2018).

Another key site included in the narco-tours offer was shut down in 2018. A museum dedicated to Pablo Escobar, run by his brother Roberto, the former treasurer of the Medellin cartel, was housed in a former hideout of the drug kingpin (Naef, 2018a). This informal museum was closed during a joint action of the municipality, the Vice-Ministry of Tourism and the Department of Migration. This measure was presented as temporary and Roberto Escobar, at the time of writing, promptly announced his intention to reopen his museum as soon as possible. Yet, like the destruction of the Mónaco building, the action taken against this controversial museum demonstrates the will of official bodies to control the narrative related to Medellin's narco-past. It is indeed the image of the city that is at stake; representations of a boomtown of innovation and progress are in conflict with the portrayal of a city doomed by violence and narcos (Figure 2.4).

Finally, action to counter the promotion of Medellin's "narco-heritage" (Naef, 2018a) beyond Colombia's borders was also undertaken. After spending close to 30 years in jail, Jhon Jairo Velásquez – alias "Popeye" – moved into Medellin's tourism scene through the organization of private and pricey narco-tours (Palomino, 2017). In 2016, the Medellin sub-secretary of tourism lodged a complaint against a Puerto Rican tourist operator, who proposed a four-day trip in Colombia, labelled "Medellín during Halloween," where one of the highlights was a dinner with Popeye (Restrepo, 2016). Pointing out that the offer did not comply with National Tourism Registry norms, the sub-secretary forced the tour operator to remove his advertisement. In the context of the global tourism offer, it was argued that action had to be taken against undermining

1 Jhon Jairo Velásquez – alias "Popeye" – was Pablo Escobar's main hitman. He died on February 6th, 2020, from cancer.

Figure 2.4: In a souvenir shop of Medellin, the portrayal of Pablo Escobar (right) contrasts with the representation of the outdoor electric stairs (left), symbol of the transformation of Medellin. (Source: Photo by the author, 2017).

of the international image of Medellin. In a similar vein, several individuals in the Colombian community, located in France, have taken action against a Parisian cocktail bar – Medellin Paris – inspired by the narco-related past of the second city of Colombia. One week after the opening (where bags of flour were exposed everywhere in the bar as fake packs of cocaine) a demonstration was organized by Colombians expatriates asking for its closure. The bar was still open when this chapter was written, but the manager has changed certain elements of his decoration: a fake tomb of Pablo Escobar has been removed and an inscription on the wall was added: "Pablo no es la Colombia" (Pablo is not Colombia) (Loisy, 2019).

2.6 Tourism and Post-Conflict in "Locolombia"

Magical realism, as used in Colombia tourism promotion, describes an attractive exoticism for potential tourists. At the same time, the narco-world, where boundaries between myth and reality are often blurred, also seems to fit into the frame of magic realism. The spectacular life of Pablo Escobar and other drug lords are often described as reality surpassing fiction. Some tour guides offering narco-tours present the development of the narco-trade in Medellin as possible only in "Locolombia" (a contraction of "loco," which means "crazy" and "Colombia"). The Netflix show

Narcos cashed in on this dark magical realism and undoubtedly helped to spread a picture of Medellin and Colombia which is very different from the one of peace and innovation promoted by official bodies. This dark representation of the city has provoked some criticisms, some specifically commenting on the fact that magical realism was part of the narrative of the series. The Colombian philosopher Armando Silva explained how audiovisual popular culture contributes to a kind of fascination and tolerance for narcos in Latin America. He added, "the major outrage of its directors [of *Narcos*] was to put all this criminal enterprise in a context of 'magical realism', justifying it by the fact that both could be born only in Colombia. Here the fiction yields to the ideology of its directors" (Silva, 2017, n.p.). The blogger Bernardo Aparicio García also stated. "The wonderful strangeness that García Márquez had brought to life in the realm of fiction, Pablo Escobar and the mágicos imposed on the real world through their ruthless ambition" (2015, p. 3). While magical realism was promoted in the tourism national narrative and associated mainly with natural wonders, other actors, from Netflix media to local guides in Medellin, tied the country's narco-past to the realm of magical realism.

Behind designations such as "locolombia," "narcolombia," or "medellinovation," diverse and sometimes dissonant images are diffused. In terms of governance, the diversity of actors acting and interfering in the branding and the tourism sector of a country like Colombia pose serious challenges for stakeholders implicated in the construction of the country's post-war image. Action undertaken to remove sites, such as the Mónaco building and Roberto Escobar's museum from Medellin's touristscape illustrates these challenges. After announcing the destruction of the Mónaco building, Medellin's mayor Federico Gutiérrez (2016–2019) insisted on the importance of developing a tourism that "adds value to the city," qualifying narco-tours as "predatory tourism." "I invite people to Medellin, obviously, to learn our history, but our history told by us as an institution. Not by some of those who run these narco-tours, who were part of these narco-terrorist structures" (Amrani, 2018).

2.7 Conclusion

In this chapter, the objective was to illustrate the diversity of actors influencing the construction of Colombia's post-war image, specifically within the tourism sector. Through the lens of tourism governance, it demonstrates some of the challenges tourism stakeholders face when promoting the image of a country which is slowly achieving peace after decades of violence. In the tourism and memorial arena of Colombia, various actors – government officials, tour guides, former narcos, international tourism operators, media producers, etc. – all participate in diffusing different and often competing representations of the country, ranging from the exploitation of a violent past and the promotion of perspectives of peace. While Colombia is promoted as a "magical" place, it conjures up different repertories of images, where the strangeness related to the narco-world can also fit in.

Medellin serves as an enlightening case study for exploring the tensions these dissonant representations can create, between the hyper-visibility of Pablo Escobar's memorabilia and the municipality's objective to portray a city that has moved forward and is now a symbol of innovation and resilience. It also demonstrates that actors influencing tourism governance in the second city of Colombia range far beyond the sole official bodies. This might be even more significant after the reopening of Colombia to international visitors, following the severe lockdown that cities like Medellin endured as part of the COVID-19 pandemic. Before the sanitary crisis, Medellin started experiencing a real boom in tourism arrivals and the authorities struggled to control the huge increase of guides moving into the city touristscape. The commune 13 and its internationally famous outdoor electric stairs, briefly mentioned above, saw more than 400,000 tourists in 2019 and became the city's main tourist attraction. With the return of tourists after the lockdown, this strategic area will certainly be at the center of new governance challenges. In 2019, more than 400 guides, most of them without an official license, were active in this specific area. While most were Colombian, an increasing number of foreigners (mainly from Venezuela and Argentina) started to narrate the history of the conflict to foreign visitors, causing discomfort and anger within the local population. To remedy the absence of any state action, the community reactivated the grassroots "Tourism network of the commune 13." The main objective was to coordinate the many actors involved in tourism activities in this district of Medellin. Furthermore, in tourist areas of the city, such as the commune 13 and the historical center, local street-gangs were also using tourism to empower themselves, through their racketing of tour guides and an increasing drug market targeting foreign tourists.

Colombia, a country rapidly repositioning itself on the world tourism map, is thus facing many issues, related to the dissemination of a positive and attractive image, as well as to the control of the many new actors integrating this expanding sector. As these lines are written and as Colombia is slowly reopening its tourist sites after several months of lockdown, the future seems open to many potentialities, but also many challenges in terms of governance. For tourism stakeholders and government officials concerned with the branding and promotion of Colombia, the most important of these challenges will be to find a middle path between oblivion and the acknowledgment of its dissonant and violent past, as well as to provide a stronger institutional frame for further tourism developments.

References

Amrani, I. (2018). Narcotours: Netflix fans uncover the real life of Pablo Escobar. The Guardian 4 October. Retrieved from: https://www.theguardian.com/world/video/2018/oct/04/narco tours-narcos-netflix-fans-uncover-the-real-life-of-pablo-escobar-video. Date accessed: 20th October 2020.

Anholt, S. (2013). Beyond the nation brand: The role of image and identity in international relations. *Exchange: The Journal of Public Diplomacy*, 2(1), 1–7.

Ashworth, G.J., Graham, B., & Tunbridge J.E. (2007). Place, identity and heritage. In B. Graham, G. J. Ashworth, & J. E. Tunbridge (Eds.), *Pluralising pasts: Heritage, identity and place in multicultural societies*. (pp. 54–67). London: Pluto Press.

Aparicio Garcia, B. (2015). I grew up in Pablo Escobar's Colombia. Here's what it was really like. Vox 21 October. Retrieved from: http://www.vox.com/2015/10/21/9571295/narcos-pablo-escobar-colombia. Date accessed: 20th October 2020.

Borges, M. d. R., Eusébio, C., & Carvalho, N. (2014). Governance for sustainable tourism: A review and directions for future research. *European Journal of Tourism Research*, 7: 45–56.

Casey, N. (2018). 25 Years After Escobar's Death, Medellín Struggles to Demolish a Legend. The New York Times 22 September. Retrieved from: https://www.nytimes.com/2018/09/22/world/americas/medellin-colombia-pablo-escobar.html Date accessed: 20th October 2020.

Causevic, S. & Lynch P. (2011). Phoenix Tourism: Post-Conflict Tourism Role. *Annals of Tourism Research*, 38(3):780–800.

Debarbieux, B. (2016). *L'espace de l'imaginaire*. Paris: CNRS Edition.

Flores, A. 1955. Magical realism in Spanish American fiction. *Hispania*, 38(2), 187–192.

González N. & Álvarez, P. (2018). Colombia en mutación: del concepto de posconflicto al pragmatismo del conflicto" *JANUS.NET e-journal of International Relations*, 9(2), DOI: https://doi.org/10.26619/1647-7251.9.2.6

Guilland, M-L. (2012). "Colombia, the only risk is wanting to stay" – From national tourism promotions to travelling in the Sierra Nevada: uses and refutations of risk. Via [Online], 1, 2012, DOI: https://doi.org/10.4000/viatourism.1248

Guilland, M-L. & Naef, P. (2019). Les défis du tourisme face à la construction de la paix en Colombie. Via [Online], 15, DOI: https://doi.org/10.4000/viatourism.3637

Gutiérrez, E. (2019). A un lugar secreto son llevados los escombros del edificio Mónaco. El Tiempo. 4th March. Retrieved from: https://www.eltiempo.com/colombia/medellin/a-un-lugar-se creto-son-llevados-los-escombros-del-edificio-monaco-de-medellin-333520 Date Accessed: 20th October 2020.

Gravari-Barbas, M. & Grabrun, N. (2012). Tourist imaginaries. Via [Online], 1, 2012, DOI: http://journals.openedition.org/viatourism/1180

Hudson, R.A. (2010). *Colombia: A country study*. Washington, D.C.: Federal Research Division.

Linde, P. (2017). Tengamos el turismo en paz. El Pais 6 February. Retrieved from: https://elpais.com/elpais/2017/02/03/planeta_futuro/1486150155_364987.html. Date accessed: 20th October 2020.

Loisy, F. (2019) Polémique autour du Medellin, le club parisien qui surfe sur la série «Narcos». Le Parisien 3 January. Retrieved from: http://www.leparisien.fr/paris-75/polemique-autour-du-medellin-le-club-parisien-qui-glorifie-pablo-escobar-03-01-2019-7980528.php. Date accessed: 20th October 2020.

Marquez, G.G. (1967). *One Hundred Years of Solitude*. New York: Harper Perennial.

Martin, G. (2009). *Gabriel Garcia Marquez: Une vie*. Paris: Grasset.

Medellín Abraza Su Historia. (2020). Retrived from: https://www.medellinabrazasuhistoria.com/ Date accessed: 20th October 2020.

Naef, P. (2016). Touring the communa: memory and transformation in Medellin, Colombia. *Journal of Tourism and Cultural Change*. 16(2), 173–190. DOI: https://doi.org/10.1080/14766825.2016.1246555
Naef, P. (2018). The commodification of narco-violence through popular culture and tourism in Medellin, Colombia. In: C. Lundberg & V. Ziakas (Eds.). *The Routledge Handbook of Popular Culture and Tourism*. (pp. 80–92) London: Routledge.
Naef, P. (2018a). "Narco-heritage" and the Touristification of the Drug Lord Pablo Escobar in Medellin, Colombia. *Journal of Anthropological Research*, 74(4), 485–502. DOI: https://doi.org/10.1086/699941.
Ndlovu, J. & Chigora, F. (2019). The moderation effect of branding on destination image in a crisis-ridden destination, Zimbabwe. R. K. Isaac, E. Çakmak & R. Butler (Eds.) *Tourism and Hospitality in Conflict-Ridden Destinations*. (pp.118–136). London: Routledge.
Organisation for Economic Co-operation and Development (OECD). (2018). OECD Tourism Trends and Policies 2018. Retrieved from: http://www.oecd.org/cfe/tourism/oecd-tourism-trends-and-policies-20767773.htm. Date accessed: 20th October 2020.
Palomino, S. (2017) La cara b del 'narcotour' de Pablo Escobar desmitifica al capo de Medellín, El Paìs, 23 June. Available at: https://elpais.com/internacional/2017/06/23/colombia/1498172784_237146.html. Date accesses: 20th October 2020.
ProColombia. (2013). The Ministry of Commerce, Industry and Tourism and ProColombia presented the new Colombian international tourism campaign abroad. ProColombia. Retrieved from: http://www.procolombia.co/en/news/colombia-magical-realism. Date accessed: 20th October 2020.
Restrepo, V. (2016) Agencia que vendió paquetes turísticos para ver a alias "Popeye" será investigada, el Colombiano 29 September. Retrieved from: http://www.elcolombiano.com/antioquia/agencia-turisticos-de-puerto-rico-vende-paquetes-para-visitar-a-alias-popeye-en-medellin-CA5075602. Date accessed: 20th October 2020.
Rivera, L. A. (2008). Managing 'Spoiled' National Identity: War, Tourism, and Memory in Croatia. *American Sociological review*, 73(4): 613–634. DOI: https://doi.org/10.1177/000312240807300405.
Salazar, N.B. (2012). *Envisioning Eden: Mobilizing Imaginaries in Tourism and Beyond*. New-York: Berghahn Books.
Salazar, N.B. & Graburn, N. (Eds.) (2016). *Tourism Imaginaries: Anthropological Approaches*. New-York: Berghahn Books.
Scott, N. & Marzano, G. (2015). Governance of tourism in OECD countries. *Tourism Recreation Research*, 40(2) 181–193
Silva, A. (2017). Narcos, drama sin fin. El Tiempo. Retrieved from: http://www.eltiempo.com/opinion/columnistas/armando-silva/narcos-drama-sin-fin-en-colombia–143754.
Slemon, S. (1988). Magic realism as post-colonial discourse. *Canadian Literature*, 116, 9–24.
Shirley, G., Wylie, E. & Friesen, W. (2018). The Branding of Post-Conflict Tourism Destinations: Theoretical Reflections and Case Studies. IN: A. Neef & J.H. Grayman, (Eds.) *The Tourism–Disaster–Conflict Nexus*. (pp. 119–139) (Community, Environment and Disaster Risk Management, Vol. 19), Bingley: Emerald Publishing Limited. DOI: https://doi.org/10.1108/S2040-726220180000019007.
Vitic, A. & Ringer, G. (2008) Branding Post-Conflict Destinations, *Journal of Travel & Tourism Marketing*, 23:2–4, 127–137, DOI: 10.1300/J073v23n02_10.
World Tourism Organization (UNWTO). (2009). Colombia. Back on the Map of World Tourism. UNWTO. Retrieved from: https://www2.unwto.org/agora/colombia-back-map-world-tourism. Date accessed: 20th October 2020.
World Tourism Organization (UNWTO). (2011). *Governance for Sustainable Tourism Development*. Algarve: Universidade do Algarve/UNWTO.

Amir Gohar

3 Egypt

3.1 History and Background

Egypt has been known throughout history as a destination for travelers, particularly since it was visited by Herodotus during ancient times and he wrote about his surprises of the vast differences between Egypt and his homeland. During the Roman Empire numerous festivals and leisure events were held, creating a need for domestic travel between cities in the Roman province of Egypt (Lindsay, 1965). Roman roads opened up major cross-border routes that established trade routes and newly conquered areas. One notable example is the Roman road network connecting the Red Sea to the Nile River, which still retains its historical value as a contemporary tourist route. Krzywinski (2000) and Snyder (2003) assert that this Roman road facilitated mobility between the Far East and Europe before the construction of the Suez Canal. During the Medieval times, Egypt was a part of major pilgrimage routes connecting north African travelers to Mecca and all of Africa to Jerusalem. Later, the Grand Tour brought the flow of European tourism toward Egypt leading to increased establishment of guesthouses, resorts, and private estates. Hotels multiplied and Europeans bought estates in many places in Egypt, including Fayoum City.

The industrial revolution and the invention of the railway resulted in capital accumulation and real estate investment in Egypt (Gregory, (2001). In fact, until World War II, members of the British aristocracy spent a majority of the winter season in Egypt (Abdelwahab, 1996). In terms of broader tourism industry investments, since the mid-19th century, Thomas Cook chose Egypt as a major destination for his third group of tour outside Great Britain (the first two were to Europe and the United States). Thomas Cook's first trip to Egypt, in 1869, took customers via steamboat down the Nile from Cairo to Aswan. Over time, Cook engaged more local people to provide help; he also improved ships and the related infrastructure so that the trip was smoother, quieter, and safer (Gregory, 2001; Hunter, 2004). Thomas Cook eventually added train excursions on the Nile connected to his steamer travel packages.

The commercialization of the automobile in the 1890s brought further changes. According to Refaat (1997) and Al-Aswany (2015), the first car brought into Egypt was a French De Dion-Bouton belonging to Khedive Ismail's grandson, Prince Aziz Hassan. In 1904, accompanied by two friends, the prince made a historic 210-kilometer journey from Cairo to Alexandria in over ten hours, despite the hundreds of difficulties resulting from the absence of roads and bridges. At the end of 1905, there were approximately 110 motorized vehicles in Cairo and 56 in Alexandria, as well as 50 motorcycle sidecars and two Dietrich-type omnibuses belonging to the newly formed

https://doi.org/10.1515/9783110638141-003

Cairo Omnibus Company. The car began to shape tourism and day-use destinations in Egypt after the launch of the Touring Club D'Egypte.[1] This group encouraged local day trips, domestic travel, and supported international tourism by conveying people from railway stations to destinations in Egypt's main cities.

During WWI and the British colonization of Egypt, travel from Europe to Egypt further increased. Although there was fledgling air service between London and Paris that became popular by the 1920s, the journey to Egypt remained time consuming. Before long distance air travel was available for the entire route to Egypt, tourists could either journey overland across Europe and then board a steamer to cross the Mediterranean, or they could simply make the entire trip by sea, a journey that took around two weeks (Fletcher, 2011). Upon arriving in Alexandria or Port Said, travelers would take the train to Cairo, and then continue either by train or by ship to upper Egypt to see antiquities and archaeological sites. The spectacular discovery of the tomb of Tutankhamen in 1922 brought tourists to Egypt in droves, and as Luxor became a greater tourist magnet, the Valley of the Kings was thronged with visitors hailing from all over the world, each wanting a glimpse of the latest treasure to be removed from the tomb. To accommodate these waves of tourists, hotels, such as the Hotel Cecil in Alexandria, began to develop around attraction areas. In response to the increase in tourism during the British imperial era, the tourist destinations changed, especially in numbers of hotels around popular attractions. This change in the built form extended to include main mobile hubs, such as train stations and airports. These establishments' architecture was influenced by European architecture, especially British architecture. In this era, the travel modality was still undergoing the same rapid transformation as in the preceding era; it primarily developed in response to the colonial demand for non-touristic transportation, but the Nile Cruises, which specifically and exclusively served travelers to upper Egypt, developed in response to tourism's demands.

During WWII and the Jet Era that followed, tour packages continued to frequent Egypt's main attractions near urban centers; as such, tourism boomed in Cairo, Luxor, Aswan, and Alexandria. Because travelers used, and their hosts maintained, urban structures built in previous eras, tourists made use of local transportation such as the tram, the local railway, and the Nile cruises (Refaat, 1997; Towner, 1995; Towner & Wall, 1991; Wolf, 1996). With the rise of ecotourism, there has been some initiatives and attempts to implement ecotourism as an alternative approach to mass tourism. There are few examples of tourism facilities designed around environmental principles and ecologically sound development practices, though they are not typical. Before the Camp David Accords of 1978, Egypt's mainstream tourism was concentrated in large cities close to Egypt's rich antiquities, the Nile, and sites

1 Tour d'Egypte is a professional road cycling stage race held each February in Egypt. Tour d'Egypte is part of the UCI Africa Tour.

of ancient ruins and civilizations. Beach activities were restricted to coastal cities such as Alexandria. Following the peace treaty, many areas in the Red Sea were opened to exploration, including natural and cultural heritage sites that offered germinal opportunities to expand the tourism market.

Today, tourism is one of the most important sectors in Egypt's economy. More than 12.8 million tourists visited Egypt in 2008, providing revenues of nearly US$11 billion. The sector employed about 12% of Egypt's workforce. Coastal tourism is one of the country's main tourism components. The government has been developing coastal areas that invited mass tourism and put huge pressure on environmental and cultural resources. Some of these have been measured and others are ignored externalities. El-Sherbiny et al (2006) showed through empirical case studies that environmental degradation in Egypt due to tourism construction is significant and that the entire construction operation needs to be addressed in the EIA process.[2]

3.2 Tourism Governance in Egypt

Effective governance is a key requirement for implementing successful tourism development (Bramwell, 2011; Bramwell & Lane, 2011; Connelly, 2007; Erkus, 2011; Yüksel et al., 2005). The expansion of tourism development on Egyptian coasts and specifically along the Red Sea has had well-documented effects from the construction of resorts, hotels, diving centers, tourism services and related infrastructure (Shaalan, 2005; Sherbiny et al., 2006), including solid waste, coral reef destruction, mangrove degradation, building in flood plains, displacement of wetlands, changes in the shoreline, and threats to turtle nesting sites (El-Gamily et al., 2001; Frihy, 2001; Salas, 2014; Sherbiny et al., 2006)

Such impacts are often blamed on the stakeholders who are responsible for the final shaping of this built environment: designers, planners, owners, and managers of tourism resorts. However, these stakeholders operate in the final stages of a lengthy process that is premised on a strong centralized governance system. To diminish ill effects from future tourism development, and to effectively redefine the pattern of future development, it is crucial to highlight the role of the entire governance system that defines tourism's built environment.

In describing the Egyptian government, Sims (2012) uses the following phrases: "strong regime," "weak state," "political vegetables," "lame leviathan," "neglectful rule," and "soft state." This variety of terms shows how difficult the system is to unpack and understand. Figure 3.1 shows the three scales that this chapter will examined in relation to the major institutions shaping tourism development.

2 E.I.A. is the Environmental Impact Assessment and it is a mandatory step to gain approval for any tourism project in Egypt.

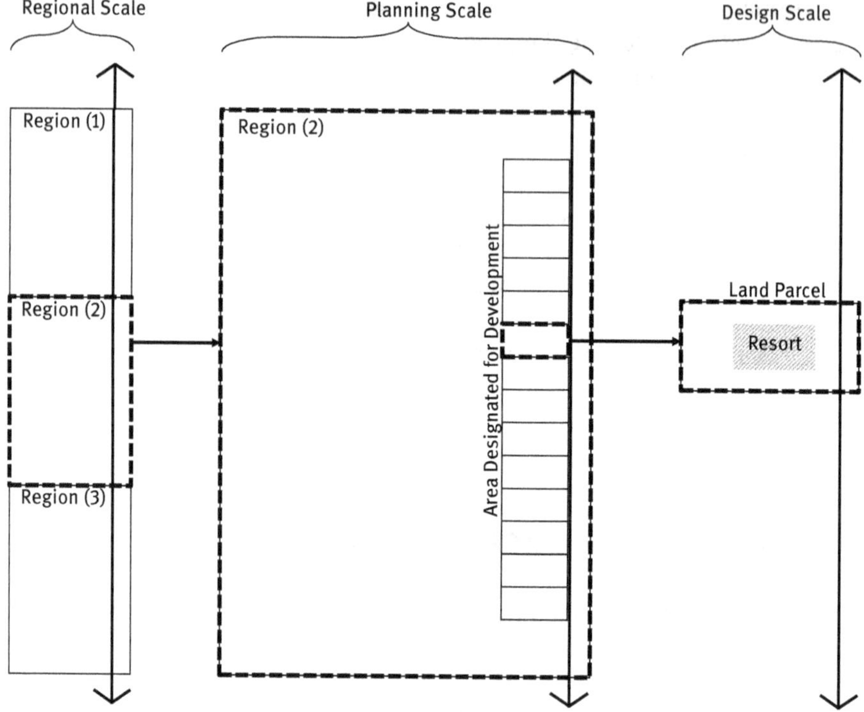

Figure 3.1: Three scales of tourism development processes in the Red Sea. (Source: Diagram by the author).

3.3 Governance and Tourism Development in the Red Sea

3.3.1 Unpacking Tourism Vis-à-Vis Central Government

In the comparison given earlier in the introduction, tourism is represented in central government in various ways. In highly centralized countries like Egypt, tourism falls under the purview of a Ministry of Tourism that plans and implements tourism operations and development throughout the country with little reliance on local government (Abdelwahab, 1996). In somewhat centralized countries like the Netherlands, no specific ministry of tourism exists. Instead, recreational activities fall under the Ministry of Agriculture, Nature Management, and Fisheries, which is responsible for setting policies that are implemented by local municipalities (Ashworth & Dietvorst, 1995). In extremely decentralized countries like the United States, there may be no centralized tourism governance at all; in the US, the now-defunct United States ravel and Tourism Administration (USTTA), which once used to operate the official global travel and tourism offices, has been dissolved. There currently exists no ministry responsible for tourism or tourism planning. Instead, relevant policies are managed at the state level for the private sector reasons previously cited.

Figure 3.2 compares these various systems of tourism oversight at the country scale. To further appreciate how the government shapes and influences the built environment along the Red Sea (i.e., coastal hotels and resorts), one needs to further understand the Egyptian model of governance. The diagram in Figure 3.3 makes clear which authorities influence tourism development in Egypt. The institutions in grey boxes are directly under the main tourism authority, and the ones with bold black borders are institutions that influence tourism development despite falling under different authorities with alternative mandates. The tourism authorities' mandate is to build more resorts and increase the number of rooms, and they are often criticized for this aggressive tourism development.

3.3.1.1 Tourism Development Authorities

Egypt's Ministry of Tourism, which is part of the Cabinet of Egypt, is the leading authority responsible for the development, promotion, and branding of tourism across the entire country. The Tourism Development Authority (TDA), a subordinate under the Ministry of Tourism, is the main organization that is responsible for planning and zoning designated regions for tourism with specific focus on coastal areas. According to Egyptian law, the Tourism Development Authority (2005) is the authority responsible land for development, land parceling, and land allocation for developers. It has regional offices along the Red Sea to supervise and follow up on tourism projects' construction.

3.3.1.2 Non-tourism Authorities: Shaping Tourism by Antagonism

The primary central governmental authority that intersects with this clear tourism development mandate is the Ministry of Environment. Formed in 1997, this ministry oversees the national parks and is specifically responsible for protecting environmentally sensitive areas, such as mangroves, coral reefs, salt marshes, and other fragile ecosystems. It functions through its implementing agency: the Egyptian Environmental Affairs Agency (EEAA). Egypt's legal basis for the environmental impact assessment (EIA) requirement was established by Law No. 4 of 1994, the Law on Protection of the Environment. This law is implemented through Executive Regulations issued by Prime Ministerial Decree No. 338 of 1995. These regulations came into full force in 1998 (Manchester University EIA Centre, 2000). According to an interview with Assem El-Gazzar, a former EIA specialist with various environmental and tourism authorities, the Egyptian Environmental Affairs Agency was established in 1997 and Law No. 4 was implemented in environmental departments with the help of other authorities. Tourism developers, designers, and

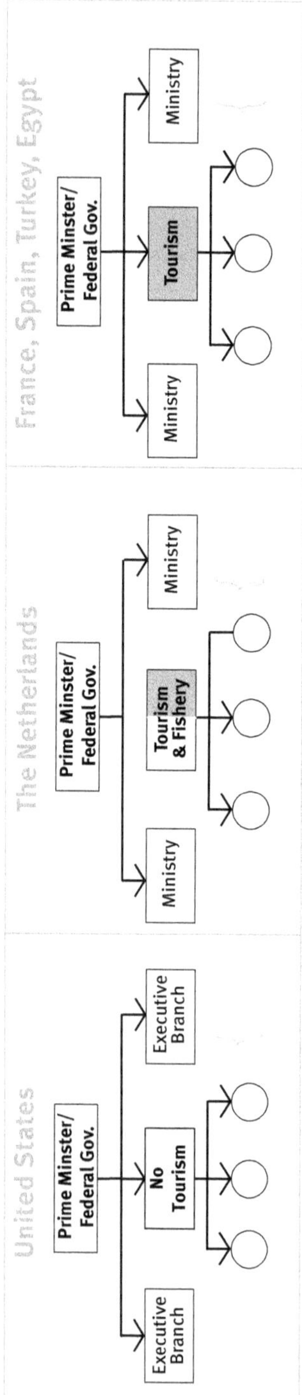

Figure 3.2: Conceptual diagram comparing three different possibilities: (1) no government, (2) mixed/shared tourism, in government, & (3) stand-alone tourism authority in central government. (Source: Diagram by the author).

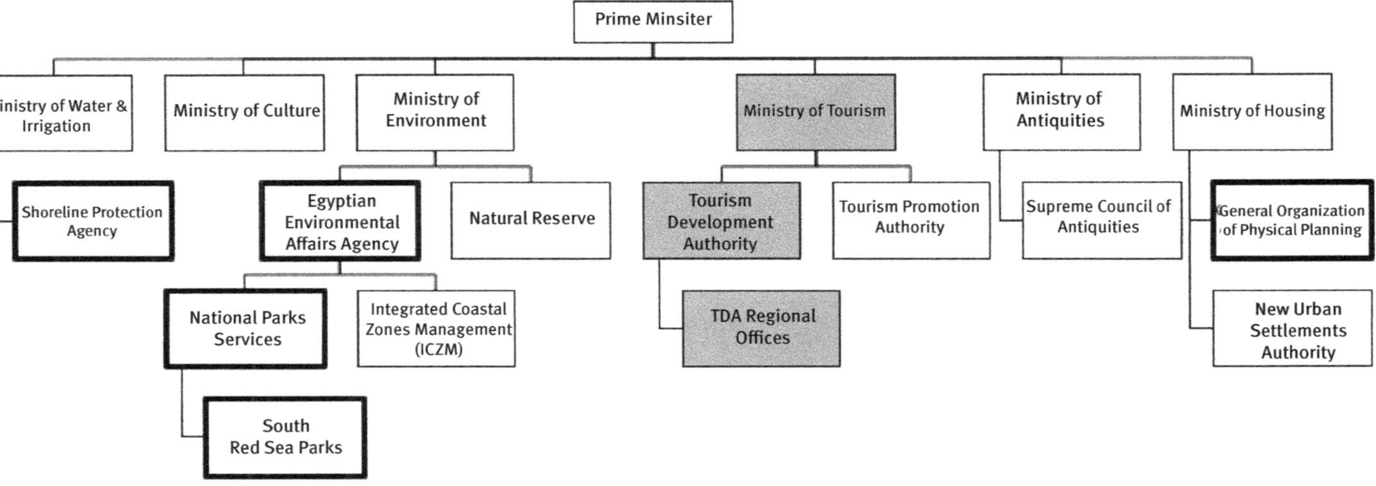

Figure 3.3: Part of the Egyptian government structure showing the tourism and environment authorities that influence/shape tourism development. (Source: Diagram by the author).

planners must obtain environmental approvals for their coastal tourism projects. There are two other central non-tourism governing authorities that are key stakeholders: the Shoreline Protection Authority, which falls under the ministry of water irrigation and controls, and which monitors and maintains the shoreline integrity and the beach buffer zone (setback) on which no construction is allowed; and the General Organization for Physical Planning (GOPP), which falls under the Ministry of Housing and Infrastructure and is the main organization responsible for infrastructure for tourism development within municipal areas.

3.3.2 Tourism in Local Government

Within the municipal boundaries, the actual day-to-day tourism on the ground is governed locally. Whether it is a city or a village, resorts within municipal boundaries fall into the governor's authority. The diagram in Figure 3.4 lays out a model of municipal boundaries (cities and villages).

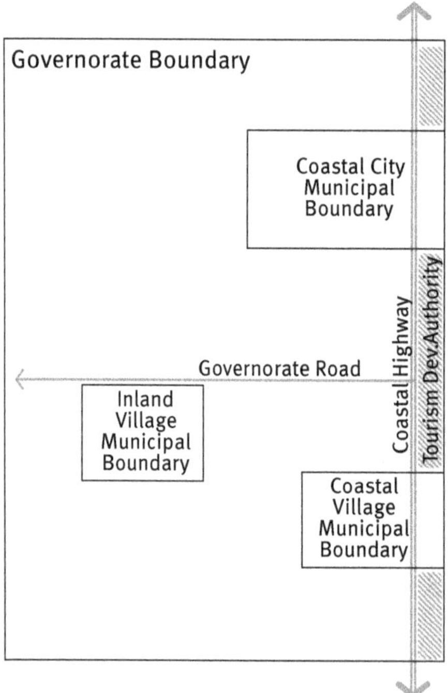

Figure 3.4: Village and city municipal boundaries are within the governorate authority and land jurisdiction. (Source: Diagram by the author).

Unlike the central government tourism authorities, which attend only to tourism activities, the local government is responsible for and concerned with all activities and land uses in the area. The master planning of the city or village is conducted to accommodate housing, urban services, infrastructure, urban facilities, and tourism. Table 3.1 lists the major differences between tourism resorts that fall within the municipal boundaries of city or village and tourism resorts that fall outside the municipal areas and are thus under the direct supervision of the central government, such as the TDA land in the Red Sea. An interview with Dr. Fahmy, a tourism-planning specialist in the Egyptian TDA for two decades, revealed the fundamental differences between parcels allocated for tourism development within the TDA and the Municipal Governorate.

Table 3.1: Main Differences between Tourism Resorts Under the TDA and Under the Red Sea Governorate (Municipal Boundaries).

	Resorts under local government	**Resorts under central government**
Land jurisdiction	Municipality jurisdiction	TDA jurisdiction
Lead agency	City mayor	TDA local office manager
Connection to utilities	Resort connected to grid	Not connected to grid (use diesel generator)
Connection to water	Using water pipes	Using local desalination units
Price per square meter	Higher price per m2 (40 EGP)	Bidding processes
Setback from the beach	50 meters	200 meters
Time to develop the project	According to approved documents	3 years for projects less than 500Km^2 and 10 years for projects over 1 million m^2
Number of rooms	More than 30 rooms per feddan	Less than 25 rooms per feddan
Built up area	More than 3 floors	Up to 3 floors
Thematic functions	Hotels	Resorts and waterfront

3.3.3 Non-governmental Influence (Community and Private Sector)

In a rather cartelized governance system, the role of non-governmental organizations becomes minimal; however, the NGOs' influence is easily distinguishable from that of government organizations. Fieldwork in the southern region of the Red Sea shows that the influence of non-governmental organization on the built environment is primarily exerted through the roles played by non-profits and local communities and by private and/or corporate interests.

3.3.3.1 Non-profits and Local Community

The few non-profits in the area focus on mangrove protection, cultural resources, and solid waste management and have little influence on the tourism market. The most influential one is Hurghada Environmental Protection and Conservation Association (HEPCA), a non-profit based in the capital of the governorate with branch offices in Southern Red Sea Egypt. HEPCA's trucks collect waste daily along a 200-kilometer stretch of coastline and transport it to the material recovery and recycling facility in the center of Marsa Alam. The waste is separated into organic and non-organic waste and processed appropriately.

Outside the municipal boundaries of cities and villages, local communities have sometimes built shelters situated within the TDA land jurisdiction that are often displaced or removed because of tourism development. However, two small settlements that have built their shacks near the shoreline have been able to remain: The Wadi el-Gimal settlement and the El-Qul'an settlement. The first is located west of a tourism resort (Shams Alam Resort) and the second is located in El-Qul'an bay on the coast further south.

3.3.3.2 Private Corporate

Private corporations and investors have very little opportunity to shape the final product, which they do either by using an environmentally friendly appearance or by adopting a sustainable management style (usually as part of a corporate policy – Hilton, for example, requires this). These are adopted on a voluntary basis, however, rather than being required by central or local government mandate.

3.4 Unpacking Governance Across Scales

3.4.1 Regional Scale

The scale at which decisions are made on the ministry level usually ignores several site-specific details. In the context of the Red Sea, for example, the Ministry of Petroleum has designated a northern stretch of the Red Sea for oil extraction, an important economic activity. But oil extraction undermines the spectacular potential for adverse effects on potential tourism on the north coast – effects on both marine and terrestrial destinations. The stretch south of this, which is dedicated to tourism and led by the Ministry of Tourism, also has quarrying and mining activities seeking gold and other minerals. The third stretch, even further south, is under the administration of Ministry of Defense; it is a politically sensitive border segment and is not ready for development. These rough divisions are based on central decisions rather than on contextual attributes. Figure 3.5 shows a schematic diagram for the three stretches of coast; it magnifies the middle (tourism) stretch to show the influence of governance at the planning level.

3.4.2 Planning Scale (Land Parceling)

Decisions and policy at the planning scale are the most complex ones. The vertical structure of the government is found to have different authorities that, by design, have conflicting mandates. Similarly, the central and local governments have land jurisdiction issues. The patterns of land parceling and attributes of development (shape and form) at this scale are guided by the following authorities:

A. The TDA manages the 5-kilometer stretch from the coast and is authorized to develop land subdivisions and allocate land for investors. Therefore, the size of the tourism project, its number of rooms, and the footprint are all physical manifestations of the decisions made by the tourism authorities.
B. The Egyptian Environmental Affairs Agency (EEAA) is responsible for protecting sensitive lands such as mangroves, wetlands, and islands. These environmental resources are commonly in conflict with tourism development because they are themselves attractions.
C. The Nature Conservation Sector (NCS) influences development near and within national parks.
D. The City Municipality designates coastal areas for tourism in addition to other urban services; the overall shape and form of the built areas are different from the tourism developments that are governed by the TDA.
E. The village authorities follow the Red Sea governorate and allocate the land parcels for tourism projects according to the city and the governorate policies.

F. The Shoreline Authority is responsible for the shoreline buffer and the type of "light structures" permitted along the coastline, such as shades, pergolas, or wooden snack bars. The shoreline buffer in the Southern Red Sea region varies from 50–200m across parcels, as per the tourism development project agreements.

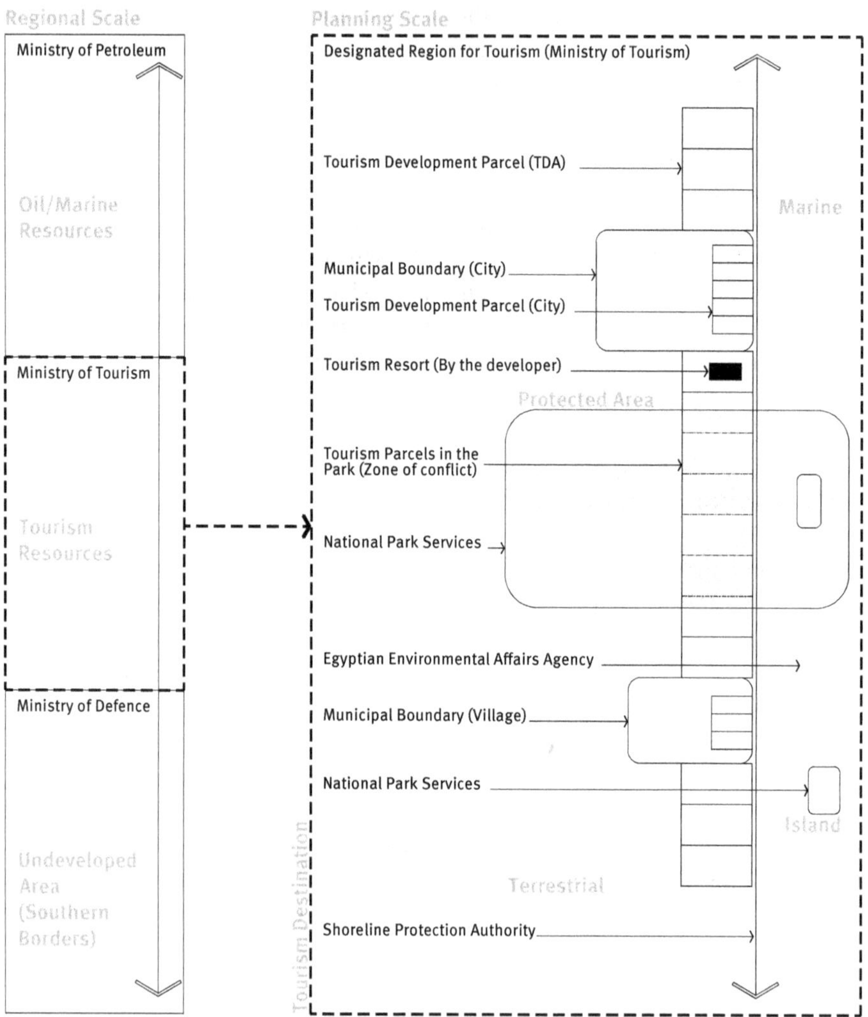

Figure 3.5: Regional and planning land designation and its influence on patterns of development. (Source: Diagram by the author).

Local community stakeholders are largely ignored in these processes, and as a result, parcel planning on this scale often comes into conflict with small local settlers on the coast. The developments of the southern Red Sea region fall into one of the following three patterns of relationship with local communities: (i) the resort is not in conflict with local tribes; (ii) the resort adjacent to a community has blocked their sea access but has compensated by providing power and water; and (iii) the resort forces local communities to relocate. The following conceptual sketches illustrate the three existing relationships between resorts and local communities (Figure 3.6).

3.4.3 Design Scale (Resort)

On the level of the land parcel, the given policy and environmental constraints leave the architect with little room to innovate design and customize because the variations in the tourism sites are not taken into account by the land subdivision system. As such, designed resorts are often in direct conflict with environmental requirements such as flood plan, coral reef, mangrove zone, or shoreline modifications. No matter who is the recrueted resort designer, the outcome of the resort shape and form will remain in direct conflict with the ecosystem.

3.4.4 Conflict Among Authorities

Tourism is shaped by tourism development forces that are guided by tourism authorities (i.e., designating lands for development and shaping them at all scales). Tourism is also shaped by other non-tourism authorities (i.e., antagonism). Government is the largest influence on tourism development in Egypt. Tourism authorities, non-tourism authorities, and even the nature of the conflicts between central and local government play a role in identifying the final pattern of tourism development.

It is not news that there is an inherited conflict between authorities in central and local government (Mayfield, 1996). There is also a financing conflict that faces government institutions during the budgeting and planning processes (EzzAlArab, 2004). Seif ElNasr (1999) asserts that the local administration can advocate that a new governor be put in place if they think he is acting against local interests. The governor's mandate can also be in conflict with that of the minister of tourism if the tourism directorate in the cities is not functioning well. Seif ElNasr also confirms that, in times of conflict, individual personalities and leadership, in addition to institutional structures, influence the locals' responses to their governorate (Figure 3.7).

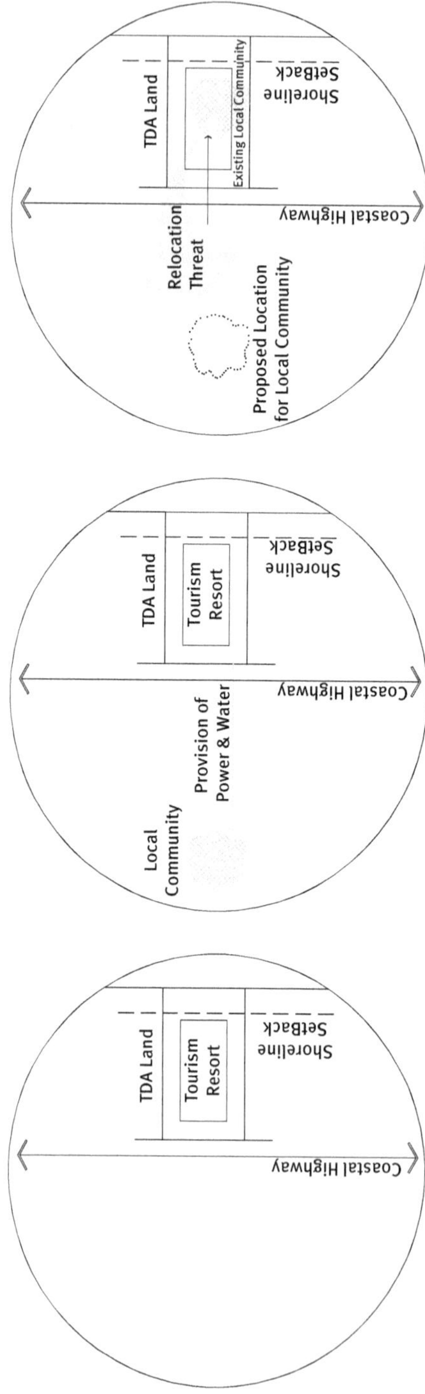

Figure 3.6: (Left) a land allocation with no local community conflict; (center) interaction with local community; (right) a threat to relocate the local community. (Source: Diagram by the author).

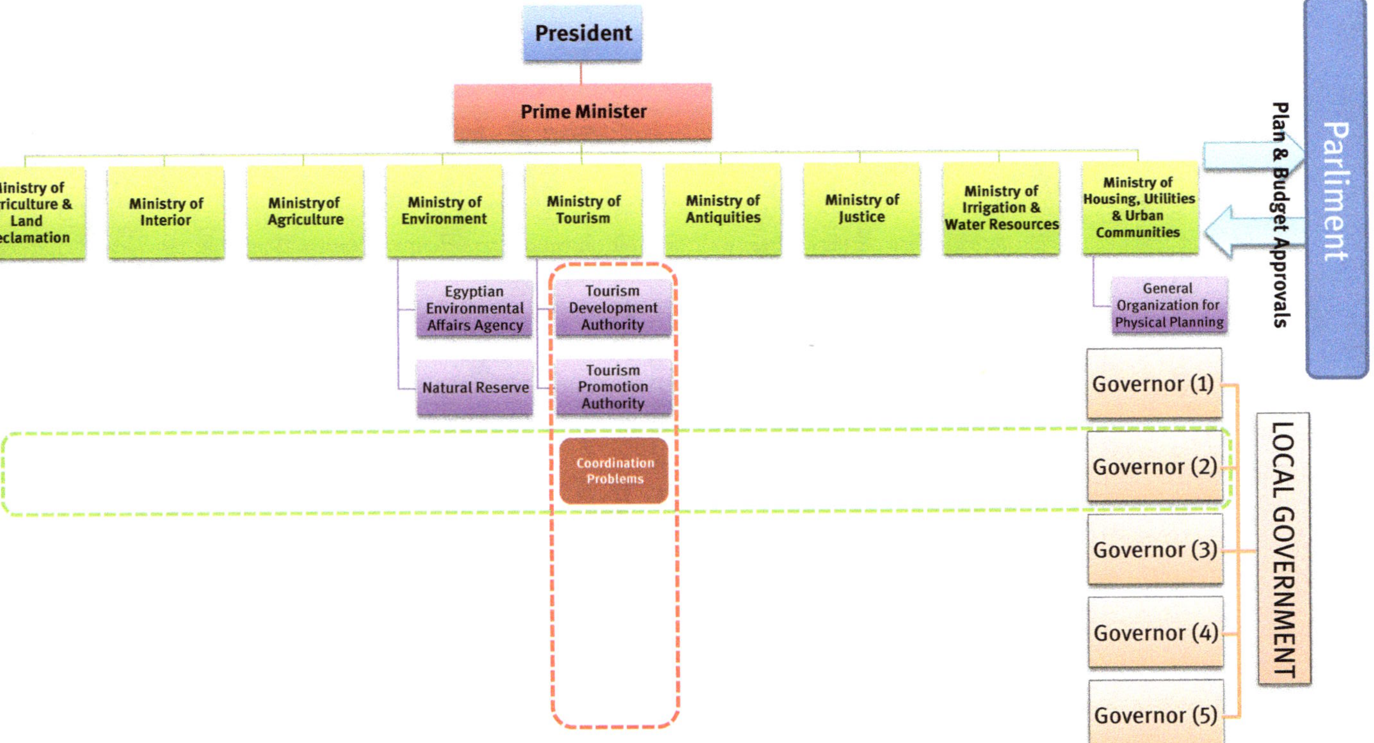

Figure 3.7: The red box represents the area of conflict among the tourism authorities in central government and the administration in the local governorate.

3.5 Conclusion and Recommendations

Tourism is better governed when all stakeholders are engaged, including both local communities and the private sector. The current top-down planning approach will compromise the integrity of tourism resources and create more environmental damage.

Governance in Egypt is fundamental to tourism development. It is almost impossible to improve tourism in the absence of good governance of such a complex phenomenon. It is also obvious that tourism authorities' institutional structure cannot be fixed outside of the context of national reform. Non-tourism authorities seem to be vital to the tourism business and can also be incorporated. The Shoreline Protection Agency, which is a key authority in influencing tourism development, can provide better planning solutions if it is consolidated with the Ministry of Tourism. Futures research should investigate the institutional implications of such restructuring upon other authorities.

Spatial planning for tourism suffers from appropriate understanding to tourism systems across scales. Unlike the typical land subdicion that takes place on flat lands in the capital, coatal areas designated for tourism incorporate complex ecosystems that needs customized planning and design rather than prototyping and/or tourism regulations that certainly can not apply on all coastal areas of the Red Sea and the Mediterranean Sea in Egypt.

The competition between dfferent authorities on how to maxemise the benefit from resourses is one of the challenges that faces the local destination reaching their full potential. The Red Sea area, both marine and terrestrial, is attractive for mining authorites, tourism authorities, investment authorities, and oil authorities. The final tourism product is shaped by the end result of these dynamics. It would achieve its full potential if (i) better harmony between different interests of centreal government and (ii) developing a metric that guide the development decision, especially that, when incorporating the economy of scale, the grouping and spatial distribution of development may influence the overall tourism planning strategies for the area.

Especially in remote areas of coastal tourism, the engagement of local communty as a stakeholder is fundamental to the success fo the tourism destination and to avoid conflicts with local ecological and cultural settings. The very few active environmental agencies have grown to become centered near the main cities only. The nomadic tribes, as the main locals impacted by tourism development are the ones need to be invited to tourism decision making on both central and local government levels.

Lastly, the understanding of tourism as a cross cutting industry that deals with many other sectores and occurs on multiple scales is crucial to govern across scales, across institutions, and across stakeholders. The diagram in Figure 3.8 summarizes and simplify the different complex relationships between government, private sector, and community non-profit orgainzations in different scales and across multiple insittutions.

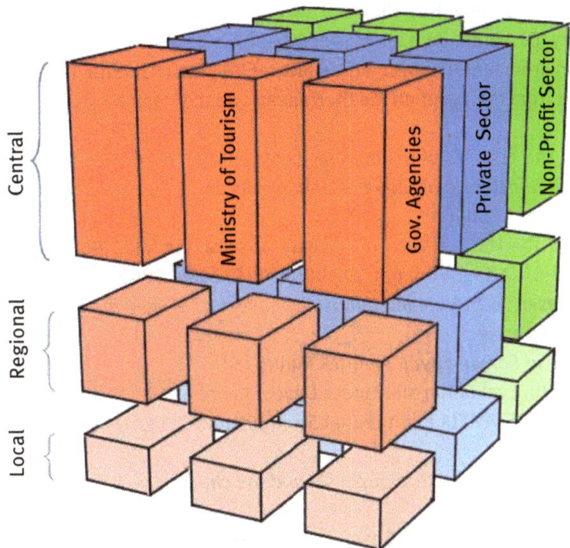

Figure 3.8: The red box represents the area of conflict among the tourism authorities in central government and the administration in the local governorate.

References

Abdelwahab, S. el-D. (1996). Tourism Development in Egypt: Competitive Strategies and Implications. *Progress in Tourism and Hospitality Research*, 2(May), 351–364.

Al-Aswany, A. (2015). *The Automobile Club of Egypt, A Novel.* Vintage Books.

Ashworth, G., & Dietvorst, A. (1995). Tourism and Spatial Transformations. (Implications for Policy and Planning).

Bramwell, B. (2011). Governance, the state and sustainable tourism: a political economy approach. *Journal of Sustainable Tourism*, 19(4–5), 459–477. https://doi.org/10.1080/09669582.2011.576765

Bramwell, B., & Lane, B. (2011). Critical research on the governance of tourism and sustainability. *Journal of Sustainable Tourism*, 19(4–5), 411–421. https://doi.org/10.1080/09669582.2011.580586

Connelly, G. (2007). Testing Governance – A Research Agenda for Exploring Urban Tourism Competitiveness Policy: The Case of Liverpool 1980–2000. *Tourism Geographies*, 9(1), 84–114. https://doi.org/10.1080/14616680601092931

El-Gamily, H., Nasr, S., & El-Raey, M. (2001). An assessment of natural and human-induced changes along Hurghada and Ras Abu Soma coastal area, Red Sea, Egypt. *International Journal of Remote Sensing*, 22(15), 2999–3014. https://doi.org/10.1080/01431160120421

Erkus, H. (2011). Modes of tourism governance: a comparison of Amsterdam and Antalya. *Anatolia – An International Journal of Tourism and Hospitality Research*, 22(3), 307–325.

EzzElArab, M. (2004). *Ministry of Local Development.* Al-Ahram Center for Political & Strategic Studies.

Fletcher, J. (2011). The "Death in Sakkara" Gallery. BBC UK. http://www.bbc.co.uk/history/ancient/egyptians/death_sakkara_gallery_02.shtml. Date accessed: 2020-07-28

Frihy, O. E. (2001). The necessity of environmental impact assessment (EIA) in implementing coastal projects: lessons learned from the Egyptian Mediterranean Coast. *Ocean & Coastal Management*, 44, 489–516.

Gregory, D. (2001). Colonial nostalgia and cultures of travel: Spaces of constructed visibility in Egypt. In *Consuming tradition, manufacturing heritage: Global norms and urban forms in the age of tourism* (pp. 111–151). Routledge London.

Hunter, R. (2004). Tourism and Empire: The Thomas Cook & Son Enterprise on the Nile, 1868–1914. *Middle Eastern Studies*, 40(5), 28–54. https://doi.org/10.1080/0026320042000265666

Krzywinski, K. (2000). *Deserting the Desert; a Threatened Cultural Landscape between the Nile and the Sea*. Alvheim & Eide Akademisk Forlag.

Lindsay, J. (1965). *Leisure and Pleasure in Roman Egypt*. Fredrick Muller Ltd.

Manchester University EIA Centre. (2000). Evaluation and Future Development of the EIA System in Egypt: A Report Prepared under the METAP EIA Institutional Strengthening Project. In World Bank (Issue December).

Mayfield, J. (1996). *Local government in Egypt: structure, process, and the challenges of reform*. American University Press.

Refaat, S. (1997). History of Motoring in Egypt. Egyptian Gazette. http://www.egy.com/historica/97-03-02.php

Salas, J. C. (2014). Evolution of the Tourism Industry on Guam. http://www.guampedia.com/evolution-of-the-tourism-industry-on-guam–2/

Seif ElNasr, R. (1999). *Concerns of Public Administration: Stories of Governors & Governorates*. Al Arabi Publishing & Distribution.

Shaalan, I. (2005). Sustainable tourism development in the Red Sea of Egypt threats and opportunities. *Journal of Cleaner Production*, 13(2), 83–87. https://doi.org/10.1016/j.jclepro.2003.12.012

Sherbiny, A. H. El, Sherif, A. H., & Hassan, A. N. (2006). Model for Environmental Risk Assessment of Tourism Project Construction on the Egyptian Red Sea Coast. *Journal of Environmental Engineering*, October, 1272–1282.

Sims, D. (2012). *Understanding Cairo: The Logic of A City Out of Control*. American University Press.

Snyder, J. (2003). Guidelines for Ecotourism Development in the Deep Range of the Red Sea Region.

TourismDevelopmentAuthority. (2005). Conditions and Regulations of Land Allocation for Tourism Development in Egypt. Ministry of Tourism.

Towner, J. (1995). What is tourism's history? *Tourism Management*, 16(5), 339–343.

Towner, J., & Wall, G. (1991). *History and Tourism*. Pergamon Press Plc and J. Jafari, 18, 71–84.

Wolf, W. (1996). *Car Mania, A Critical History of Transport*. Pluto Press.

Yüksel, F., Bramwell, B., & Yüksel, A. (2005). Centralized and decentralized tourism governance in Turkey. *Annals of Tourism Research*, 32(4), 859–886. https://doi.org/10.1016/j.annals.2004.09.006

Kristina Svels & Laura Puolamäki

4 Finland

4.1 History and Background

In order to analyze and grasp the width of the Finnish tourism industry we start with laying down some facts about geographical boundaries, the background of the construction of a welfare society and how seasonality shapes tourism. Finland belongs to the Nordic Countries as a part of the geographical area called Fennoscandia. It shares borders with Sweden, Norway, and Russia. Finland became an autonomous part of Russia in 1809 after being part of the Swedish Kingdom from the Middle Ages. During the Russian era the first national organisation of tourism, Turistföreningen i Finland (in Swedish) later renamed to Finland's turistförbund/ Suomen Matkailuliitto (in Swedish/Finnish) was established in 1887.

Since the beginning of the 20th century the country has developed along the Nordic welfare model. As a forerunner nation, women in Finland gained eligibility to the Parliament and the right to vote in 1906. Finland got its independence from Russia on December 6th, 1917. A new political and juridical era began when the country joined European Union (EU) on January 1st, 1995.

Finland has strong cultural, political, and economic ties to all Nordic countries. Due to its historically important linkage to Sweden, the country is constitutionally a bilingual country, guaranteeing equal constitutional rights to the Finnish (87.9%) and Swedish (5.2%) speaking populations (Statistics Finland Population, 2017). The total population in Finland has more than 5 million inhabitants, forming a scarcely populated nation with 16.3 persons/km2. The population in the capital area surrounding Helsinki in the south of Finland exceeds 1 million (Statistics Finland, 2018a).

Finland possesses a total area of 338,448 km2 of which water accounts for 10.2%. While called "The Land of the Thousand Lakes," the actual number of lakes in Finland exceeds the thousand; more precisely estimated there are about 168,000 lakes (over 500 m2) and 187,000 islands (over 100 m2). The Finnish coastline extends a total stretch of 14,018 km shoreline (Statistics Finland, 2018b) and the country has the largest forest areas in Europe, which covers 86% of the country.

Tourism in Finland depends on, and at the same time profits from four distinctly different seasons; winter (December to February, average temperature –4 to –6 °C), spring (March to May, average temperature –2 to +10 °C), summer (June to

https://doi.org/10.1515/9783110638141-004

August, average temperature +14 to +15 °C), and autumn (September to November, average temperature +10 to 0 °C). The country has a temperate climate that, in rare occasions, can reach down to −40 °C and up to +30 °C, also temperatures and timing of seasons may vary between the South and the North (Ilmatieteenlaitos, n.d./a). The daylight is an important imperative for all Nordic countries. Since 25% of Finland's area is above the Arctic Polar Circle, the amount of daylight changes drastically throughout the year. There is at least one day per year the sun does not set or rise above the Arctic Polar Circle. The Nordic light, Aurora Borealis, is seen in the whole of Finland, though with greater likelihood in the more northern parts of the country (Ilmatieteenlaitos, n.d./b) and has become an important tourism attractor during the last years, especially among Asian visitors (Suvanto et al., 2017).

Nature and rural recreational landscapes create the tourism foundation of Finnish tourism. In the late 1800s the Finnish Arctic explorer Erik Adolf Nordenskiöld, inspired by the North American national park movement, enhanced the development towards protection of natural resources and landscapes. There have since then been purposeful protective and governing measures, as in the case of national parks and less imposed managerial control on the natural reserves (Perttula, 2006). In 1938, the first national parks Pallas-Ounastunturi (now Pallas-Yllästunturi) and Pyhätunturi (now Pyhä-Luosto) were established alongside the natural reserves Pisavaara and Malla. In 2019, there are 40 national parks with a total area of 10,022 km2. As in the other Nordic countries, tourism is extendedly based on a public access regime. The legally secured Right of Public Access is strongly rooted in the Nordic tradition of outdoor recreation, allowing free access and movement in nature, and is generally taken for granted among the public (Fredman et al., 2013). This customary right has developed into comprising practically all-natural areas with some restrictions concerning commercialization of (tourism) services on the land of others.

In the next sections, the state's impact on the development and support of the Finnish tourism sector is presented as it stands at the end of 2019. The aim with the forthcoming sections, is to provide an analytical and descriptive understanding of how society can support an industry in development within its societal context, its institutionalization, and its strategic lineups.

4.2 Finland's Governmental Strategy

Finland is an independent parliamentary republic. The government structure lies within the framework of representative democracy, meaning that all citizens over 18 years old have the right to vote in parliamentary elections. The President is head of state and elected every 6th year and the Prime Minister, is head of the government (President of the Republic of Finland; Finnish Government, n.d.). The 200-seat unicameral Parliament is elected every 4th year and exercises legislative power and proclaims its governmental program. This brings new strategic goals and policy visions for the tourism sector at least every fourth year when a new program is implied.

One important implication on the national tourism structure is the Finnish economy, ranked the 4th of knowledge economies in Europe. Tourism belongs to the largest sector of the classified Finnish economy sectors economy, service and administration (72.2%), positioned before the manufacturing and refining cluster (31.4%), and primary production (2.9%). Although tourism belongs to the service sector other industries precede, e.g. manufacturing within the largest industries as electronics (21.6%), machinery, vehicles and other engineered metal products (21.1%), forest industry (13.1%), and chemicals (10.9%). For the tourism industry, Finland draws benefits from having been a member of the EU monetary euro-zone since January 1st, 1999, and the currency used is the Euro (€).

The tourism industry is included within the scope of enterprises among the focus areas of the Ministry of Economic Affairs and Employment (MEAE) where priorities of the Finnish tourism policy are set. Other national focus areas, except for tourism, are energy, competition and consumers, working life, and regions. Growth and network-oriented tourism enterprises with expanding international operations have priority when public subsidies are distributed. Funding is directed at areas providing the tourism industry with stronger operating prerequisites, such as transport connections, improvements in energy-efficiency, and the maintenance of national parks and hiking routes.

The EU membership and yearly payment of membership fees to the union makes Finland eligible for European Funds. Public subsidies for the development of the tourism industry are provided through such multi-funding channels as the European Regional Development Fund (ERDF), European Social Fund (ESF), European Agricultural Fund for Rural Development (EAFRD), European Maritime and Fisheries Fund (EMFF), Business Finland programs and direct grants provided by several Ministries, e.g. Ministry of Agriculture and Forestry, Ministry of Environment and Ministry of Education and Culture.

There are no explicit tourism laws regulating tourism activities in the Finnish legislation. General consumer laws apply to the travel, hospitality, and tourism sectors. Other laws influencing the tourism sector are laws regulating the use of products and services, construction sector, corporate businesses, transportation, taxation, employment, health, and security are all separately constituted in the Finnish law. This applies to all sectors of tourism, for example transport of visitors, construction of accommodation facilities, and other recreational venues. EU legislation, including the directives on free movement within the union as well as on tourism, is applicable in Finland.

When looking at tourism as a part of governmental industry, the process of reshaping the Finnish tourism structure advanced in 2001, while the tourism organization originating from 1887 was transformed into a governmental office for promoting tourism, Matkailun edistämiskeskus (MEK, in Finnish). In 2015, MEK became a part of the Finnish foreign trade office and in 2018 Business Finland was established as a part of the Team Finland network bringing together state-funded internationalization services for domestic and international customers in marketing, business development and funding.

Business Finland's unit for tourism, Visit Finland (VF), operates on a governmental level focusing on destination management and developing Finland's travel image, helps Finnish travel companies to internationalize and to develop and to sell and market high-quality travel products. VF cooperates with travel destination regions, businesses, export promoters, and embassies. It also produces tourism research and statistics and hosts familiarization trips for foreign tour operators and tourism influencers and travel bloggers. Operations by VF are aimed to implement the strategic goals of Finland's tourism strategy "Achieving more together – sustainable growth and renewal in Finnish tourism. Finland's tourism strategy 2019–2028 and action plan 2019–2023" which has been developed and implemented by MEAE. It envisions Finland as the most sustainably growing tourist destination in the Nordic countries. From 2015 to 2018 funding from MEAE and other supportive measures for tourism implementation were channeled through three different growth programs called Finrelax, StopOver Finland, and Finnish Archipelago. These governmental growth programs were designed in central administration, but during the implementation phase operations were channeled through Visit Finland to regions, local level, SMEs, and micro level as development projects. Stakeholder participation was not included in the design phase making the action a top-down tourism development process.

In 2017 the government granted €16 Million for an additional growth program for international growth and renewal called Tourism 4.0 (2018–19), implemented jointly with MEAE and the Ministry of Agriculture and Forestry, Ministry of Environment and Ministry of Education and Culture. It targeted marketing of larger brand regions and themes like Lapland, Lakeland, Archipelago, and Helsinki, nature tourism, digitalization, and digital accessibility. Additionally, seasonality under selected tourism themes of nature, wellbeing, food, culture, and sustainability are targeted.

4.3 Current Tourism Profile

4.3.1 Tourism Volume

According to Visit Finland's Visitor Survey (2018), Finland received 8.3 million foreign visitors in 2017. Of these travelers 5 million were overnight visitors and 3.3 million were day-trippers. A total of 21.9 million overnight stays were registered, of which 31% were foreign visitors. The foreign visitors brought in a total €3 billion to Finland. The multiple effects of tourism on other sectors is €1 to 56 cents. Tourism investments in total were €1.7 billion. Russia was the most important market, while the biggest growth was reported from the Chinese market (+35.3%) (Visit Finland, 2018). In 2019, the growth was 3% among foreign visitors (Visit Finland Visitor Survey, 2019).

4.3.2 Tourism Distribution

The year 2017 was a marker for international tourism in Finland (Visit Finland, 2020). Overnight stays of international visitors increased by 14% (+813,000 stays) of a total of 6.6 million and especially in Lapland the increase was high (22%). The other areas with growth in international tourism were the Helsinki region (+14%), Lakeland region in Central Finland (12%), and the coastal areas and archipelago in the South and West of the country (+8%) (Business Finland, 2018).

One reason for growth in the Helsinki region is increased cruise line traffic, with stops in Helsinki on tours of the Baltic Sea. The number of cruise ships visiting was 303 bringing 603,500 cruise passengers to the capital in 2019 (Port of Helsinki, 2019). Docking is arranged by Port of Helsinki, which is a company owned by the City of Helsinki. Another reason for a noticeable increase in incoming visitors might have been Finland's centennial anniversary of independence in 2017, which gained extensive attention globally.

4.3.3 Tourism as Part of the Economy

The growing tendency of an expansive tourism industry worldwide also shows in Finland. The tourism industry generates a growing share of the total Finnish economy and is predicted to increase during the forthcoming years. Tourism's contribution to the GDP during 2011–2016 has been on a stable level of 2.5%. Tourism consumption in 2016 was €13.8 billion, which equates to €4.6 billion of the GDP. The total of export earnings by international tourism in 2017 was €4.4 billion Foreign travelers accounted for 27%, or €3.66 billion of total tourism consumption (Business Finland, 2019). Several sectors are influenced by the tourism industry, such as building and construction, transport, and commerce. Tourism has become a major economic driver, particularly in remote areas in Eastern and Northern Finland.

The central government is able to directly influence the growth and viability of tourism enterprises by increasing the industry's price-competitiveness, improving the accessibility of tourist destinations, and promoting all-season tourism. Nevertheless, raising the value-added taxes would harm the tourism industry, according to tourism actors, and introducing passenger tax on flights or road, kilometer or congestion charging would hinder the growth of tourism according to the industry representatives (Ministry of Economic Affairs and Employment of Finland, 2018).

Total employment in the tourism sector was 140,000 people in 2016 (of which in F&B 49%, passenger transport 25%, cultural, sports- and recreational industry and travel agencies 15%, and accommodation 11%). This counts for 5.5% of the total employment base in Finland (Business Finland, 2019).

4.4 Shaping Tourism Development

4.4.1 Finland – Made of Nature and Culture Between East and West

Securing income continuity for local sustainability is the key goal in Finnish tourism development. National governance promotes tourism in international markets and offers resources for research, development, capacity building, and certification. These resources are channeled through Business Finland. Tourism in Finland is generally built upon a well-functioning infrastructure where accessibility within, to, and from the country is partly provided by publicly financed operators e.g. airports, roads, railroads, and ports. Finland had belonged to the European Union's Schengen Area since 2001. This mobility-zone functions as a single jurisdiction for international travels and has a common visa policy making the free movement of persons effortless within the zone (Schengen Visa Info, n.d.)

According to the Organization for Economic Co-operation and Development (OECD), the comparative strengths of Finland's tourism lies in the contrasts between its modern culture and its nature-based cultural heritage, the meeting of east and west, technology, the Finnish way of life, and creativity (OECD, 2018). The framework for the Finnish tourism industry is based upon an open, secure, and equal society well-suitable for receiving, catering for, and making a living out of tourism. The country has during the last decades been listed and scored as the top countries in several world rankings, due to its safety, reliability, and non-corruptive qualifications that makes the country attractive and plausible for tourism development and in general mirrors the level of services provided (Toolbox Finland, 2018).

The Finnish tourism industry is complementary to other industrial sectors. A stable and resilient society combined with a greening economy is the basis on which the development across sectors is targeted on a national level. This keeps national and tourism development of main attractions on the same track. The main tourist destination is the capital Helsinki. The tourism dependent region, Lapland, is the second next popular destination, yet vulnerable to international economic trends and unforeseen challenges as extensive investments in resorts and infrastructure have been pursued for the tourism industry. For example, the Santa Claus tourism during winter seasons has been affected by decreased economic situations from the largest visitor markets: the UK and China.

4.4.2 Tourism in Central Government Versus Local Government

The Ministry of Economic Affairs and Employment (MEAE) across other ministries and operators, have the responsibility of the tourism industry which is divided on several levels: national, regional, and local as depicted in the OECD diagram (2018, see Figure 4.1). In regard to tourism governance in Finland, the government creates, upholds, and partially implements national policy, strategies, and funding programs for tourism, gathers statistics, and manages surveys. The Finnish tourism system is a combination of governmental bodies and national, regional, and local tourism organizations. Regional government authorities implement national strategies, develop regional strategies and programs, coordinate, and grant funding. Funding of the tourism sector, development, and marketing is channeled through Centers for Economic Development, Transport, and the Environment (ELY Centers) and Regional Councils. EU funding for tourism is sourced from the European Regional Development Fund, European Social Fund, and European Agricultural Fund for Rural Development.

Regional tourism organizations show different mixtures of public and private participation. The regional destination management organizations focus on marketing, networking, and implementation of regional strategies and work directly with tourists on a local level and through service clusters composed by municipalities, private companies, and authority representation. Municipalities may own attractions, such as recreational areas, islands, or lighthouses, but they hand the commercial activities over to tourism entrepreneurs. World Heritage Sites are managed by steering groups led by a mixture of public and private actors, bringing together various actors from the fields of cultural heritage, tourism, nature conservation, and local governance. Nature parks are managed by the governmental environmental organization Metsähallitus (Metsähallitus, n.d.), which certifies its business partners by co-creating services for visitors in line with their conservation goals. Official Tourism information bureaus are diminishing and being replaced by regional and local initiatives, often run by some NGOs or SMEs. There are local tourism marketing and selling pursued by municipalities and some public funded libraries serve as tourism information points.

4.4.3 Finnish Tourism Economy Based on Freedom of Sharing and Networking

Finland is a high-tech country where 89% of Finns, aged 16 to 89, are daily Internet users and 76% use it several times a day (Statistics Finland, 2018c). Combined with traditions of technology development and innovations, Finland has a strong basis of entrepreneurial skills. It is predicted that 84% of the total Finnish population have a mobile phone and the 4G network is well developed and composes up to 83% of the total national mobile coverage (2017), which makes the use of social

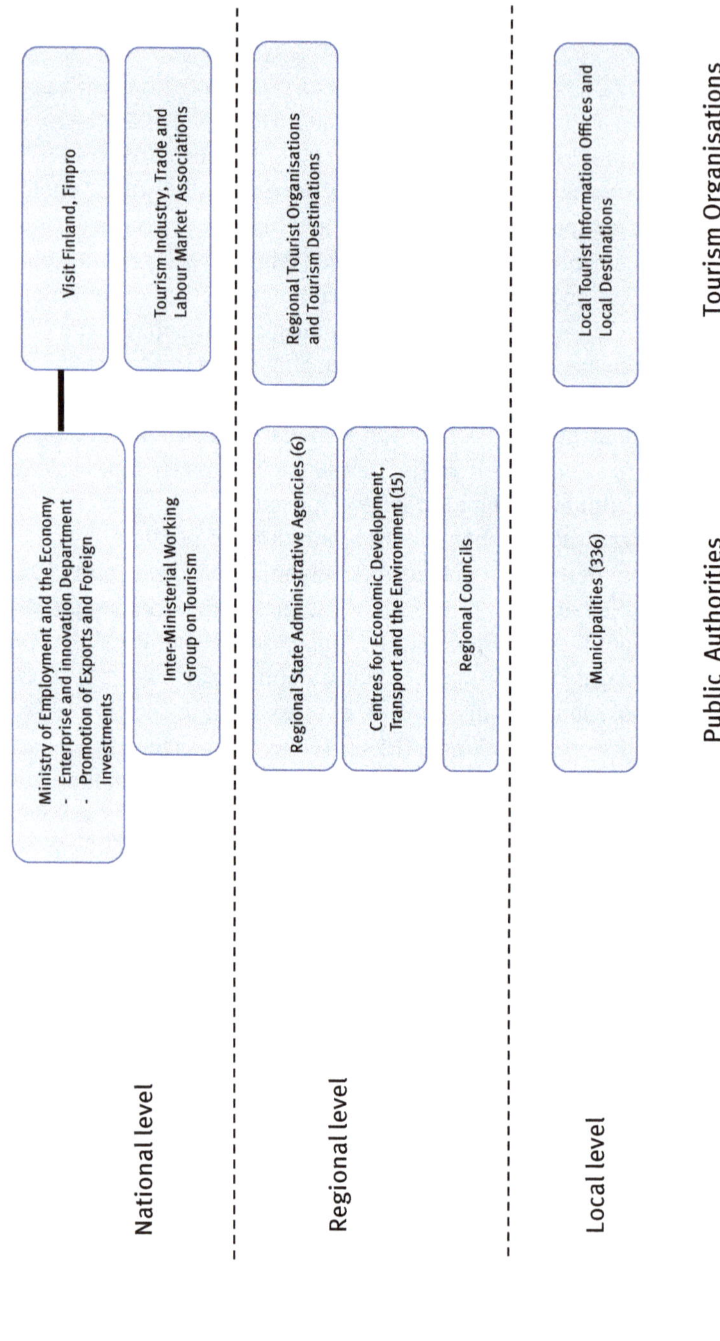

Figure 4.1: Organizational chart of tourism bodies in Finland. (Source: Adapted from OECD diagram).

media in the network economy-based tourism industry easy and efficient. Also, online marketing and booking of Finnish tourism products is one of the key areas to be developed governmentally. In 2017, the Finnish government predicted the Finnish sharing economy to grow 1200% by the year 2020. In 2016 there were approximately 300,000 providers of shared services within tourism; 10% were within the accommodation sector and 2% in transportation. National users of shared services are estimated to 250,000 of which 21% in accommodation and 21% in transportation (Economic Affairs and Employment of Finland, 2017). Besides international sharing-platforms like Airbnb, Wimdu, Couchsurfing, Eatwith, and Uber, new tourism products originally from the Finnish sharing-economy have emerged, see Table 4.1.

Table 4.1: Example of Finnish sharing economy.

Food	Transport	Local activities	Local activities
The Restaurant day concept with pop-up restaurants, established in 2011 as a quarterly food carnival. Anyone can during this day establish a restaurant or cafe, at home or in a public space, can provide food services for remuneration to the public. www.visitnordic.com/en/attraction/restaurant-day	Private boat owners can rent their boats to other private people or enterprises with enough experience from boating on **Skipperi** www.skipperi.com	**Doerz** bridge visitors and local communities together with experience-based tourism products and services through digital and peer marketing. The goal is to produce tourism-based income directly to local communities. www.doerz.com	**Cosy Finland** offers since 2005 home visits for tourists in private homes mainly in the Helsinki area. Its business model intertwines with experiences of local culture and mundane life with local people. Cosy Finland can be seen as a forerunner of network economy actors. www.cosyfinland.com

Selling tourism packages in Finland is based on Finnish legislation (Finlex, 2017) and EU directives making the practice controlled and safe for the buyer, though may discourage smaller entrepreneurs from developing their own services and products on a smaller scale as legislation obliges businesses to set a deposit for securing customer compensations. Some small-scale businesses, which often operate inside the network economy, may find this restricting and even financially hindering.

4.5 Finnish Tourism Dynamics and Realities

Visit Finland manages strategic development of tourism and is supported by tailored funding to marketing activities, and from international to national level markets. National tourism support is provided by Visit Finland Academia that offers detailed business coaching, Quality1000-programme certification of companies, services and products, implementation of the "Roadmap for growth and renewal" strategies and it organizes seminars and conferences. It also monitors various official tourism statistics for tourism development and actualizing of scenarios. Based on the tourism strategy "Roadmap for growth and renewal 2015–2025," Visit Finland presented the tourism analysis report "Know your customers." The groundwork of the previous strategy and report supports Finland's new tourism strategy for 2019–2028, with accompanying action plans for 2019–2023. The aim of the new strategy is for tourism to be developed as a responsible and growing service business that generates welfare and creates jobs across the country focusing on five priorities: supporting sustainable development, responding to the digital transformation, improving accessibility, taking into account the needs of the tourism sector, and ensuring an operating environment that supports competitiveness.

In recent years cooperation with the film industry has become a significant marketing tool especially in Lapland, but also in other regions like the West Coast (West Finland Film Commission, 2018) as Finnish natural resources have become sought after by filmmakers worldwide. A working group of the MEAE published the report "Small risk – large opportunities; a proposal for an incentive system for audio visual production" (Ministry of Education and Culture, 2016, p.18). The working group's proposal suggested providing incentives to increase economic effects of audio-visual production in Finland, a reduction of at most 25% of eligible expenses. One of the productions granted with cash rebate in 2017 was Ailo's Journey, a movie expected to attract international family tourists to Lapland, attracted by natural scenery and wildlife. Teasers from Ailo's Journey were shown on VF's web page to connect the destination images to the media production (see Figure 4.2). Finland has also been utilized for mirroring sets and replicating former Eastern bloc societies in film productions.

4.6 Other Tourism Stakeholders

4.6.1 Non-Profit Tourism Influencers

Non-profit organizations have a strong connection to the democratic society in Finland. Though not directly combined with tourism, there are some exceptions. For the past decades, the owner of Linnanmäki amusement park in Helsinki, the Children's Day Foundation, has been maintaining and developing the park in order to raise funds for Finnish child welfare work. In 2017, the foundation raised a total of €4.5 million (Children's Day Foundation, n.d.).

Another non-profit movement that influences tourism development and innovation is Slush, a student-driven non-profit movement, originally founded to change attitudes toward entrepreneurship. The objective of Slush is to facilitate founder and

Figure 4.2: Incentives for audio visual production in Finland is a part of a national strategy for utilizing nature as a resource in the filmindustry. Ailo's Journey is one of the films produced with the support of the audio-visual incentive system. (Source: Image source MRP Matila Röhr Productions).

investor meetings and to build a world-wide startup community. In 2018, Slush was organized in Tokyo, Shanghai, Singapore, and Helsinki. In 2018, a total of 3,141 start-ups from over 87 countries gathered at Slush in Helsinki where 10,000 pre-booked investor start-up meetings were held and thousands of other encounters took place during the Slush Week (Slush, 2018). This event is therefore seen as promoting Finnish tourism both indirectly and directly.

4.6.2 Transportation by National Corporate Companies and Private Companies

The government in Finland has contributed to tourism development through financing and supporting state-owned enterprises, mainly within transportation and infrastructure. Through the national airline Finnair, a part of the OneWorld alliance and co-owned by the Finnish state, Finland has become a northern hub between Asia and northern Europe. Hence, long-term investments and commitments have been taken by the national airline. Finnair's carriers labelled "The Land of the Midnight Sun," "The Land of the Nordic light," and "The Land of Moomins" are promoting Finland as a travelling destination worldwide.

The privately-owned ferry companies Viking Line and Tallink-Silja Line, among others, carry passengers, goods, and vehicles across the Baltic Sea while promoting tourism with their marketing activities and customer programs. This enables fixed prices on their products and creates road trip packages. In 2017, the busiest ferry line was the route between Helsinki and Tallinn (Estonia), with 48.1% of ferry passengers, while ferry lines between Helsinki and Turku to Stockholm (Sweden) carried 46.1% of passengers.

4.6.3 Finnish Tour Operators and Bloggers

The Association of Finnish Travel Agents (SMAL/AFTA) is a consortium of over 170 travel agencies, tour operators, and incoming agencies that drives its members' interests in relations with public authorities, legislators as well as in the field of domestic and international organizations. Total sales of SMAL/AFTA member companies in 2015 amounted to over €1.9 billion, which is about 95% of the sector's total sales. They have started to organize The Nordic Bloggers' Experience connected to MATKA Nordic Travel Fair. In 2017, there were 71 bloggers writing in 18 languages, from 27 countries with more than 2 million monthly blog readers, 1 million Instagram followers, 1 million Twitter followers, 1.2 million Facebook followers, and 150,000 followers on YouTube (Kaukokaipuu, 2015).

4.6.4 Quality Control Companies and Corporate Social Responsibility (CSR)

Several national and international labels and certificates are in use. Many international labels are tailored for accommodation and products, while ISO 14000, EMAS, and EU Ecolabel are suitable for services as well. The European level program European Tourism Indicator System Toolkit for Sustainable Destinations (ETIS) is designed for destinations. Quality labels and certificates steer sustainable choices of tourism businesses through control programs and audits. The systems simultaneously frame sustainability and quality with the indicators used. Recently it has been noticed that socio-cultural sustainability is especially challenging to measure and monitor, and it is currently touched upon only slightly in quality labels or surveys. Prime Minister's office, with its Government's analysis, assessment, and research activities office, has financed a study for creating a toolbox for steering tourism and socio-cultural sustainability in cultural environments (Veijola et. al. 2020; Prime Minister's Office 2020). A new governmental sustainability label for tourism, Sustainable Travel Finland, has certified the first enterprises and destinations in 2020 (Business Finland, 2020).

Within the Ministry of Economic Affairs and Employment (MEAE) the Committee on Corporate Social Responsibility (CSR) is responsible for national strategies of CSR together with the Ministry of Foreign Affairs (external economic relations, development policy), the Prime Minister's Office (State's ownership policy), and the Ministry of the Environment (sustainable development). Advanced CSR is currently not used in marketing of tourism services in Finland, or in many cases even seen as a marketing advantage. This may be due to the long traditions in compliance with labor legislation, occupational safety and health legislation, and environmental legislation, or history in gender equality.

4.7 Tourism Impact

4.7.1 Ecological Impact

Natural parks are in general easy to access, well-guided, offer considerable information in their visitor centers, and have safe infrastructure. In some places the number of visitors is at times too high, and problems such as uncontrolled widening of paths, insufficient capacity for parking, and loss of quietness and serenity begin to occur. The popularity of nature tourism among domestic and international visitors causes pressure in some of the national parks and other recreational areas. Nevertheless, there are no immediate threats of mass tourism in Finnish nature destinations. The most obvious consequences of increase in nature tourism are overuse of paths and littering. These manifest an increased need for development and construction of maintenance infrastructure.

4.7.2 Socio-Economic Impact

Social and economic impacts due to tourism are seen favorably in Finland (WTTC, 2017). Regional development is pursued in many regions via tourism development, as it introduces new, external sources of monetary income, public funding, and local job opportunities (Saarinen, 2003). In Lapland increase in winter tourism has caused positive social impacts in the form of tourism generated income to local people, improved infrastructure in the most popular areas, better connections between North and South, and vital higher education and research. On the other hand, the hub destination of Lapland in the city of Rovaniemi, has the emerging challenge of rising rents, especially in winter season, when hotel capacity is full and people are keen to find alternative accommodation. This may cause lack of rental apartments of reasonable price for local people, seasonal employees, and students.

4.8 Types of Tourism

Finnish tourism is highly based on interconnected experiences from nature and culture, hence built, commercial attractions are not imposingly visible. The general tourism flow in Finland is privately operated and noticed in everyday surroundings such as in rural and recreational landscapes, cultural settings where nature and culture merge. Hospitality and accommodation establishments would in some historical cases have a central place in the urban centers, though operated by private owners. Tourist attractions managed by government authority affiliation are the national parks, including the natural World Heritage site Kvarken Archipelago (Svels, 2017) and some ancient cultural buildings.

4.8.1 Community-Based Tourism

Community-based tourism in Finland uses living heritage as a resource for reviving and conserving both tangible and living heritage (Haanpää et al., 2018; Puolamäki, 2017). One common type of community based, voluntary, non-profit work is called “talkoot.” In tourism activities “talkoot” is less visible, though appear within local communities especially during the summer months, as activity supporting local history museums and village festivals. This generates profit both for local associations, communities and for municipalities. One example is the St. Olav's waterway, the pilgrimage route from Turku in Finland to Trondheim in Norway that was built and is maintained in cooperation with local communities, municipalities, and NGOs. Community participation is engaged in parishes along the route, especially in the Åland archipelago, between Finland and Sweden in the Baltic Sea. There are several churches from medieval and historic periods because the route is an ancient waterway across the Baltic Sea. Churches are focal points along the route, and services for pilgrims are produced by local people, village societies and enterprises. Another example is the Old Rauma World Heritage Site, where the Old Rauma Association, the resident NGO organizes annual free access events, e.g. open yards, flea markets, concerts (Puolamäki, 2020) and living Christmas calendar in house windows. Most of the houses are privately owned and function as homes or businesses.

4.8.2 Rural Tourism

Finland is a rural country by character (Eurostat, 2017), meaning that 95% of the total area is classified as rural spaces. Since the 1970s “green wave”, rural tourism has become a natural part of the multifunctional rurality. Rural companies have combined other business activities, such as farming, animal husbandry, and forestry, with tourism to stabilize their economy. Rural tourism, including cottage holidays, farm holidays, bed & breakfast lodging, farm visits and group catering, organized activity services, and holiday villages is one of the traditional forms of tourism. Well-being tourism has during the last years become one of the most popular forms of rural tourism in Finland (Tervo-Kankare and Tuohino, 2016).

The vision of Finnish rural tourism for 2020 includes several aspects of sustainability (Ministry of Economic Affairs and Employment of Finland, 2006). Rural tourism enterprises in Finland are often family-owned, micro-sized companies utilizing traditional and local food, culturally valuable surroundings, and serving experiences in the clean Finnish nature. Many entrepreneurs are handed down the rural tourism company from previous generations (Komppula, 2007). There are approximately 4900 rural tourism entrepreneurs in Finland (Niemi & Ahlstedt, 2013). The Finnish Association of Rural Tourism Entrepreneurs (SMMY) was established in 1995 and is a special interest organization for the 180 rural entrepreneurs in the network.

4.8.3 Second Homes and Resorts

Seasonal changes mainly define leisure tourism in Finland, meaning that possibilities for tourism vary due to natural changes in landscape, temperature, and weather (ice/snow or open waters). A challenge for the tourism industry has always been to prolong the tourism seasons. Peak seasons vary geographically; in Lapland the peak season is winter, when in other parts of the country it is the summer season. Traditional second homes, situated by the lakes and the seashore, are also traditional locations for Finns to spend summers at. The constructions vary from traditional modest cabins to refurbished year-long accommodation. In total there are approximately 500,000 second homes in Finland (Czeslaw et al., 2015; Mökkibarometri, 2016).

Holiday resorts and time-sharing accommodation is common in Finland. In Lapland the holiday resorts of Levi and Ylläs are the most popular. Seasonality has been strong in Lapland in general, but Northern resorts have invested in the summer season with new golf courses and other activities, and investments paid off. In South Karelia Region resorts in Lappeenranta offer large accommodation capacity with spas. In this area the holiday seasons in Russia create peaks in visitor numbers, but possibilities for tax-free shopping are also significant for Russian tourists. Exchange rate of the Russian ruble impacts tourism near the border between Finland and Russia.

4.8.4 Sports Year-Round

The main tourism activity during the winter in Finland is skiing (both Nordic and alpine skiing), winter safaris with husky or reindeer, and the search for the Northern lights. During the summer season recreational fishing, camping, hiking, kayaking, rowing, and biking are activities that attract visitors, and in many cases large national audiences. Skiing centers have been seasonally challenged, however, to be able to extend their main product assortment with e.g. MTB, hiking, golf, rowing, and fishing activities during the summer months.

Athletic competitions are a set part of the Finnish national identity. Sports events that create tourism attractions are ice-hockey, Nordic skiing, ski jumping, motorsports, and orienteering. Today an increase in new sport events like triathlon, running/marathon/trail, road biking/MTB can be noticed. Also, atypical Finnish events like Wife Carrying World Championship and Swamp Soccer World Championship attracts both national and international visitors.

4.8.5 Cultural, Historic, and World Heritage Tourism

Cultural tourism, both tangible and intangible forms, are important parts of the Finnish tourism product assortment. The Ministry of Education and Culture governs cultural infrastructure related to tourism, like the Finnish Heritage Agency, museums, World Heritage Sites (WHSs), nomination processes in Finland for inscription to the list of UNESCO's Intangible Cultural Heritage, fields of art and culture and the creative industries. UNESCO WHSs have established a set scene for protection of cultural and natural heritage within Finnish tourism. There are six WHSs designated on cultural values (Fortress of Suomenlinna [1991], Old Rauma [1991], Petäjävesi Old Church [1994], Verla Groundwood and Board Mill [1996], Sammallahdenmäki [1999], and Struve Geodetic Arc [2005]) and one designated on natural merits (Kvarken Archipelago [2006]).

The mandate and responsibility for overseeing the Finnish WHSs are given the Ministry of Education and Culture for the cultural sites and the Ministry of Environment for the natural sites. Finland's WHSs cooperate through their NGO, The Association of World Heritage Sites in Finland (AWHSF). AWHSF, established in 2016, implements the National World Heritage Strategy of the Ministry of Education and Culture with the funding distributed through Matka 4.0 program. With this funding, AWHSF has completed a visitor survey in Finland's WH sites in 2017–2019 (Association of World Heritage Sites in Finland, 2019) and a cultural tourism project for new tourist products for the sites and improved online access to information about Finnish WHSs (Association of World Heritage Sites in Finland, 2018).

4.8.6 Festivals

Year round, but especially during the summer months, festivals are a popular element in Finland. The NGO Finland Festivals works to improve operating conditions for festivals and to influence government policy in its sector by establishing close ties with parties that formulate relevant public policy. The Finland Festival operation also maintains statistics. In 2017, members of Finland Festivals sold 822,593 tickets and counted 2,272,338 visitors with free access events and invited guests. The most popular festivals in Finland in 2017 were Ruisrock (Turku), Ilosaarirock (Joensuu), Pori Jazz (Pori), Helsinki Festival (Helsinki,) and Savonlinna Opera Festival (Savonlinna). There is no statistic of domestic and international visitors in Festivals, but festival tourism is one of the areas where Visit Finland estimates growth internationally (Finland Festivals, n.d.).

4.8.7 Wellness Tourism

Hot springs, minerals, and salty waters are not generic in Finland as in other Nordic countries (e.g. Sweden, Norway, and Iceland). Instead, the leading star of relaxation is the Finnish Sauna, the place for cleaning, health, and relaxation. Today the sauna has an unquestionable place in the Finnish family life, as well as in tourism. Sauna has also become a brand and provides a multitude of tourist attractions around the country. Visit Finland has produced a sauna campaign for Japanese markets, presenting 100 different saunas in Finland (Visit Finland, n.d.). In construction, a recent phenomenon is combining intangible cultural heritage, Finnish modern architecture, and the historical Finnish building material wood with tourism. A forerunner is Löyly in central Helsinki, a combination of Scandinavian wooden architecture and tradition of sauna bathing (Löyly, n.d.).

4.9 Conclusion and Recommendations

As suggested initially, the description of the Finnish tourism sector is viewed from an analytical and descriptive evaluation perspective rather than from a more critical point of departure. The multiple layered Finnish tourism sector provides the assessment an opportunity to reach out and present even smaller, yet important parts of the industry. As times and society change rapidly, so does the needs and postulates for the industry. As a proof, Finnish entrepreneurs have proven to be flexible and adjustable. The portrait provided in this chapter is therefore to be perceived as a still of the time.

Finland is between tourism strategy phases portraying Finland's tourism governance structures merging public and private sectors in service production and accessibility, as well as merging of public economy, the sharing economy, and private business. This makes tourism an interface between public and private, global and local, virtual and real life. Tourism operators from local to national level have started to cooperate with sharing economy-based small actors, embedding their services on web (landing) pages and encouraging innovation of new tourism products with competitions and funds. There is no research data available, but observations suggest the new, agile forms of governance may have enhanced growth in the tourism industry.

A fast-changing society makes a state anchored tourism governance style rigid though stiff. The Finnish governance approach may be perceived as well-anchored due to a profound and covering legislation, though somehow stiff in practice. The multi-governance level of tourism systems is rather top-down steered, nonetheless, accepted by most stakeholders as normative for the industry (Åberg & Svels, 2017) and moving towards a more open and inclusive system. The governmental acceptance of the sharing economy is changing the rules of the game among tourism actors in Finland. Contested and challenged companies as Airbnb and Uber have introduced a new way of organizing tourism services providing fair market shares; yet, new Finnish enterprises have followed and are now a part of the growing sharing-economy. Supported by development of social media the network-based society makes tourism vulnerable, fast, and transparent for all.

In the Nordic countries, distances are a challenge for transportation and visitors often are conveyed by airplane. Finland has actively taken part in international negotiations towards fighting climate change and striving towards a low emission society. The report of IPCC on Global Warming (IPCC, n.d.) inspired the debate about climate change and the need to increase air traffic taxes. The Ministry of Transport and Communications responded with a six-point program aiming for carbon-free transport, notably not referring to passenger tax on flights as one way towards change. Aspirations of growth in tourism and objectives aiming for carbon-free transports are quite complex to combine politically and practically. More likely these objectives interact in consumer discourses, facilitating trends of slow travel and "staycation," and result in new products and services within the sharing economy, to be mainstreamed later in national tourism strategies and marketing activities. The new tourism strategy though sets "Finland as the most sustainably growing tourist destination in the Nordic countries" to its vision by year 2023 (Ministry of Economic Affairs and Employment, 2018).

Tourism governance in Finland is quite agile, effective, and reflective in its own frame, but the lack of cross-sectoral development may cause problems in the coming years. Co-creation of tourism between governance sectors and levels of governance jointly with stakeholders might reveal invisible hindrances for development and growth, produce innovations, and enhance sustainability. It could also harness new kinds of expertise to the tourism industry, if network economy models created and tested in new services will be implemented in governance structures. National tourism strategies and supportive growth programs linked with targeted tourism marketing and research implemented by Visit Finland have been successful. Also, familiarization trips for bloggers and tour operators would be too demanding for individual tourism actors, but joint efforts are within the limits of mid-size and small operators. Enhancing the sharing economy and networking of tourism actors of all sizes in network-based online services fit in well in the modern tourism business based on peer marketing and use of social media. Here, Finnish tourism governance is on the right track into the future.

A clear suggestion to the government is not to decrease funding to the tourism sector. This would supply local businesses resources to actively take part in the local and regional level tourism sector development. One example of concrete government action during the COVID-19 challenges, its attempts looking for means to enhance inbound tourism from within and outside the Schengen area, in order to ease the tourism economy in the most vulnerable parts of the country, Lapland being the number one target area. As of October 1st, 2020, special groups and charter or group tours were planned to be permitted to enter Finland even from high-incidence countries, subject to certain conditions (Finnish Government, 2020). Yet, the unstable state of the pandemic, both nationally and globally, challenged the tourism industry and permanent government regulations even for one season were unable to set, and the country maintained closed during the pandemic.

Enhancing cross-sectoral governance would improve the possibilities for sustainable growth in nature tourism, especially in state-governed areas like national parks. Maintenance budget of Metsähallitus has decreased for several years, and the organization is struggling with sufficient management of waste, firewood, and structures in the most popular parks. Nature tourism is seen as a potential sector of growth but tourism in nature is expensive to sustain. Cutting in public funding to organizations who maintain accessibility to and within protected areas creates a discrepancy between use and maintenance. A similar situation is in WHSs and cultural attractions run by the National Heritage Agency, which has been forced to cut the opening hours in some of the cultural sites, or even close them. Simultaneously, the government is increasing funding for marketing of nature and culture tourism, which are seen as a potential sector for growth. Therefore, a recommendation to tourism funding actors is to keep the public funding and resources to natural- and cultural tourism maintenance elevated and prioritized.

References

Association of World Heritage Sites in Finland. (2019). Visitor Survey project 2017–2019. Retrieved from: https://www.maailmanperinto.fi/en/visitor-survey-project-2017-2019/. Date accessed: 19.11.2019

Association of World Heritage Sites in Finland. (2018). Cultural Tourism project 2018–2019. Retrieved from: https://www.maailmanperinto.fi/en/cultural-tourism-project-2018-2019-2/. Date accessed: 19.11.2019

Business Finland. (2020). Kestävän matkailun puolesta. https://www.businessfinland.fi/suomalaisille-asiakkaille/palvelut/matkailun-edistaminen/vastuullisuus/sustainable-travel-finland#stored. Date accessed: 19.11.2019

Business Finland. (2019). Tourism in Finland stays on record level. https://www.businessfinland.fi/en/whats-new/news/2019/tourism-in-finland-stays-on-record-level. Date accessed: 3.10.2019

Business Finland. (2018). Kaikkien aikojen matkailuvuosi – ennätys kasvu teki Suomesta Pohjois-Euroopan kiinnostavimman matkailumaan. News 15.02.2018. https://www.businessfinland.fi/ajankohtaista/uutiset/2018/kaikkien-aikojen-matkailuvuosi/ Date accessed: 15.11.2019

Czeslaw, A., Vepsäläinen, M., Strandell, A., Hiltunen, M., Pitkänen, K. Hall, M., Rinne, J., Hannonen, O., Paloniemi, R., and Åkerlund, U. (2015). Second home tourism in Finland – Perceptions of citizens and municipalities on the state and development of second home tourism. Reports of the Finnish Environment Institute 22en/2015. Retrieved from: https://helda.helsinki.fi/handle/10138/155090. Date accessed: 15.10.2018

Children's Day Foundation. (n.d.). Retrieved from: https://www.linnanmaki.fi/en/children-s-day-foundation/. Date accessed: 16.12.2018

Country Economic. (2017). Quarterly GDP improves in Finland in third quarter. Retrieved from: https://countryeconomy.com/gdp/finland?year=2017. Date accessed: 7.12.2018

European Commission. (n.d.). European Tourism Indicator System Toolkit for Sustainable Destinations. Retrieved from: https://ec.europa.eu/growth/sectors/tourism/offer/sustain able/indicators_en. Date accessed: 7.12.2018

European Union. (n.d.). Overview Finland. Retrieved from: https://europa.eu/european-union/about-eu/countries/member-countries/finland_en. Date accessed: 7.12.2018

Eurostat. (2017). Eurostat regional yearbook. https://ec.europa.eu/eurostat/documents/3217494/8222062/KS-HA-17-001-EN-N.pdf/eaebe7fa-0c80-45af-ab41-0f806c433763. Date accessed: 10.10.2020

Finavia. (2018). Air traffic passenger statistics. Retrieved from: https://www.finavia.fi/en/about-fi navia/about-air-traffic/traffic-statistics/traffic-statistics-year. Date accessed: 5.12.2018

Finland festivals. (n.d.). Tilastot. Retrieved from: http://www.festivals.fi/tilastot/#.XBdpBSBS82x. Date accessed: 5.12.2018

Finlex (922/2017). Laki matkapalveluyhdistelmien tarjoajan valvonta- ja maksukyvyttömyyssuojamaksusta. 922/2017. Retrieved from: https://www.finlex.fi/fi/laki/alkup/2017/20170922. Date accessed: 15.11.2018

Finnish Government. (2020). Retrieved from: https://valtioneuvosto.fi/en/information-on-coronavi rus/current-restrictions. Date accessed: 22.10.2020

Finnish Government. (n.d.) Retrieved from: https://valtioneuvosto.fi/en/frontpage. Date accessed: 7.12.2018

Fredman, P., Stenseke, M., Sandell, K. and Mossing, A. (Eds.). (2013). Friluftsliv i förändring: resultat från ett forskningsprogram, slutrapport, Naturvårdsverket, report 6547. Retrieved from: http://miun.diva-portal.org/smash/get/diva2:689842/FULLTEXT01.pdf Date accessed: 1.11.2019

Haanpää, R., Puolamäki, L. and Raike, E. (Eds.) (2018). Toolkit for co-creation of cultural heritage databank. In Living with cultural heritage. Sharing experiences and knowledge around the Baltic Sea. Kulttuurituotannon ja maisemantutkimuksen julkaisuja 53, University of Turku. Retrieved from: https://liviheri.files.wordpress.com/2018/04/liviheri_web.pdf. Date accessed:7.12.2018

Haila, K., Rannikko, H., Valtakari, M. and Nyman, J. (2018). AV-tuotantokannustimen väliarviointi. Retrieved from: https://www.businessfinland.fi/globalassets/finnish-customers/news/news/2018/av-tuotantokannustimen_valiarviointi__impact_brief_business_finland.pdf Date accessed: 7.12.2018

Ilmatieteenlaitos (n.d./a). Seasons in Finland. https://en.ilmatieteenlaitos.fi/seasons-in-finland Date accessed:15.12.2018

Ilmatieteenlaitos (n.d./b). Northern Lights. http://en.ilmatieteenlaitos.fi/northern-lights Date accessed: 15.12.2018

IPCC. (n.d.). The Intergovernmental Panel on Climate Change. Global Warming of 1.5 ºC. https://www.ipcc.ch/sr15/ Date accessed: 15.11.2019

Kaukokaipuu. (2015). Nordic Bloggers' Experience – reissuvideo 1/3 https://kaukokaipuumatkablogi.net/nordic-bloggers-experience-reissuvideo-13/

Komppula, R. (2007). Developing rural tourism in Finland through entrepreneurship. In R. Thomas, & M. Augustyn (Eds.) *Tourism in the new Europe. Perspectives on SME policies and practices. Advances in tourism research series*, 123–134. Oxford: Elsevier.

Löyly. (n.d.). https://www.loylyhelsinki.fi/en/ Date accessed: 19.10.2020

Metsähallitus. (n.d.). https://metsa.fi Date accessed: 5.12.2018

Ministry of Economic Affairs and Employment of Finland. (2018). Collaborative Economy in Finland – Current State and Outlook . Publications of the Ministry of Economic Affairs and Employment MEAE reports 9/2017. http://urn.fi/URN:ISBN:978-952-327-196-8

Ministry of Economic Affairs and Employment of Finland. (n.d./a). Focus areas Tourism https://tem.fi/en/tourism Date accessed: 15.11.2019

Ministry of Economic Affairs and Employment of Finland. (n.d./b). Public Subsidies. https://tem.fi/en/public-subsidies-for-developing-tourism Date accessed:10.12.2018

Ministry of Economic Affairs and Employment of Finland. (2019). Achieving more together – sustainable growth and renewal in Finnish tourism: Finland's tourism strategy 2019–2028 and action plan 2019–2023. https://julkaisut.valtioneuvosto.fi/handle/10024/162136. Date accessed: 22.10.2020

Ministry of Economic Affairs and Employment of Finland. (2006). Suomen matkailustrategia vuoteen 2020.

Ministry of Education and Culture. (2016). Small risk – large opportunities; a proposal for an incentive system for audiovisual production. 18/2016. http://urn.fi/URN:ISBN:978-952-263-406-1 Date accessed:

Mökkibarometri. (2016). Saaristoasiain neuvottelukunta. Maa- ja metsäalousministeriö. https://mmm.fi/documents/1410837/1880296/Mokkibarometri+2016/7b69ab48-5859-4b55-8dc2-5514cdfa6000 Date accessed: 15.11.2019

National Parks. (n.d.). History of the Finnish National Parks. https://www.nationalparks.fi/nationalparks Date accessed: 10.12.2018

Niemi, J., and Ahlstedt, J. (2013). Finnish Agriculture and Rural Industries. MTT Taloustutkimus 114a. https://jukuri.luke.fi/bitstream/handle/10024/481072/jul114a_FA2013.pdf?sequence=1&isAllowed=y Date accessed: 4.12.2018

OECD. (2018). OECD Tourism Trends and Policies 2018. Paris: OECD. https://www.oecd-ilibrary.org/economics/oecd-economic-surveys-finland_19990545 Date accessed: 15.11.2019

Perttula, M. (2006). Suomen kansallispuistojärjestelmän kehittyminen 1960–1990-luvuilla ja U.S. National Park Servicen vaikutukset puistojen hoitoon. Metsähallituksen luonnonsuojelujulkaisuja. Sarja A 155. https://julkaisut.metsa.fi/assets/pdf/lp/Asarja/a155.pdf Date accessed: 17.11.2019

Port of Helsinki. (2019n.d.). New record for international cruises in HelsinkiRisteilykaudesta odotetaan värikästä ja huippuvilkasta. News 2811.1004.20198. https://www.portofhelsinki.fi/en/port-helsinki/whats-new/news/new-record-international-cruises-helsinki https://www.portofhelsinki.fi/helsingin-satama/ajankohtaista/uutiset/risteilykaudesta-odotetaan-varikasta-ja-huippuvilkasta Date accessed: 17.1012.202018

President of the Republic of Finland. (n.d.) https://www.presidentti.fi/en/ Date accessed: 15.11.2019

Puolamäki, L. (2017). Sustainability, Heritage and Tourism in The Three Historic Towns. Conference proceedings, Heritage, Tourism and Hospitality International Conference HTHIC 2017, Pori, Finland. http://urn.fi/URN:NBN:fi-fe2017122156012

Puolamäki, L. (2020). We Are Opening the Gate with a Clef. Encounters Between People, Space, and Objects Through Chamber Music. Conference proceedings, Heritage, Tourism and Hospitality International Conference HTHIC 2020.
Saarinen, J. (2003). The Regional Economics of Tourism in Northern Finland: The Socio-economic Implications of Recent Tourism Development and Future Possibilities for Regional Development, *Scandinavian Journal of Hospitality and Tourism*, 3:2, 91–113, DOI: 10.1080/15022250310001927
Schengen Visa Info. (2018n.d.). Application Requirements Finland. Retrieved from: https://www.schengenvisainfo.com/finland-visa/. Date accessed: 1.12.2018
Slush. (2018). https://www.slush.org/why-attend/startups/ Date accessed: 7.12.2018
SMAL/AFTA. (n.d.). http://www.smal.fi/fi/topmenu/In-English Date accessed: 7.12.2018
SMMY. (n.d.). https://www.smmy.fi/in-english/ Date accessed: 10.12.2018
Statistics Finland. (2018a). Kunnat asukasluvun mukaan. http://www.stat.fi/tup/alueonline/y_kunnatasukas.html Date accessed:7.12.2018
Statistics Finland. (2018b). Environment and Natural Resources Geographical data. http://www.stat.fi/tup/suoluk/suoluk_alue_en.html Date accessed: 10.12.2018
Statistics Finland. (2018c). Use of information and communications technology by individuals. http://www.stat.fi/til/sutivi/index_en.html Date accessed: 10.12.2018
Statistics Finland. (2017). Population. http://www.stat.fi/tup/suoluk/suoluk_vaesto_en.html Date accessed: 7.12.2018
Suvanto, H., Sudakova, L., Kattai, K., Grīnberga-Zālīte, G., Bulderberga, Z. (2017). Japanese tourists in Finland, Estonia and Latvia – a literature review. University of Helsinki Ruralia Institute. http://hdl.handle.net/10138/229444. Date accessed: 26.10.2020.
Svels, K. (2017).World Heritage management and tourism development: A study of public involvement and contested ambitions in the World Heritage Kvarken Archipelago. Doctoral thesis, Åbo Akademi University, department of Social policy. http://www.doria.fi/handle/10024/134803
TEM. (n.d.). Roadmap for growth and renewal in Finnish tourism for 2015–2025. https://tem.fi/en/roadmap-for-growth-and-renewal-in-finnish-tourism-in-2015-2025 Date accessed: 10.12.2018
Tervo-Kankare, K. and Tuohino, A. (2016). Defining 'rurality' for rural wellbeing tourism – Halfacree's conceptual triad of the production of rural space in practical-level tourism development in Northern Europe. Nordia Geographical Publications 45: 2, 37–52. https://nordia.journal.fi/article/view/64928
Toolbox Finland. (n.d.). Finland in International Rankings and Comparisons. https://toolbox.finland.fi Date accessed: 7.12.2018
Trading Economics. (2018). Finland Unemployment Rate https://tradingeconomics.com/finland/unemployment-rate Date accessed: 7.12.2018
Veijola, S. and Kyyrö, K. (2020). Multidisciplinary Measurement Methods for Sustainable Growth of Tourism in Cultural Environments. Publications of the Government's analysis, assessment and research activities 2020:26. Prime Minister's Office. http://urn.fi/URN:ISBN:978-952-287-934-9 Date accessed: 6.11.2020.
Visit Finland. (2020). Towards responsible travel industry. https://www.businessfinland.fi/en/do-business-with-finland/visit-finland/sustainable-travel-finland-label/ Date accessed: 6.11.2020.
Visit Finland. (2018). Visitors' survey 2017. HYPERLINK "https://protect2.fireeye.com/v1/url=k=31323334-501d0a38-31357b2d-454441504e31-5955b2e8df5fe018&q=1&e=9a4fdec5-cefb-4506-a84e-d853b42db189&u=https%3A%2F%2Fwww.businessfinland.fi%2Fglobalassets%2Fjulkaisut%2Fvisit-finland%2Ftutkimukset%2F2018%2Ft2018-visit-finland-visitor-survey-

2017.pdf" https://www.businessfinland.fi/globalassets/julkaisut/visit-finland/tutkimukset/2018/t2018-visit-finland-visitor-survey-2017.pdf. Date accessed: 3.12.2018

Visit Finland. (2020). Finnish tourism year 2017 – key facts and figures. HYPERLINK "https://protect2.fireeye.com/v1/url?k=31323334-501d0a38-31357b2d-454441504e31-92f165356a688504&q=1&e=9a4fdec5-cefb-4506-a84e-d853b42db189&u=https%3A%2F%2Fwww.businessfinland.fi%2F4a711e%2Fglobalassets%2Fjulkaisut%2Fvisit-finland%2Ftutkimukset%2F2020%2Fmatkailutilinpito_raportti_2017_2018" https://www.businessfinland.fi/4a711e/globalassets/julkaisut/visit-finland/tutkimukset/2020/matkailutilinpito_raportti_2017_2018.pdf. Date accessed: 1.3.2020

Visit Finland. (n.d.). Join the Finns in the Sauna. https://www.visitfinland.com/article/join-the-finns-in-the-sauna/ Date accessed: 5.12.2018

West Finland Film Commission. (2018). https://wffc.fi/

Wikipedia. (n.d.). List of political parties in Finland. https://en.wikipedia.org/wiki/List_of_political_parties_in_Finland Date accessed: 5.12.2018

World Economic Forum. (2016). Networked Readiness Index. http://reports.weforum.org/global-information-technology-report-2016/networked-readiness-index/ Date accessed: 3.12.2018

WTTC. (2017). Finland 2017 Annual research: Key facts. https://wttc.org/Research/Economic-Impact Date accessed: 10.12.2018

Åberg, K.G., and Svels, K. (2017). Destination development in Ostrobothnia: great expectations of less involvement. *Scandinavian Journal of Hospitality and Tourism*, 17, 1–17. http://dx.doi.org/10.1080/15022250.2017.1312076

Bertram M. Gordon

5 France

5.1 History and Background

5.1.1 Early History of Tourism in France

France has been a center for tourism since the creation of the modern French state in the early modern period and, in some ways, even before. Caves at Chauvet and Lascaux drew migratory visitors, some gazing undoubtedly with tourist curiosity, for religious pilgrimages and the paintings on their walls. Re-discovered in 1940, the Lascaux paintings are approximately 15,000 years old. During the Middle Ages, Christian pilgrims on the road to Santiago de Compostela trekked across much of southwestern France, stopping at inns and surely gazing at sites along the way. Most significant for France was the publication in 1552 of the Guide des Chemins de France, which also discussed sites to visit and, in many ways, marks the beginning of French travel publications. As other Renaissance guidebooks, it followed the model of the Description of Greece, written in the 2nd century A.D. by the Greek traveler Pausanias. The first French road guide, the Guide outlined 283 itineraries in France and described local foods in a manner that Antoni Maczak later called "the prototypes of dishes later given star ratings in the Michelin guides" (Maczak, 1995, p. 25).

The emergence of the modern French state with the construction of better roads and canals together with improved domestic security during the 17th and 18th centuries, especially after the end of the religious wars in 1589, facilitated increased travel. Among the privileged who wrote accounts of their journeys, were François Rabelais (1494–1553) and Michel de Montaigne (1553–1592) (Montaigne, 1983). Following the end of the religious wars with the accession of Henry IV in 1589, the new king and the Duc de Sully, named high commissioner of highways and public works, began planning to modernize French roads, a process put into effect under Louis XIV (Gerbod, 2002). State intervention in the economy actively pursued by Louis XIV's Finance Minister, Jean-Baptiste Colbert, and since associated with him as Colbertisme, set the model for later government involvement in the area of tourism.

5.1.2 The Emergence of France as a Tourist Destination

The development of a vocabulary of tourism from the late 17th through the early 19th centuries was capped in France by Stendhal's Mémoires d'un Touriste, published in 1837–1838. In addition to publicizing his travels through France en route to Italy, this book helped establish the usage of the word "touriste," that had appeared in French in the early 19th century (Trésor de la Langue Française). The

https://doi.org/10.1515/9783110638141-005

importance of tourism in France was recognized as early as 1827 with the establishment of an information office for travelers by the Journal des voyageurs et des étrangers in Paris. It offered information about hotels, public roads, and hotels, among others (Venayre, 2012). An early example of governmental intervention to promote tourism, in this case gastronomic, came in 1855, when the Bordeaux Chamber of Commerce established a classification and rating system of place names for local wines, the appellation contrôlée, to be used in the Paris World's Fair of that year. What Eric Hobsbawm (1983) referenced as the "invention of tradition" contributed to France as a tourist destination.

5.1.3 State Promotion of Tourism in France

In 1910, a state Office National du Tourisme was established under the Minister of Public Works to promote it. This was the first national administrative agency created with its own budget to deal with issues of tourism. State sanctioned paid holidays (congés payés) in 1936 stimulated domestic tourism as did the developing youth hostel movement (Auberges de jeunesse), which emphasized touring in and becoming acquainted with the various regions of France. State involvement in the administration of tourism was extended during World War II with the establishment by the regime of Marshal Pétain of governmental regional committees for tourism throughout the country (Gordon 2018). Tourism, both domestic and foreign, has continued to grow during the second half of the 20th century and into the 21st (Mesplier 1986).

Following a brief look at France's long history of governmental involvement with the economy dating back to pre-revolutionary times, this chapter examines the relationships between the French state, a highly centralized republic with a strong executive, and the tourism industry in contemporary France. It addresses the Code de tourisme, enacted in 2006 and revised since, by which the state regulates many activities related to tourism, then turns to the role of central and local governments in administering the Eiffel Tower, as well as local government action by the city of Lille to have that city designated a "European Cultural Capital." Also discussed is Atout France, the state tourism agency, which publishes annual studies of tourism trends and makes suggestions for the development of tourism in France itself. The chapter then addresses government and private partnerships in administering Disneyland Paris and concludes with concerns for security following the November 2015 attacks and the 2019 Yellow Vest violence leading to an even greater state role in tourism in France.

5.2 The Parameters of State Involvement with Tourism

5.2.1 Tourism and the Public Domain in France

Historically, not all aspects of life were open to public curiosity, particularly in the form of tourism. Restrictions invariably limit official definitions of tourism to the public domain incorporating the state, church, and public landscape. State symbols include sites such as the Château de Versailles and the Arc de Triomphe and are generally limited to museums and other legally established monuments dedicated to the propagation of an official cult. Religious symbols function in a similar manner. They include the churches themselves, such as Nôtre-Dame in Paris, as well as marriages and often religious rituals, and usually reflect a combination of church and state values. Public landscape sites involve forests and trees, seashores and beaches, and lakes. These sites are normally managed, often as national or regional state parks. As elsewhere, publicly sanctioned tourism in France focuses on sites of state, church, and public landscape and it is in this sense that tourism should be understood when examining its governance. The French government on both national and local levels continues to play a pivotal role in the organization of tourism in the country. In an article on the state and the development of Disneyland Paris, Anne-Marie d'Hauteserre (2001) noted that the central government's role focused on three facets: regulation, financial support, and land development. She wrote:

> Exterior and Interior ministries define trade and commerce regulations. The Social Affairs and Work ministry determines employment conditions. Rural tourism is dependent on the Agriculture ministry. Major infrastructural investments for tourism projects are often made as parts of National Plans by ministries other than the tourism one. In the Xth Plan, the Ministry of Tourism was responsible for only 27% of all funds allocated by the state to tourism. (p. 126)

d'Hauteserre's analysis was based largely on the role of the state with respect to the Eiffel Tower and although there have been occasional changes in the functioning of the various governmental offices, the state continues to be very much involved in French tourism. Its promotion of tourism in France, if anything, has intensified in recent years. A meeting of the Interministerial Council for Tourism [Conseil interministériel du tourisme (CIT)] in January 2018 announced a policy designed to advance tourism to France in two major ways, first by promoting tourism to France from abroad and, second, by investment in sites in France to make them readily accessible to visitors. Actions to promote tourism from abroad included faster delivery of visas, shortening the lines at airports, and improving the roads in France. French gastronomy, placed on the UNESCO Heritage List in 2010, was now promoted by the staging of "Good France/Goût de France," an annual event to bring chefs from France to demonstrate their skills in many countries throughout the world in the expectation that this would attract more visitors to France. Regarding investments, the CIT also facilitated loans to the hotel industry, as well as helping coordinate local security agencies to better protect tourists (Conseil interministériel du tourisme (2018).

5.2.2 Measuring Tourism in France

In 1970, France ranked third in the number of tourist visits measured by the United Nations World Tourism Organization (UNWTO). The French Revolution Bicentennial celebrations helped move France into first place in the UNWTO statistics, where it has remained since 1988. France's tourism figures, however, are undoubtedly inflated in that many visitors counted as tourists in France are travellers from northern Europe on their way to Spain or Italy. These visitors stop again in France on their way home so, consequently, they are counted twice or more but their stays are often shorter than in other countries. In tourism income, France has ranked consistently below several other countries, in fifth place in 2016, trailing the United States, Spain, Thailand, and China, according to the UNWTO (2017) statistics. Reductions in tourism income totalled 22.9% between 2014 and 2015 and an additional 5.3% from 2015 through 2016. The declines appear to have been related to terrorist attacks in 2015, which slowed tourism growth especially in Paris. The attacks of 2015, followed by more in 2016, meant that as of late 2017, terrorism had become the primary preoccupation, passing unemployment, for people in general in France ("Le terrorisme" 2017). UNWTO figures for 2016 showed France continuing to attract the largest number of international visitors but with a slight decline from the previous year. While helpful in drawing a picture of international tourism generally, these figures do not reflect domestic tourists in a given country.

5.2.3 Paris as a Tourism Magnet

Tourism has become an important segment of the French economy, referenced in one study as the largest economic sector in France, creating some 1.3 million jobs, many held by young people. It produces more income than the automobile industry or agriculture in France (Barnu and Hamouche 2014). The city of Paris is a major tourist attraction. The Eiffel Tower, for instance, built as part of the 1889 International Exposition staged to celebrate the centennial of the French Revolution, remains a leading urban tourist site, as is the Palais de Chaillot in Trocadéro, built for the 1937 International Exposition. A study of France's 43 most visited destinations by both French and international visitors in 2006 gathered by the ONT (Office National du Tourisme) listed the Notre-Dame Cathedral in Paris in first place with some 13.5 million visitors, followed by the Fontainebleau forest with 13 million,

Disneyland Paris with 12.8 million, the Saint-Ouen flea market with 11 million, and the Sacré-Cœur Basilica in Montmartre with 10.5 million (Sites touristiques, 2011).

More recently, the Musée du Quai Branly was opened in 2006 and it, too, attracts visitors (Palmares 2009). The Paris region leads the world in the number of international congresses. Its main drawing card is its integration in international networks of exchanges and contacts, an integration strengthened by the local articulation of air links with rapid trains moving passengers both within France and across Europe, as well as the Réseau Express Régional (the RER), an extensive network of suburban trains linking destinations within the metropolitan area. State direction and support in the construction of the rail networks made Paris a major center for transport and tourism (d'Hauteserre 2001, 147).

5.3 France: A Highly Centralized State

5.3.1 A Republic with a Strong Executive

As of 2016, France ranked sixth in the world in the value of its GDP and tourism represented seven per cent of it. France is a republic, with a president and legislature elected every five years. Historically, the French state has been relatively centralized dating back to the reign of Louis XIV if not before. Recently, efforts have been made to give more autonomy to regional administrations, especially in regard to tourist sites. In addition, the French government is subject to European Union regulations, including open borders, which also serves to promote tourism. The role of the central government, however, remains key. In the words of d'Hauteserre (2001):

> France has a unitary, centralized government, which decides tourism policy at the national level, even if efforts have been made to decentralize government activity. Implementation of French public policy is highly complex in that it involves many actors, different levels of action and authority, and many types of policy instruments. Much tourism development is subject to regulatory bodies created independently of tourism concerns, such as aviation and transport, labor relations, hotel construction and management. (p. 146)

The Fifth Republic, in place since 1958, is a system with a strong central government, headed by a powerful executive, the President, elected by universal suffrage since 1962. Presidents are elected for five-year terms. Emmanuel Macron, the current President, was elected in May 2017. His political party, La République En Marche!, is relatively new, having been created in 2016 and is generally seen as pro-European with a focus on "modernizing" France.

5.3.2 The Macron Government's Tourism Initiatives

Reflecting the longstanding practice of significant government involvement in tourism, on July 26th, 2017, Édouard Philippe, newly appointed Prime Minister, announced a desire to increase the numbers of visitors to France from the then 83 million to 100 million by 2020. As mentioned above, by actions of the CIT, waiting times in France's airports were to be reduced, as was the time needed to acquire visas from countries including Russia, one of the largest sources of visitors to France. Philippe's goals were on track with increases in tourism to France in 2018 and 2019 but were sidetracked by the COVID-19 pandemic of 2020, whose long-term effects remain to be seen (Rose 2020). His initiative followed a similar move by the state, which in 2014 initiated a program to make visas available within 48 hours in China. A study, released on April 8th, 2016, indicated that in response to the measures taken to ease visa requirements for Chinese and other Asians, the proportion of visitors from Asia had increased by 22.7% from the previous year. Included in the measure easing visa requirements were provisions for the defense and protection of French brands in international websites, assuring access to the Internet in tourist zones, and the training of tourist professionals in effective computer use. The 2014 program appears to have helped increase the numbers of Chinese tourists in France. Philippe also proposed in 2017 increasing the number of apprenticeship contracts in coordination with those in the travel industry ("Objectif: 100 millions de tourists," 2017).

5.3.3 The State and the 2006 Code du Tourisme

The complexity together with the many ties between the state at various levels and tourism in France makes it frequently difficult to clearly demarcate the public from the private sector. Governmental agencies in France have recognized the importance of developing and encouraging tourism to enhance the national economy but also as a factor in protecting the environment, as well as the rights of consumers, and in preserving and enhancing France's cultural heritage (patrimoine). Analogous to the Napoleonic Law Code of the early nineteenth century, a Code du Tourisme was created by the state in 2006 putting into one place a wide variety of government dispositions and regulations concerning tourism (Py, 2007). The Code is continually revised and was updated most recently in January 2018. It codifies regulations concerning the "general organization of tourism," (see Figure 5.1 below) the business and professional aspects of tourism, and tourism facilities such as restaurants, hotels, and campsites." It also sets fiscal regulations regarding tourism as well as government support in promoting vacation benefits, together with spelling out the functions of the national government, the region, the department (département or county), and the commune ("Code du tourisme," 2018).

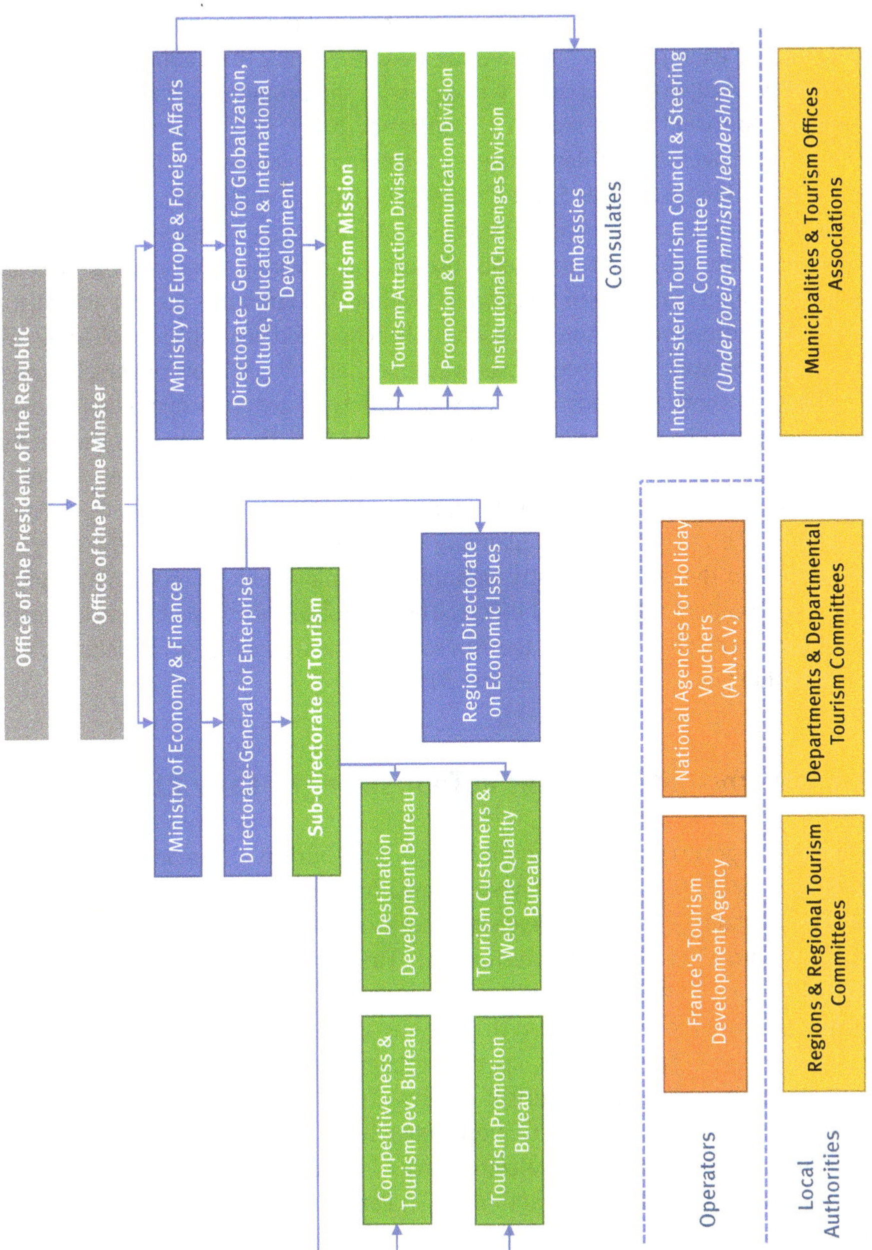

Figure 5.1: France: Organizational chart of tourism bodies. (Source: OECD, adapted from the Directorate-General for Enterprise, 2018).

The Code spells out in detail the spheres of the different governmental levels, specifying, for example that "a commune may, following the deliberations of the municipal council, create an agency tasked with the promotion of tourism, called an office of tourism" (Sous-section 1, Legifrance 2018). Special provisions within the Code also cover French dependencies including Guadeloupe, Guyana, Martinique, and Réunion, as well as Saint-Pierre-et-Miquelon and Mayotte. The Code also governs licensing of tour operators, as well as a cluster of regulations concerning the renting of facilities on a part-time basis, presumably covering those renting space to tourists and others via Airbnb (Section 8: Contrat de jouissance, Legifrance 2018). Regulations addressing museums, seashores, beaches, and boating are spelled out as are the rights of workers to receive paid vacations, and the controls of vehicles used for tourism. In short, the Code covers a wide variety of activities related to tourism that are governed by state agencies, national and local. These activities frequently cross into the domain of government branches not specifically designated for tourism, such as the ministries of Labor, Interior, Foreign Affairs, and others. As of 2014, for example, the government was planning to help train an additional 100,000 persons for jobs in the tourist industry through its employment centers (Pôle emploi) and in collaboration with regional offices. The goal of the government was to create an additional 500,000 jobs in the tourism sector to capture 5% of an anticipated additional billion tourists between then and 2030.

5.3.4 Government Initiatives in Tourism in the Early 21st Century

Analyzing the governmental structure in regard to tourism at the beginning of the 21st century, Py (2007) noted three types of initiatives for its development: the competition among state agencies, the diversity among the kinds of state involvement, and the putting into place of relevant institutions. The Ministry of Tourism was charged in 1974 with the promotion of tourism in France and the coordination of the work of all other ministerial departments in this sphere. There was a significant increase in the powers of local governments, specifically departments and communes, which had had little involvement in tourism matters. The significant increase in tourism from the 1950s on, however, led to more extensive local government involvement, heightened under a state decentralization law of March 2nd, 1982, by which funds were transferred from the central authorities in Paris to the local administrations. On the local level, the communes have been granted the authority to create "zones" for tourist and leisure activities, especially significant in areas of nature tourism such as mountains and seashores. The departments were now authorized to spend funds allotted to them by the national government for infrastructural projects including the modernization of hotels in the countryside, together with the setting up of hiking trails. The regional authorities, administering larger areas, have the authority to manage tourism infrastructure with funds transferred from the national government, which may be used to help set up regional nature parks, canals, and water ports. As Py (2007) points out, however, the local government spheres of activity are all part of a larger government agenda in which the specifics of tourism are rarely, if ever, mentioned. Whereas provisions for local government roles in the administration of tourism are written into the legislation, they invariably appear in the context of larger economic policies, rarely is tourism singled out for special attention.

5.3.5 Local Government and Regional Initiatives to Promote Tourism

On the local level, Comités régionaux du Tourisme (regional tourism committees), created in 1942 and reorganized by a 1987 law, work through studies, planning, management, and infrastructure to promote tourism abroad to their regions, sometimes in collaboration with other regional committees. The SNCF (French National Railway System) was to begin a renovation project in 2014, with the Eurostar section to be renovated beginning in 2015. Paris city authorities were studying plans for re-routing traffic around the Gare du Nord, a leading railway station in Europe, comparable only to Saint-Pancras in London.

5.4 The Eiffel Tower

5.4.1 Local Government and Tourism in France: The Eiffel Tower

An example of the tourism role of local governments is the administration of the Eiffel Tower, one of the best-known tourism landmarks in the world and owned by Paris, its locality, as both a city and a department. The Council of Paris (Conseil de Paris) is the deliberative body responsible for the governing of Paris. It possesses both the powers of a Paris Municipal Council (Conseil municipal) and those of a General Council (Departmental Council) for the Département de Paris, as defined by the so-called PLM Law (Loi PLM) of 1982 that redefined the governance of Paris, Lyon, and Marseilles. Paris is, in effect, the only territorial collectivity in France to be both a commune (commune or municipality) and a department and this arrangement has been a fact even longer, since the passage of the law of July 10th, 1964, which totally reorganized the Paris region. The mayor of Paris presides over the Council of Paris and therefore holds in her hands the powers of mayor and of president of the departmental council. The Eiffel Tower is often indicated as France's most visited destination after Disneyland Paris. French statistical publications show the Tower as consistently a leading attraction there, listed as the most popularly visited "cultural" attraction in the Annuaire Statistique de la France in 1952 and vying with the Château de Versailles as the most popular cultural site through the following years. By the second decade of the 21st century, the Tower, now classified in a more general listing of sites, was in fourth place with some 5.9 million visitors in 2016 (Gordon 2011; 7e Partie, "Les sites touristiques en France," 2017).

5.4.2 Administering the Eiffel Tower: SETE (Société d'Exploitation de la tour Eiffel)

The Tower is owned by Paris jointly as a department and a city. It is administered by the SETE (Société d'Exploitation de la tour Eiffel) a local public organization, on behalf of both the City of Paris, which owns 60% of the Tower, and the Department of Paris, the owner of the other 40%. Created in 2005, the SETE is tasked with keeping the Tower's installations up-to-date, and maintaining the welcoming spaces and services for visitors, while at the same time also maintaining "a high level of security." It employs some 340 workers and with more than 6 million visitors each year, earned €77.8 million in 2016. Altogether, nearly 600 people keep the Tower open and accessible each day with half employed by the SETE and the other half by various concessionaries and tenants (Société d'Exploitation de la Tour Eiffel, 2018).

5.4.3 Concessionaires at the Eiffel Tower

The concessionaires include SCSC, a branch of Lagardère Travel Retail, which manages nine souvenir stands at the Tower; a joint Sodexo-Alain Ducasse company that administers three buffets and two restaurants, the "58 Tour Eiffel" on the Tower's first level and the "Jules Verne" on the second. The Wika Company, which makes precision measuring instruments, is also located in the Tower, whose long-distance views are useful in making its products. Sub-tenants (sous-occupant) include public agencies such as the police, fire fighters, television and radio broadcasting company, the Institute for Nuclear Safety, and Météo (Weather) France. A company, Airparif, which monitors the air quality in France, is also located in the Tower (Société d'Exploitation de la Tour Eiffel, 2018).

5.4.4 Renewal of the Contract with SETE: Administering the Eiffel Tower Today

On November 1st, 2017, the City of Paris renewed its contract with the SETE for another fifteen years. SETE's mission, elaborated in its website is "to put the visitor at the heart of its preoccupations" along three basic axes:

1) Improving the visitor's experience, through the sales of tickets online, the maintenance of the gardens around the Tower and the maintenance and improvement of the Tower's infrastructure, including making Wi-Fi available to visitors.
2) Maintaining the security and accessibility of the Tower, redoing the second level, and painting and renewing its lights that sparkle at night and can be seen from almost anywhere in Paris.
3) Reinforcing and spreading the renown of the Tower by participating in events such as the Olympic Games and Universal Expo scheduled for Paris, and organizing its own events as well as maintaining and enhancing the presence of the Tower on the Internet (Société d'Exploitation de la Tour Eiffel, 2018).

Since 2008, the SETE has also administered the observatory of the clientele, enabling it to better gauge visitor satisfaction with their experiences at the Tower. Its findings in 2014 indicated that 77% of the visitors were generally satisfied with their experiences at the Tower with 92.5% indicating that they would recommend a visit to others (Société d'Exploitation de la Tour Eiffel, 2018).

5.4.5 National and Local Governments: Divisions of Responsibilities for Tourism in France

In general, it is the national government that is charged with establishing professional regulations to protect consumers as well as tourism professionals, including the classification of hotels and inns and the laws relating to the organization of travel and vacation stays. The communes determine land use regarding tourism as part of their urban planning, codified in their more general Plans locaux d'urbanisme (PLU, or local urban plans). They take charge of campgrounds, caravanning, and leisure colonies, while local mayors are responsible for police services. Commune authorities are also responsible for information services for tourists, whether they plan to travel in France or abroad; helping to stage fairs and salons; promoting tourism among journalists both French and international; and putting reservation services into place when needed. In general, the state authorities cooperate to manage tourism facilities and collect statistics related to tourism. Statistics gathered by the national government are disseminated to those in the industry. National and regional governments advertise to promote tourism around specific themes. Contrats de destination (destination contracts) bring everyone, public and private, in the tourism sector together around a brand to better promote a "new international clientele" ("Tourisme" 2016).

5.4.6 "Destination France 2010–2020"

These contracts, generated by Atout France with the support of the French government, are part of the plan called "Destination France 2010–2020" to increase the visibility of France's tourism sites in the international marketplace. The key actors for a locale join together in a "Destination France" contract to market their site by improving local transport, hotel accommodations, restaurants, and cultural and leisure activities over a period of several years. The group works together to target specific international markets. Those who sign these contracts are generally local authorities in charge of economic development together with representatives of their tourism organizations, tourism offices, and transport officials, all of whom are placed in contact with the targeted markets ("Contrats de destinations" 2018). Once again, the tourism infrastructure and marketing are underpinned with funds and expertise from the government. The national and communal authorities also work together with the French National Railway System and through the Caisse des dépots et consignations (Deposits and Consignments Fund), an investment agency of the French state under parliamentary control that has invested funds in companies such as the Accor Hotel chain and Club Méditerranée (Py 2007; "Summary" 2018).

5.4.7 The Secretariat of State for Foreign Trade, Promotion of Tourism, and of French Abroad; and "Atout France"

An elaborate and frequently shifting network of national state agencies involved with the promotion and management of tourism includes the current national cabinet office in charge of tourism, the Secretariat of State for Foreign Trade, Promotion of Tourism, and of French Abroad (Secrétariat d'État chargé du Commerce extérieur, de la promotion du Tourisme et des Français de l'étranger), continuing the pattern of combining responsibilities for tourism with other state functions. ("Renforcer l'attractivité" 2018; Boudet, "Gouvernement Philippe" (2017). This position, lower than cabinet minister, combines the supervision of tourism development with other functions, working with international trade in general seeking to expand French exports. Specific to tourism, the Secretariat of State supervises Atout France, which was created by a merger of the Maison de la France (House of France), the promotional agency for foreign travel to France, and ODIT France (Observation, développement et ingénierie touristiques France; Tourism Observation, Development, and Engineering France), the tourism administrative office. As of March 2018, Atout France was the sole national organization for the tourism sector funded by the Ministry of Tourism ("The Essential Guide" 2018). The Maison de la France included a network of offices in 29 countries in 2007 (Py 2007).

5.4.8 "Atout France" and Shifting Government Organizations

Arguing for state policies more directly focused on the promotion of tourism, Julien Barnu and Amine Hamouche pointed out that during the 40-year period from 1974 through 2013, 10 different cabinet and subcabinet-level ministries were assigned to manage tourism affairs (Barnu and Hamouche 2014). Indeed, the creation of Atout France is the product of a long series of shifting government organizations involved with the promotion and management of tourism in France. Legislation passed during the 1990s gave more authority to the local administrative units, notably the communes, now allowed to manage welcoming services and tourist offices, often jointly with town municipal councils, granted the authority to set up local tourism offices by a law of 1992. These offices are administered by a variety of people with differing stakes in the tourism trade. Communes that attract large numbers of tourists, for example to their seashores, mountains, or spas, but with fewer than 2,000 inhabitants may now receive extra funds from the state. Mountain communes are now allowed to tax ski lifts and to derive royalty fees for cross-country ski trails. Those with casinos may tax the games and receive back taxes from the state (Py 2007).

5.4.9 Tourism Promotion: The Case of Lille

Among the examples of local government action to promote tourism is the program undertaken by the city of Lille in northern France to make its residents more tourist-friendly. A program was begun there in 1999 to turn local residents into "ambassadors of their city." Martine Aubry, the Mayor, argued that the city's citizens needed to be encouraged to be proud of their city and to regard those coming there as "visitors" rather than "tourists." In 2004 amid celebratory events the city was named a "European Cultural Capital." The events of 2004 drew some 9 million visitors and 17,000 "ambassadors" (Barnu & Hamouche 2014; "Histoire de Lille" 2018).

5.4.10 Government and Private Initiatives in Tourism

In terms of shaping tourism development, the impact of the public as opposed to the private sector has been mixed. A decree of October 1st, 2014, under the umbrella of promoting French exports abroad, gave to the Secretary of State in Charge of Foreign Commerce (secrétaire d'Etat chargé du commerce extérieur) the task of promoting tourism to France in other countries (Legifrance 2014). A conference held on December 17th, 2014, included both government and industry representatives and focused on using computer technology in the hotel industry. Emphasized were improvements in the tourism website and the development of digital tourism applications.

5.4.11 Government and Private Initiatives in Tourism: Collaborating Rather than Competing

Additional cooperation between the state and private interests occurs in the collection and dissemination of tourism statistics. The central and regional administrations advertise to promote tourism to France around specific themes. Contrats de destination (destination contracts) bring everyone, public and private, in the tourism sector together around a brand to better promote a "new international clientele." Examples include heritage, as in the case of Mont Saint-Michel and its bay, wine and gastronomic tourism in Burgundy and Bordeaux, mountains such as the Alps and the Pyrenees, and sport and relaxation as in golf in Biarritz. Since 2009 Atout France has been the sole state tourism agency. Atout France publishes studies annually of tourism trends in France and internationally together with suggestions for development of tourism in France itself. Special attention was being paid in 2017 and 2018 to the coasts, secondary residences or vacation homes, and leisure parks. Atout France was also studying the potential market for tourism in France among Mexicans and Colombians ("Chiffres clés et études" 2018). The last two national groups were singled out because of their economic growth and strong economies, which, according to an Atout France study, made them potential sources for increased tourism to France, as was the case in other emerging economies such as China and Brazil ("Le Potentiel touristique" 2017). Publications consulted so far argue that the various agencies involved in tourism are collaborating more than competing. The private sector plays a large role in tourism exemplified by hotel chains including Hilton, Holiday Inn, and Novotel. The same is true of tour companies.

5.4.12 Government and Private Initiatives in Tourism: Disneyland Paris

An example of the complex interrelationships between the national government and local involvement in tourism is the administration of Disneyland Paris. Since its opening in 1992, Disneyland Paris has been among the most visited tourist sites in France in the various statistical counts compiled there. The 2009 Memento du Tourisme divided the most visited tourist sites in France into "cultural" and "non-cultural" sections and counted paid entries to arrive at its figures. The Louvre was listed as the number one "cultural" site and Disneyland Paris as the leading "non-cultural" attraction. None of the 30 leading non-cultural sites was located in Paris, whereas 23 of the 29 cultural destinations, largely museums, were situated in the capital (Gordon 2011; Palmares 2009). Disneyland Paris led all tourism sites in France with 13.4 million visitors in 2016, according to Memento du Tourisme statistics (7e Partie, "Les sites touristiques en France," 2017).

As d'Hauteserre wrote in her study of the site, "The implementation of the Disneyland Paris project in France would have been impossible without the planning culture and structures set up by the French government" (2001, p. 123). In addition to the French government, the supra-national European Economic Community, predecessor of the European Union, also played a role in the placement of Disneyland in France by enforcing a uniform interest rate across the Continent, thereby preventing Spain, for example, from offering lower interest rates to the Walt Disney Company (d'Hauteserre, 2001). The contract signed in 1987 was for a public–private partnership, meaning that the French government assumed specific obligations and the Disney Corporation others. Exemplifying the complexity of the French state in its administration of tourism, the signatories included the central government, the Île-de-France regional authority, the Seine-et-Marne department, the Paris transportation authority, and the EPA (Etablissement Public d'Aménagement) Marne and EPAFrance, governmental bodies established to monitor the execution of the agreements (d'Hauteserre, 2001). Once again, the development of a major tourist site involved significant government action but was also subsumed in a larger project, in this case the development of the New Town of Marne-La-Vallée.

5.4.13 Disneyland Paris: A Public-Private Partnership

D'Hauteserre points out that as a "New Town," Marne-La-Vallée was large enough to be divided into four sectors with Disneyland in sector IV. A new EPA, EPAFrance, was created to direct the development of sector IV and thus of the Disney park. During construction in the 1990s, the state invested some 2.7 billion French francs in the development of Disneyland with a total private investment of 23 billion French francs. The connections with rail transport were highlighted by the development of the RER suburban railway network, rapid speed trains (TGV), and highway exits to reach the Marne-La-Vallée New Town. Additional infrastructure including improved water supply and sewers were built to service the Disneyland locale. In addition, a cost/benefit analysis and an environmental impact study were done for the government (d'Hauteserre, 2001). The main characteristic of the Disneyland project noted by d'Hauteserre is that it is a public–private partnership in which the state and the private company assume specific risks for which they are fully responsible in contrast to mixed economies in which state and private companies assume risks and responsibilities together. The EPA's trustees were largely elected officials, once more blurring the distinction between state and private action in the French economy (d'Hauteserre 2001).

5.4.14 Additional Government Promotion of Tourism in France

To promote investment in the tourism industry and the training of its professionals across the country, the government's Service du tourisme, du commerce, de l'artisanat et services (STCAS; Tourism, Commerce, and Artisan Service) collects and disseminates relevant information and data. Another state agency, STCAS operates under the Direction générale des entreprises (General Director of Businesses) within the Ministry of the Economy and Finance ("Service du tourisme" 2018). STCAS operates under a chain of command at the top of which is the Ministry of the Economy and Finance. A bac technologique Hôtellerie-Restauration (secondary level degree in hotel and restaurant management) was introduced in 2015 to focus on technical as well as foreign language training for those in the hotel industry.

5.5 Conclusion: A Continuing Governmental Presence in French Tourism

In conclusion, the state both on national and more recently local levels maintains and even expands its historical role in the guiding and informing the tourism industry in France. With its strategic investments, its licensing procedures, and its information gathering, it is difficult to overestimate the role of the government in shaping tourism in France, especially in regard to the promotion of tourism into the country from outside France. By the early 1990s, however, the central government was moving away from investment in "heavy infrastructure for tourism" because an economic recession had limited the availability of funds. Investment in tourism development was now left more in the hands of local communities. Commenting on governmental involvement in Disneyland Paris, d'Hauteserre (2001) wrote:

> Three centuries of centralized [sic] bureaucracy, reinforced by the tight control required to manage Fordist production mean that, even in its most liberal hour, French capitalism remains in the hands of the state. It has been reluctant to adopt post-Fordist principles of development until forced by recessionary economic circumstances. (p. 138)

The national government's role in the creation and administration of Disneyland Paris is but one example of a continuing state presence in France's tourism. The French state continues to guide and inform the tourism industry in France. With its strategic investments, its licensing procedures and its information gathering, it is difficult to overestimate the role of the government in shaping tourism in France, especially in regard to the promotion of tourism from outside France as in the case of the Contrats de destination mentioned above.

Continuing examples of the important role played by the French government in tourism include its support for the establishment in 2014 and 2015 of five distinct tourism offices to publicize France's sites to audiences abroad. These offices focused on

wine tourism, mountain tourism, eco- and "slow-tourism," artisanal and luxury tourism, and urban or nocturnal tourism. By 2020, 20 destinations supported with Contrats de destination were divided among the five major categories with the regions indicated below ("Les Contrats de Destination" 2020):

A. Heritage [L'Offre Patrimoniale]
   1. Mont Saint-Michel and its Bay
   2. The Loire Valley
   3. Normandie Paris Ile-de-France – The World of Impressionism
   4. The Lens Louvre and surrounding area
   5. Paris, the greater city
B. Ecotourism, Good Living and the Discovery of Natural and Heritage Sites [Ecotourisme, Bien Vivre et Découverte de Sites Naturels et Patrimoniaux]
   1. Brittany
   2. Lifestyle in Provence
   3. Corsica
   4. Dordogne Valley
   5. Guyana
C. Wine Tourism and Gastronomy [Oenotourisme et Gastronomie]
   1. Champagne
   2. Burgundy
   3. Bordeaux
   4. Lyon – City of Gastronomy
D. Mountains and Health [La Montagne et le Ressourcement]
   1. Jura Mountains
   2. Vosges Mountains
   3. Travel in the Alps
   4. Auvergne
   5. Pyrenees
E. Sport and Relaxation [Sport et Détente]
   1. Biarritz – Golf destination

Since the attacks of November 2015, the government, with a budget of €2.5 million, has gathered those in the tourism industry to plan the promotion of their industry outside France. Atout France was given the task of putting the program into operation. An internet campaign also targeted sixteen strategic markets in Europe together with emerging and more distant countries, comprising 83% of international stays in France. The goal was to attract some 100 million tourists during the year 2020 (Laine, 2016). Despite political tensions surrounding the Yellow Vest movement in 2019, France's tourism trade rebounded with some 90 million visitors in 2019, with a drop expected because of the COVID-19 pandemic in 2020. Much depends on the possibility of more violence and security concerns among the public, as well as the effects of the COVID-19 pandemic, which the government has attempted to mitigate through information gathering and making financial grants available to restaurants and tourist accommodation providers to help them during the pandemic and create more sustainable tourism models ("Tourism in France" 2020).

Clearly, it seems fair to say that the long-term model of state involvement in the governance of tourism in France will continue into the foreseeable future. The state may assume an even larger role as protector of sites in Paris, where increased numbers of armed guards are visible at tourist locales and other popular places elsewhere in the country. Given the increases in tourism in France during the past two centuries, which with the development of steamships, railways, and automobiles, has seen mobility increase among the population, especially among the middle and working classes, it seems probable that with changes as the market develops, tourism to and within France will continue to develop, even if outstripped ultimately by larger countries such as China. Speculation about improvements is difficult as there are many factors involved. In keeping with historical modeling of a strong state centered in Paris, much of the state power over tourism resides in Paris. Whether those who, such as Barnu and Hamouche, call for even greater focus and concentration on the promotion of tourism in France will win out remains to be seen. Much of the information consulted to date comes from state sources, statistics may be questioned, and the tourism studies may be interpreted in a variety of ways. The term "tourist" continues to imply negative connotations in France as elsewhere, although the numbers of tourists has continued to increase, even if many of them prefer to use other terms, such as "traveler," to describe themselves. One direction, for reform in the immediate future, however, might be to attempt to minimize the costs, especially in high-priced areas such as Paris.

References

Ayrault, J.M. & Fekl, M. (2016). "Innover pour retrouver de la croissance: La stratégie pour un tourisme français leader mondial," Gouvernement.fr. Retrieved from http://www.gouvernement.fr/action/la-strategie-pour-un-tourisme-francais-leader-mondial.

Barnu, J. & Hamouche, A. (2014). *Industrie du tourisme, Le mythe de laquais* Paris: Presse des Mines.

Boudet, A. (2017). "Gouvernement Philippe: Qui va s'occuper du tourisme, du logement et de tous ces autres ministères disparus?" Huffpost, Retrieved from https://www.huffingtonpost.fr/2017/05/18/gouvernement-philippe-qui-va-soccuper-du-tourisme-du-logement_a_22096743/.

Boyer, M. (1996) *L'invention du tourisme* Paris: Découvertes Gallimard/Art de Vivre.

Boyer, M. (2000). Les séries des guides imprimés portatifs, de Charles Etienne aux XIXe et XXe siècles. in G. Chabaud et. al. (Eds.). *Les Guide Imprimés du XVIe au XXe Siècle, Villes, paysages, voyages*, pp. 339–352). Paris: Belin.

"Chiffres clés et études," Atout France, France.fr, (2018). Retrieved from http://www.atout-france.fr/services/chiffres-cles-et-etudes.
"Code du tourisme, Version consolidée au 19 janvier 2018," Legifrance, La service publique de la diffusion du droit, République Française. Retrieved from https://www.legifrance.gouv.fr/affichCode.do?cidTexte=LEGITEXT000006074073.
Comités régionaux du Tourisme (CRT). (2015). Veille info tourisme, DGE, Direction Générale des Entreprises, 4 November 2015. Retrieved from: http://www.veilleinfotourisme.fr/comites-regionaux-du-tourisme-crt–92146.kjsp.
"Contrats de destinations: Des outils innovants et opérationnels pour accélérer le développement des destinations touristiques," (2018). Atout France, France.fr. Retrieved from http://atout-france.fr/content/contrats-de-destinations.
Conseil interministériel du tourisme (2018). "Assurer le succès touristique de nos territoires – 19 Janvier 2018." Retrieved from https://www.gouvernement.fr/sites/default/files/contenu/piece-jointe/2018/01/dossier_de_presse_-_conseil_interministeriel_du_tourisme_-_assurer_le_succes_touristique_de_nos_territoires_-_19_janvier_2018.pdf.
d'Hauteserre, A-M. (2001). "The Role of the French State: Shifting from Supporting Large Tourism Projects like Disneyland Paris to a Diffusely Forceful Presence," *Current Issues in Tourism*, 4:204, 121–150.
Gerbod, P. (2002). *Voyager en Europe (Du Moyen Age au IIIe millénaire)* Paris: L'Harmattan.
Gordon, B.M. (2011). The Evolving Popularity of Tourist Sites in France: What Can Be Learned from French Statistical Publications? *Journal of Tourism History*, 3:2, 91–107.
Gordon, B.M. (2018). *War Tourism: Second World War France from Defeat and Occupation to the Creation of Heritage Ithaca*, New York: Cornell University Press.
"Histoire de Lille," Lille Office de Tourisme et des Congrès (2018). Retrieved from https://www.lilletourism.com/histoire-de-lille.html.
Hobsbawm, E. (1983) *The Invention of Tradition* Cambridge, U.K.: Cambridge University Press.
Lainé, L. (2016). "Dossier de l'été 2016: les chiffres clés du tourisme," l'Echo touristique. Retrieved from: http://www.lechotouristique.com/article/dossier-de-l-ete-2016-les-chiffres-cles-du-tourisme,83771.
"Le Potentiel touristique des voyageurs mexicains et colombiens, Pour l'Europe et la France," Atout France, France.fr, (2017). Retrieved from http://atout-france.fr/publications/le-potentiel-touristique-des-voyageurs-mexicains-et-colombiens.
"Le terrorisme devient la première crainte des Français," Le Point (7 December 2017). Retrieved from http://www.lepoint.fr/societe/le-terrorisme-devient-la-premiere-crainte-des-francais-07-12-2017-2177892_23.php?&m_i=CFoCDFZHaP9R6mg%2BorHKgS3aUAca8FMDcnkGZ5z_ZWR5Q7Sduotef9s9CfvInnzyb%2Brbz4UAduOgJ_xq3gQX%2Bx2z2MADCt&boc=347791&M_BT=57272340052#xtor=EPR-6-[Newsletter-Mi-journee]-20171207.
Legifrance. (2014). Décret n° 2014–1105 du 1er octobre 2014 relatif aux attributions déléguées au secrétaire d'Etat chargé du commerce extérieur, de la promotion du tourisme et des Français de l'étranger, Legifrance – Le service public de l'accès au droit, Retrieved from: https://www.legifrance.gouv.fr/affichTexte.do?cidTexte=JORFTEXT000029527270&dateTexte=&oldAction=dernierJO&categorieLien=id.
"Les Contrats de Destination" (2020). Le portail de la Direction générale des Entreprises, Ministère de l'Économie, des Finances et de la Relance, entreprises.gouv.fr. 3 March 2020. Retrieved from https://www.entreprises.gouv.fr/fr/tourisme/developpement-et-competitivite-du-secteur/contrats-de-destination.
Maczak, A. (1995). *Travel in Early Modern Europe* Cambridge, U.K.: Polity Press.
Mesplier, A. (1986). *Le Tourisme en France: Etude Régional, 2nd ed. Montreuil*, France: Bréal.

Montaigne, M. de. (1983). *Travel Journal (1580–1581)*, trans. by D. M. Frame. San Francisco: North Point Press.
"Objectif: 100 millions de touristes, Les premières mesures proposées par le Premier ministre," La transformation du pays est En Marche! (2017). Retrieved from https://en-marche.fr/articles/actualites/objectif-tourisme.
"Palmares des 30 [sic] premiers sites culturels (entrées comptabilisées)', Statistiques et études économiques, in 'Memento du Tourisme," (2009).
Peyroutet, C. (1995). *La France Touristique*, Paris: Nathan.
Py, P. (2007). *Le tourisme, un phénomène économique*, Paris: La documentation Française.
"Renforcer l'attractivité et le rayonnement de la France," (2018) Tourisme, France Diplomatie, 17 April 2018. Retrieved from https://www.diplomatie.gouv.fr/fr/politique-etrangere-de-la-france/tourisme/.
Rose, M. (2020). "France unveils 18 bln euro plan for 'crown jewels' tourism sector," Reuters. Retrieved from https://www.reuters.com/article/us-health-coronavirus-france-tourism/france-unveils-18-bln-euro-plan-for-crown-jewels-tourism-sector-idUSKBN22Q1JP.
Section 8: Contrat de jouissance d'immeuble à temps partagé, in the "Code du tourisme, Version consolidée au 19 janvier 2018," Legifrance, La service publique de la diffusion du droit, République Française (2018). Retrieved from https://www.legifrance.gouv.fr/affichCode.do;jsessionid=06727A65C6F40A91AF077D188427A615.tplgfr36s_1?idSectionTA=LEGISCTA000020897186&cidTexte=LEGITEXT000006074073&dateTexte=20180509.
7e Partie, *"Les sites touristiques en France," in Memento du Tourisme 2017* (Paris: Direction Générale des Entreprises, Ministère de l'Économie et des Finances, 2017).
"Service du tourisme, du commerce, de l'artisanat et services (STCAS)," Direction de l'information légale et administrative (Premier ministre) (2018). Retrieved from https://lannuaire.service-public.fr/gouvernement/administration-centrale-ou-ministere_181072.
"Sites touristiques, Fréquentation des sites touristiques," Atout France, (2011). Retrieved from http://www.atout-france.fr/frequentation-sites-touristiques.
Société d'Exploitation de la Tour Eiffel, 2018. Retrieved from https://sete.toureiffel.paris/fr/la-sete.
Sous-section 1: Dispositions communes applicables aux offices de tourisme, Article L133-1, in the "Code du tourisme, Version consolidée au 19 janvier 2018," Legifrance, La service publique de la diffusion du droit, République Française. Retrieved from https://www.legifrance.gouv.fr/affichCode.do;jsessionid=06727A65C6F40A91AF077D188427A615.tplgfr36s_1?idSectionTA=LEGISCTA000006175501&cidTexte=LEGITEXT000006074073&dateTexte=20180509.
Souyris J-D. & Delage, B. (1979). *Voyage et Troisième Age, Talence*, France: Maison des Sciences de l'Homme d'Aquitaine.
"Summary: Caisse des Depots et Consignations," S&P Global Ratings, (2018). Retrieved from http://www.caissedesdepots.fr/sites/default/files/medias/relations_investisseurs/sps_rating_direct_analysis-12may2016.pdf.
"The Essential Guide for Everything French in London," FranceInLondon,com (2018). Retrieved from http://www.franceinlondon.com/en-Business-in-London-645-Atout-France-Maison-de-la-France–Atout-France-french-information-line.html.
"Tourism in France," France Diplomacy, updated October 2020, Ministère de l'Europe et des Affaires Étrangères. Retrieved from https://www.diplomatie.gouv.fr/en/french-foreign-policy/tourism/.
"Tourisme," DGE Direction Générale des Entreprises, updated 7 July 2016. Retrieved from http://www.entreprises.gouv.fr/tourisme/contrat-destination-outil-developpement-l-attractivite-touristique.

Trésor de la Langue Française. Retrieved from http://atilf.atilf.fr/dendien/scripts/tlfiv5/advanced.exe?8;s=401128260.

UNWTO (2017) "International tourism in 2016 – Key trends and outlook" in Tourism Highlights 2017 Edition. Retrieved from https://www.e-unwto.org/doi/pdf/10.18111/9789284419029.

Venayre, S. (2012). *Panarama du Voyage 1780–1920, Mots, figures, pratiques*. Paris: Les Belles Lettres.

Priyanka Ghosh

6 India

6.1 Introduction

In the past two decades, tourism has emerged as a strong sector of India's economy. Globally it is also the fastest growing industry (Goodwin & Chaudhary, 2017). Tourism can be defined as movement of people to places where they stay for a temporary period of time (Goodwin & Chaudhary, 2017). The meaning of tourism could be different to different people. Travel, relaxation, recreation, and exploring different cultural traditions all can fall under the purview of tourism (Bhatia 1985). World Tourism Organization (WTO), a United Nations specialized agency, defines tourism as activities of visitors "traveling to and staying in places outside their usual environment not more than one year for leisure, business, and other purposes" (UNWTO, 2019).

For developing countries, tourism industry has already proved to be a significant source of foreign exchange (Wunder, 2000). The growth of India's tourism sector is related to the emergence of a growing urban middle class who can afford to visit India's varied landscape, cultural, and religious heritage. In 2013, there were 1.14 billion of domestic tourists in India as compared to 270 million domestic visitors in 2002 (Draft National Tourism Policy, 2015). Being a labor-intensive industry, tourism has immense opportunity to create direct employment for 23 million people in India (Draft National Tourism Policy, 2015). As per the Travel and Tourism Competitive Index (TTCI), India is ranked 34th among the world's 140 most attractive countries for tourism (Business Standard, 2019). According to the TTCI report 2019, India scores well in terms of having natural resources (14th) as well as cultural resources (8th) (Calderwood and Soshkin, 2019).

Tourism sector in India contributed to 12.38% of total jobs in 2017–2018 out of which 5.40% was direct employment (Ministry of Tourism, Annual Report, 2018–2019). Despite such possibility of expansion and growth, India's tourism sector does not play a major role in world tourism (Travel and Tourism Competitive Report, 2011). Considering India's limited role in the global tourism industry, the Ministry of Tourism has developed a new policy in 2015 in which the government wants to promote India as "Must EXPERIENCE" and "Must REVISIT" destination. This new draft policy also highlighted several objectives of government to strengthen tourism sector which includes increasing India's share in arrival of world tourists from current 0.68% to 1% by 2020 and to 2% by 2025 (Draft National Tourism Policy 2015). However, this goal has been set prior to COVID-19 pandemic situation and in the first quarter of 2020 (January to March) Asia and the Pacific region of the world had experienced a 35% decrease in tourist arrival due to travel restrictions (UNWTO May 2020). Against this backdrop, this chapter examines how tourism in India is shaped by its governance system. Additionally, it describes tourism policies in India and the diversity of tourism products India can offer to its domestic and international visitors.

https://doi.org/10.1515/9783110638141-006

6.2 Tourism Administration and India's Tourism Policies

After independence, India only had few hotels in large cities and at hill stations which provided accommodation to visitors. Also, there were few travel agencies who could cater to the needs of the tourists. Since 1956, India focused on developing tourist facilities which also coincided with the Second Five Year Plan (National Tourism Policy, 1982). The ministry of tourism was established in 1958 (Hannam, 2010). India formulated its first tourism policy in 1982. The tourism development by Indian government passed through different phases from isolated planning to integrated area development during fifth and sixth Five Year Plans (Hannam, 2010). During the sixth Five Year Plan, the government emphasized the value of domestic tourism and different states were involved to develop domestic tourism sector (Ahuja, 1999).

India's tourism sector is primarily governed by the Ministry of Tourism which is a nodal agency that formulates policies and programs and promote tourism development in the country (Figure 6.1). The ministry also coordinates among various central government agencies, state governments and union territories, and representatives of private sectors. Secretary (Tourism) is the administrative head of the Ministry of Tourism who is also the Director General (DG) tourism. The office of the secretary and the office of the DG are now combined to implement various plans and programs (Ministry of Tourism, 2019). Currently, the ministry has 20 field offices within the country, eight offices overseas, and one sub-ordinate office/project i.e. Indian Institute of Skiing and Mountaineering (IISM) or Gulmarg Winter Sports Project (GWSP) (Ministry of Tourism, 2019). The field offices within the country provide valuable information to the visitors and monitor ongoing field projects. The IISM now offers various ski and other courses in the valley of Jammu and Kashmir. The overseas offices are responsible for marketing and promotion of various tourism products (Ministry of Tourism, 2019). Under the Ministry of Tourism, there is one government undertaking named India Tourism Development Corporation (ITDC) which was established in 1966. The ITDC runs hotels and restaurants in various places of India and also responsible for production and dissemination of tourist publicity literature (Ministry of Tourism, 2019). ITDC directly reports to the Ministry of Tourism and mostly engaged in building large scale tourism projects (Hannam, 2004). Besides ITDC, there are several autonomous institutions which fall under the Ministry of Tourism: Indian Institute of Tourism and Travel Management (IITTM), National Institute of Water Sports (NIWS), National Council for Hotel Management and Catering Technology (NCHMCT), the Institutes of Hotel Management, and Indian Culinary Institute (Hannam, 2010; Ministry of Tourism, Annual Report 2018–2019).

In the first tourism policy of 1982, the government mentioned its several objectives, including providing socio-economic benefits to local communities, job creation, and earning of foreign exchange (National Tourism Policy, 1982). In the first tourism

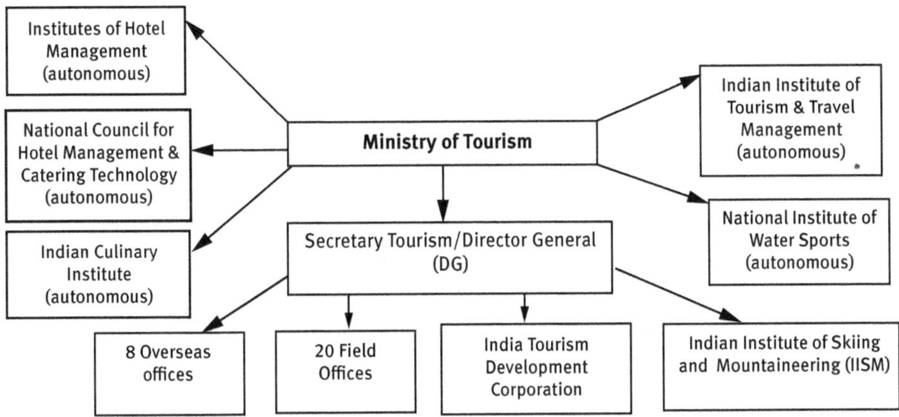

Figure 6.1: Structure of Tourism Governance in India.

policy, the government outlined a plan of action that emphasized involvement of youth in national integration. The policy also paid attention to social tourism (National Tourism Policy, 1982). The tourism policy of 1982 focused on development of selected travel circuits, change the tendency of concentration in few urban centers, diversify attractions for tourists, and opening up economically lagging areas which can attract visitors such as archaeological sites, places with scenic beauty, and places which can attract tourists for its art and craft (National Tourism Policy, 1982). The travel circuits catered to the needs of domestic tourists as compared to the international ones (Hannam, 2004).

In the 1990s, economic liberalization in the country led to special emphasis on tourism as a sector of economic investment (Hannam, 2010). In 1992, a National Action Plan was declared with an aim of improving tourism infrastructure and bringing more foreign tourists in the country. In 2002, the Ministry of Tourism introduced its new national tourism policy which is considered a very important stepping stone for development of India's tourism sector. The new tourism policy in 2002 recognized tourism as an important sector for country's economic growth and put emphasis on domestic tourism. The objectives of 2002 tourism policy were to position tourism as a national priority, enhance and maintain India's competitiveness in world tourism market, improve existing tourism products, and expand these products according to market demand. In order to expand tourism products, the policy focused on developing sustainable beach and coastal tourism resort products, especially along the western coast of India. Also, it focused on developing Kochi and Andaman and Nicobar Islands as international cruise destinations. The 2002 policy put an emphasis on ecotourism, which would eradicate poverty, create new jobs, protect biodiversity and cultural heritage of a place (National Tourism Policy, 2002). The aims of the new policy were to develop world class infrastructure and effective marketing strategies. One of the salient features of this tourism policy was that it acknowledges the important role of private sector along with the government in tourism development (National Tourism Policy, 2002).

Since 2002, India's tourism governance adopted a strategy of destination branding which has become important for wider economic development (Hannam, 2010). Destination branding helps in creating a unique identity of a place which provides it a competence in the global tourism market (Hannam, 2004). In case of India, the destination branding occurred through the campaign of "Incredible India," which portrayed diversity and exoticness of India. This campaign changed India's image as a country of poverty to a country of modern values to be explored by the tourists (Kant, 2009). According to Bandyopadhyay and Morais (2005), the incredible India focused on following themes: personal enlightenment and wellness, cultural diversity reflected in the geographical diversity, cultural richness, natural beauty, exotic wildlife, and royal treatment along with comfort.

Then, in 2005, the Ministry of Tourism launched "Atithi Devo Bhava" to complement the "Incredible India" campaign. The purpose was to sensitize local people towards preserving India's cultural heritage as well as to create a sense of responsibility towards domestic and international visitors (Indian Institute of Tourism and Travel Management, 2011). In September 2018, the Ministry of Tourism launched "Incredible India" mobile application with respect to growing influence of internet in disseminating information related to tourism service providers such as domestic tour operators, adventure tour operators, travel agents, and guides (Ministry of Tourism, Annual Report 2018–2019). Besides destination branding, other important factors which boosted India's tourism sector are deregulation in airline industry, sustained economic growth, and online marketing of tourism activities (Hannam, 2010).

6.3 Tourism Governance in India

India's geographical and cultural diversity provides various opportunities of tourism activities. On the basis of tourism products, tourism in India can be classified into cultural tourism, beach and coastal tourism, adventure tourism, ecotourism, wildlife tourism, medical tourism, ayurveda tourism, and village tourism (National Tourism Policy, 2002). Recently, the Indian government identified several other tourism products for development and promotion such as cruise tourism (both river and ocean cruise), wellness tourism, golf tourism, polo tourism, and film tourism (Ministry of Tourism, Annual Report 2018–2019). Although the Ministry of Tourism acts as a nodal agency in India, every state in India has their own tourism policy (Hannam, 2010).

6.3.1 Regional Tourism Governance: Central versus State

In terms of tourism governance in India and creating positive image in the global tourism economy, Kerala presents a successful case study. Kerala-which is branded as "God's Own Country" first initiated its tourism development by establishing a beach resort at Kovalam in 1976 and the state government of Kerala declared tourism as an industry in 1986. Since then Kerala created its own brand in global tourism market especially market in Europe (Kerala Tourism Policy, 2012; Thimm, 2017). Recently, the state has put more emphasis on attracting domestic tourists as economic recession in Europe has affected the share of foreign tourists (Kerala Tourism Policy, 2012). Kerala receives 12% of India's international tourists arrival. The major countries from which Kerala receives bulk of its international tourists are UK, USA, France, Germany, and Australia (Kerala Tourism Policy, 2012; Thimm, 2017). This section discusses regional tourism governance in India's southern state of Kerala and addresses the discord between Central and State government of India's while implementing environmental legislations in the tourism sector.

The discord between state and central governments on tourism development is often visible through the implementation of Coastal Regulation Zone (CRZ), an important environmental legislation which was first introduced in 1991 (Figure 6.2). In 1996, the Supreme Court of India urged its strict implementation. The entire coastal areas of India are divided into four coastal zones from the high tide line (HTL) to 500 meters inland (Sreekumar & Parayil, 2002). Zone I extends from the HTL to 500 meters inland and includes ecologically sensitive areas such as national parks, reserve forests, wildlife habitats, marine reserves, coral reefs, heritage areas, and breeding and spawning ground of fish. Zone II is the developed urban areas where industrial activities are prohibited. Developed and undeveloped rural areas fall under zone III. In the zone III, rural areas which fall under the HTL to 200 meters inland are considered as the 'No Development Zone' (NDZ). Most of the economic activities are prohibited in the NDZ (Sreekumar &Parayil, 2002). The CRZ notification, 1991 has been amended 34 times after its introduction and as per the 2011 notification CRZ I also includes the ecologically sensitive areas where tourism activities and infrastructure development are prohibited (Kukreti 2019).

However, the violation of CRZ can be clearly visible if we trace the historical development of coastal tourism in Kerala. For instance, in 1998, more than 28% of coastal violations in Kerala occurred due to tourism activities and more than 26% of these violations occurred in the NDZ (Sreekumar &Parayil, 2002). Nevertheless, the state government argued for a relaxation of CRZ regulations irrespective of the fact that the large NGOs like Kerala Sasthra Sahithya Parishad (KSSP) asked for strict implementation of CRZ. The decision-making process in Kerala's tourism sector is still influenced by political elites. Tourism sector in Kerala is largely controlled by the state government that does not share power to local bodies (Sreekumar & Parayil,

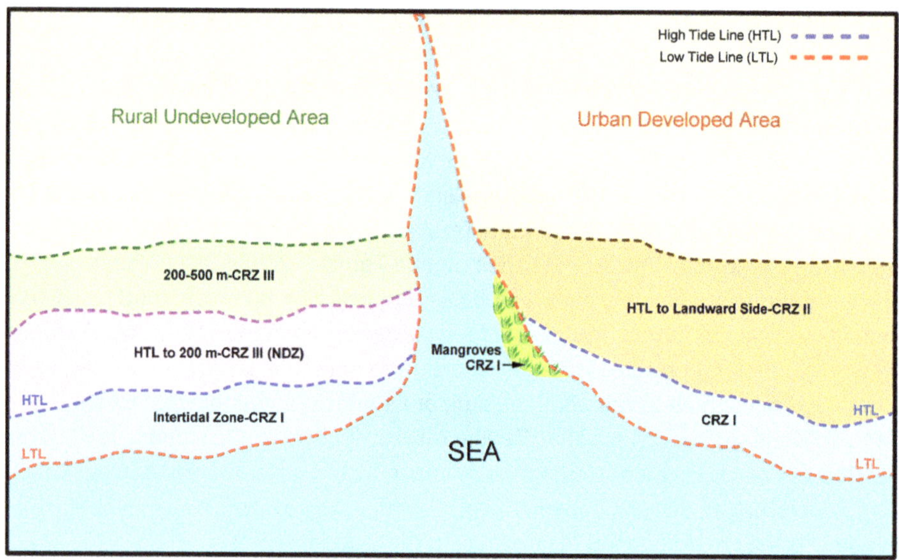

Figure 6.2: Different Zones of Coastal Regulation, 1991.

2002). The violation of CRZ regulations can also be drawn from the Sundarbans region of India.

Several federal and state government departments manage the Sundarbans region of India. For example, the Project Tiger which was launched in India in 1973 in the Sundarbans is under direct supervision of Prime Minister (Ghosh, 2018). CRZ regulation is applicable for the entire Sundarban Biosphere Reserve and in the recent past National Green Tribunal (NGT) has ordered to demolish all hotels, resorts, and lodges. The violation of CRZ notification also included expansion and remodeling of Sajnekhali Tourist Lodge located in Sajnekhali Wildlife Sanctuary without obtaining a "Consent to Operate" certificate from the West Bengal Pollution Control Board (WBPCB). The lodge falls under the CRZ 1 which is the eco-sensitive zone (Gupta, 2015). The state government appealed to the Ministry of Environment, Government of India, for an amendment of CRZ Notification 2011 and remove the word "biosphere" from the section 7.i.A(e) of CRZ notification. This would save all the constructions from demolition in this coastal wetland ecosystem (Bandyopadhyay, 2015).

The centralization of power and decision making by the state government in Kerala can explicitly be found from the case of Kumarakaran Bird Sanctuary in which hotels and resorts were built without considering the objections raised by the local panchayat and environmental activists (Sreekumar & Parayil, 2002). However, in Kerala Tourism Policy of 2012, the state government of Kerala has recognized the role of local self-governments by stating that:

> In planning and development of tourism in destinations, the local self-governments can play a decisive role. Tourism development programmes will be integrated with other developmental activities of local self-governments. We will encourage and assist local bodies to form Tourism Working Group in places of tourism importance. (Kerala Tourism Policy, 2012, p. 12)

Recently, a new Coastal Regulation Zone Notification 2018 has been proposed which would open more coastal areas for commercial activities thus leading to economic growth (Kukreti, 2019). Conservationist and Environmental Activists have argued that risk would be far greater than benefits as ecology and coastal communities will be vulnerable to rising sea level and natural hazards (Kukreti, 2019).

6.3.2 Influential Forest Management over Tourism

The state executive power in India lies with the Prime Minister and his cabinet whereas the bureaucratic power in India lies in the hands of Indian Administrative Service (Hannam, 2004). Below this hierarchy is the Ministry of Home Affairs which is responsible for the internal security and depends on its own state apparatus known as the Indian Police Service (IPS) (Hannam, 2004). The Ministry of Environment and Forest lies below in terms of hierarchy in the state apparatus and controls over one third of India's territory in the sense that one third of the India's total land area should be under forest cover (National Forest Policy, 1988). Nature-based tourism and wildlife tourism can take place in this one-third forest land which include all the national parks and sanctuaries (Hannam, 2004). Forest cover in India also falls under three categories: Protection forests, national forests, and village forests (Kumar Sanjay, 2002a). National parks fall under the category of Protection forests (Hannam, 2004). Ministry of Environment and Forests deploys its power over one-third of forest land through its own civil service called Indian Forest Service (IFS). However, this is an elite institution with a complex hierarchical bureaucratic structure which mostly remained same since the colonial period (Hannam, 1999; 2004). The Ministry of Tourism in India does not have a similar civil service like the IFS of the Forest Department (Hannam, 2004).

Tourism development in different states of India often face conflicts with the forest management especially in the case of nature-based tourism activities or ecotourism. The Department of Tourism and the Department of Forest often come into conflicts in terms of building accommodation facilities and other related infrastructure development. There are more than 800 protected areas (PAs) in India including national parks and sanctuaries (Ministry of Environment, Forest and Climate Change, Annual Report 2018–19). These protected areas are smaller in size because increasing population pressure and existing land use do not allow creation of large, protected areas in India (Kumar Suhas, 2002b). The size of protected areas in India on average is less than 300 sq. km. (Karanth & DeFries, 2010). However, the smaller PAs are not ecologically viable unless PAs are connected to nearby PAs (Kumar, 2002). Certain national parks in India are designated to protect certain animal such as tiger based on the Project Tiger launched in 1973 (Ghosh, 2014; Hannam, 2005). The Project Tiger was originally launched in nine tiger reserves in India with an objective to protect and maintain tiger population for ecological, scientific, cultural, and aesthetic value. The second objective of such a conservation project was to preserve such land for tourism and recreation. However, this second objective was not received much importance (Ghosh, 2014; Hannam, 2005). Project Tiger also prohibited hunting within the boundary of tiger reserve as well as human habitation (Hannam, 2004).

The park managers of tiger reserves in India often see tourism as obstacle rather than an opportunity (Hannam, 2004; 2005). In other words, tourism has been as subordinated to wildlife conservation, even a major environmental issue in PAs (Hannam, 2004). Even the eco-development projects which were launched around national parks only cater to reduce biotic pressure by providing livelihood to people living on the boundary of protected areas. Tourism development is not an important goal of such eco-development projects and mostly used as an incentive to relocate rural people outside the boundaries of national park (Hannam, 2005). Thus, the Ministry of Environment and Forest is dominant over the tourism department when it comes to tourism activities in a protected area such as a national park or biosphere reserve.

The dominant role of the Forest Department can be understood through different aspects of park management, such as allowing certain number of tourists, vehicles to park entry fee. For example, the Forest Department daily regulates the number of private vehicles within the boundary of Ranthambore National Park (Karanth & Defries, 2010). In the Sundarbans Tiger Reserve (STR), number of visitors are regulated through park entry fee of Rs. 40 (US $0.74) per individual (Ghosh, 2014). The number of tourists' boat is also regulated by the Forest Department, which also need to pay the park entry (Ghosh, 2014).In some protected areas such as Nagarahole and Mudumalai, the Forest Department provides bus services for the visitors (Karanth & Defries, 2010). Most of India's national parks follow a dual pricing system in terms of park entry fee. In other words, park entry fee is higher in case of foreign tourists which is often resented by reducing the duration of stay (Hannam, 2005). Foreign nationals also pay higher price when it comes to hiring a tour guide (Ghosh, 2014). For instance, in 2012–13 tourist season, the foreign tourists paid Rs. 600 (US $11.05) for hiring a tour guide as compared to Indian nationals who paid Rs. 300 (US $5.53) (Ghosh, 2014). The park managers in India sometimes acknowledge this dual pricing system and accept it in a positive note that it helps to regulate number of tourists within the park boundary. Additionally, this dual pricing policy has not economically benefitted local communities living on the edge of the parks (Ghosh, 2014). The Forest Department manages the tourism revenue in majority of the protected areas in India except Periyar where 56% of revenues are given to Periyar Foundation for ecodevelopment (Karanth & Defries, 2010).

The influential forest management over tourism also becomes prominent how the protected area managers in India view the role of tourism in conservation. Tourism is considered a mere tool in broader conservation framework rather than an avenue of gaining revenue for the local people living around the protected areas (Hannam, 2005). The conflict between the forest management and tourism development is clearly visible how local communities living around a protected area perceive tourism (Ghosh, 2018). In general, perception of tourism among people living surrounding a protected area varies in South Asia (Ghosh & Ghosh, 2018; Karanth & Nepal, 2012;). Local people living around Kanha and Ranthambore National Parks perceived that people who do not live there benefitted most from the tourism and highlighted the negative impacts of tourism on local culture (Ghosh & Ghosh, 2018). The leakage of tourism revenue either to private tourism operators, hotels and lodge owners, or to the government led people perceive tourism in negative light in India (Banerjee, 2007). This negative perception about tourism occurs due to inefficient management by the Forest Department who lack any formal training in managing ecotourism or nature-based tourism in and around protected areas (Ghosh & Ghosh, 2018). Less than 5% of revenue that is received from park entry fee in India reach to local communities living on the boundary of the protected areas (Banerjee, 2012).

As tourism development is not a priority of the park managers in India, they try to regulate tourism behavior by prescribing certain code of conduct. In protected areas of India, these codes of conduct instruct people not to carry any polythene, not to pollute any pristine habitat, not to enter tiger reserve without a permit from the Forest Department, not to disturb animals, not to stay in tiger reserve areas, not to make noise, etc. (Ghosh, 2014). The managing of visitors' behavior through such code of conduct is part of the "soft" visitor management in order to reduce the visitors' impacts on the protected areas (Mason, 2005). Nevertheless, such regulations have been mostly ineffective in case of Indian protected areas due to lack of proper implementation (Reeve & Edwards, 2002; Hannam, 2005). Instead of modifying tourists' behavior through some specific code of conduct within the park boundaries, attention can be given to improving overall visitor's experience. According to Mason (2005), this can be achieved through education which would involve interpretation and dissemination of information about a particular site. Interpretation is an important educational process which can help visitors to be more mindful towards environment by transforming their thinking and behavior (Moscardo, 1996; Mason, 2005). Many of India's national parks such as The Rajiv Gandhi National Park, also known as Nagarahole National Park, do not have effective interpretation programs (Hannam, 2005).

6.4 Conclusion

Tourism governance in India is largely shaped by the Ministry of Tourism which is responsible for formation of national tourism policies. However, every state in India can formulate its own tourism policy (Hannam, 2010). Various types of tourism activities are practiced in India such as nature-based tourism, ecotourism, wildlife tourism, and medical tourism. As compared to nature-based tourism, wildlife, or ecotourism, private stakeholders dominate the sector of medical tourism (Sen Gupta, 2008). Increasingly, India has become a favored destination for foreign medical tourists. In 2005–2006, the value of medical tourism in India was over US $310 million, and during the same period of time 1 million foreign tourists visited the country (Sen Gupta, 2008). However, in case of nature-based tourism, wildlife, and ecotourism, the Forest Department in every state over-powers the Tourism Department in terms of decision making. Forest Department sees tourism as a nuisance as tourists make noise while watching wildlife and this is often mentioned by the local tour guides of the Sundarbans Biosphere Reserve. Also, tourists are blamed for increasing pollution in and around protected areas in India (Ghosh &Ghosh, 2018). The discord between the Ministry of Tourism and the Ministry of Environment and Forest suggests the need for a more cooperative approach towards decision-making involving tourism around the protected areas of India. There is no one-size-fits-all solution when it comes to find a middle ground between environmental conservation and tourism development. Central and State governments should cooperate with each other through exchanging dialogs while making decisions regarding tourism development in and around protected area. Such decisions should consider unique geography, ecology, and local socio-economic and cultural conditions of a particular region. In other words, tourism development in and around a protected areas or forested areas should be dealt with case-by-case basis for an effective governance.

References

Ahuja, O. P. (1999). Domestic tourism and its linkage with international tourism – Indian case study. In *Domestic Tourism in India*, Eds. D. S. Bhardwaj, M. Chaudhary, and K. K. Kamra. New Delhi: Indus Publishing Company.

Bandyopadhyay, K. (2015). CRZ puts Sundarbans in jeopardy. The Times of India, Date accessed: August 03, 2015. Retrieved from: http://goo.gl/nElOVd.

Bandyopadhyay, R., & Morais, D. (2005). Representative dissonance: India's self and western image. *Annals of Tourism Research* 32(4), 1006–1021.

Banerjee, A. (2007). An evaluation of the potential and limitations of ecotourism as a vehicle for biodiversity conservation and sustainable developemnt in the protected areas of India (Doctoral dissertation). Available from ProQuest Dissertations & Theses Global database. (UMI No. 3267150)

Banerjee, A. (2012). Is wildlife tourism benefiting Indian protected areas? A survey. Current Issues in Tourism, 15(3), 211–227.

Bhatia, A. K. (1985). *Tourism Development, Principles and Practices*. New Delhi: Sterling Publisher Pvt. Ltd.

Business Standard. 2019. India ranked 34th on world travel, tourism competitiveness index: Report. Retrieved from: https://www.business-standard.com/article/pti-stories/india-moves-

up-6-places-to-34th-rank-on-world-travel-tourism-competitiveness-index-wef-report-119090400693_1.html. Date accessed: September 04, 2019.
Calderwood, L. U. and Soshkin, M. 2019. *The Travel & Tourism Competitiveness Report*. Geneva: World Economic Forum.
Ghosh, A. (2018). *Sustainability Conflicts in Coastal India: Hazards, Changing Climate and Development Discourses in the Sundarbans*. Cham, Switzerland: Springer International Publishing.
Ghosh, P. (2014). Subsistence and biodiversity conservation in the Sundarban Biosphere Reserve, West Bengal, India. Ph.D. diss., University of Kentucky, Lexington, USA. Retrieved from: http://uknowledge.uky.edu/geography_etds/26.
Ghosh, P. & Ghosh, A. (2018). Is ecotourism a panacea? Political ecology perspectives from the Sundarban Biosphere Reserve, India. *GeoJournal* 83(1), 1–22. DOI 10.1007/s10708-018-9862-7
Government of Kerala. (2012). *Kerala Tourism Policy*. Department of Tourism: Government of Kerala.
Goodwin, R. D. & Chaudhary, S. K. (2017). Eco-Tourism Dimensions and Directions in India: An Empirical Study of Andhra Pradesh. *Journal of Commerce & Management Thought*, 8(3), 436–451.
Gupta, J. (2015). NGT orders ban on all construction in the Sundarbans. The Times of India, Date accessed: August 06, 2015. Retrieved from: http://goo.gl/a5A08g
Hannam, K. (1999) Environmental management in India: recent challenges to the Indian forest service. *Journal of Environmental Planning and Management*, 42(2), 221–233.
Hannam, K. (2004). Tourism and forest management in India: The role of the state in limiting tourism development. *Tourism Geographies*, 6(3), 331–351.
Hannam, K. (2010). Governing and Promoting Tourism in India. In *Tourism and India: a critical introduction*, Eds. A. Diekmann and K. Hannam, (pp. 13–30). New York: Routledge.
Karanth, K.K., & DeFries, R. (2010). Nature-based tourism in Indian protected areas: New challenges for park management. *Conservation Letters* 00, 1–13.
Kant, A. (2009). *Branding India: An Incredible Story*. New Delhi: Harper Collins.
Kukreti, I. (2019). Coastal Regulation Zone Notification: What development are we clearing our coasts for. Down To Earth, Date accessed: February 04, 2019. Retrieved from:https://www.downtoearth.org.in/coverage/governance/coastal-regulation-zone-notification-what-development-are-we-clearing-our-coasts-for-63061
Kumar, S. (2002a). Does 'participation' in common pool resource management help the poor? A social cost-benefit analysis of Joint Forest Management in Jharkhand, India. *World Development*, 30(5), 763–782.
Kumar, S. (2002b). Wildlife Tourism in India: Need to Trade with Care. In *Indian Wildlife: Threats and Preservation*, Edited by D. Sharma. (pp. 72–94) New Delhi: Anmol Publication.
Mason, P. (2005). Visitor Management in Protected Areas: From 'Hard' to 'Soft' Approaches? *Current Issues in Tourism*, 8 (2–3), 181–194. DOI: 10.1080/13683500508668213
Moscardo G. (1996). Mindful visitors: Heritage and tourism. *Annals of Tourism Research*, 23(2), 376–97.
Ministry of Environment and Forests. (1988). National Forest Policy 1988. New Delhi: Ministry of Environment and Forests. Accessed February 9, 2022. Retrieved from: http://asbb.gov.in/Downloads/National%20Forest%20Policy.pdf
Ministry of Tourism, Government of India. (1982). National Tourism Policy 1982. New Delhi: Ministry of Tourism, Government of India. Accessed February 9, 2022. Retrieved from: https://tourism.gov.in/sites/default/files/2019-10/Tourism%20Policy%201982.pdf.

Ministry of Tourism, Government of India. (2002). National Tourism Policy 2002. New Delhi: Ministry of Tourism, Government of India. Accessed August 28, 2019. Retrieved from: https://tourism.gov.in/sites/default/files/2019-11/National_tourism_Policy_2002.pdf

Ministry of Tourism, Government of India. (2015). Draft National Tourism Policy 2015. New Delhi: Ministry of Tourism, Government of India.

Ministry of Tourism, Government of India. (2018–2019). *Annual Report.* New Delhi: Ministry of Tourism, Government of India.

Ministry of Tourism, Government of India. (2019). "The Organisation." Accessed December 8, 2019. Retrieved from: http://tourism.gov.in/organisation

Reeve, J. & Edwards, V. (2002). Visitor codes in Indian protected areas: Worth the paper they are written on? Paper presented at the Tourism and the Natural Environment International Conference, University of Brighton, October.

Sen Gupta, A. (2008). Medical Tourism in India: winners and losers. *Indian Journal of Medical Ethics*, 5(1), 4–5.

Sreekumar, T. T., & G. Parayil. (2002). Contentions and Contradictions of Tourism as Development Option: The Case of Kerala, India. *Third World Quarterly*, 23(3), 529–548.

Thimm, T. (2017). The Kerala Tourism Model – An Indian State on the Road to Sustainable Development. *Sustainable Development*, 25, 77–91.

UNWTO, (2020). *UNWTO World Tourism Barometer: Special Focus on the Impact of Covid-19.* Madrid: United Nations World Tourism Organization.

Wunder, S. (2000). *Ecotourism and economic Incentives – an empirical approach.* Ecological Economics, 32, 465–479.

Monica Pascoli

7 Italy

7.1 Background

7.1.1 The Post-War Period and the Creation of the Ministry for Tourism

As a reaction to the forceful state intervention in every sector of private life during the fascist regime of WWII, the post-war years were characterized by a clear distinction between public and private spheres. It was a time when the tourist sector had to be rebuilt, since both the transportation system and hotel facilities had been seriously damaged during the war. The tourist demand (both domestic and international) became substantial again soon after the end of the war.

The Constitution, which came into force in 1948, made a clear distinction between the role of the State in coordinating, supervising and fostering tourism, and the actions of development and promotion at a more local level (Berrino, 2011, p. 266). In 1947, the Government established a Commissariato per il Turismo (Tourism Committee), directly controlled by the Presidency of the Council of Ministers; it acted as government body while ENIT (Ente Nazionale Italiano Turismo [National Board for Tourism]) maintained an executive role. The government did not enter actively into the tourism sector, but coordinated, supported, and financed the activities promoted at a more local level. The local authorities – for example, the 92 EPTs (Ente Provinciale del Turismo – Provincial Board for Tourism) and the 187 Autonomous Agencies – were preserved, but their functions changed as they were granted more decision-making power.

The Tourism Committee managed the ERP (European Recovery Plan) funds and was active in the reconstruction of tourist facilities, by reactivating the peripheral system (Agencies and EPTs) (Berrino, 2011, p. 266). These peripheral authorities were concretely involved in the recovery of the sector and had full operational and financial autonomy. The EPTs were financed by funds granted by both Provinces and Chambers of Commerce, and their revenues included tourist and local taxes. This financing system continued throughout the 1950s, giving rise to farcical situations. In provinces with a noteworthy productive density (hence with lower tourist appeal), the local EPTs were extremely well-funded. In the South, however, which had greater needs due to the increased number of tourists arriving there, the funds were negligible. In 1958, the Italian government tried to amend this situation: it was decided that EPTs would no longer receive funds from local authorities, but from the Tourism Committee alone, so that they would directly depend on the central State. Their role became rather hazy. Sometimes they were capable of real entrepreneurial actions, but under other circumstances their initiatives were quite useless.

https://doi.org/10.1515/9783110638141-007

The establishment of a Ministry for Tourism dates back to 1959. The Minister appointed at the time was able to reorganize the sector thanks to the approval of four legislative texts aimed at promoting a centralizing policy. ENIT was reformed and was given the sole task of popularizing Italian destinations abroad; the Aziende Autonome (Autonomous Agencies) were in charge of enterprises aimed at promoting the territory, while EPTs (the acronym now stood for Ente Promozione Turistica [Tourism Promotion Board]) became the key public actors at a local level, supervising all tourism-related activities in their district. Moreover, the Minister confirmed the role of the Central Council for Tourism, with consultative and research functions.

In brief, the Ministry was the direct administrator, although according to the Constitution this role should have been covered by the regions. The central State was greatly involved in fostering tourism in Southern Italy, but with little success (in the early 1960s, only 4% of the Italian population had ever been to the South). In 1967, a law was passed for the promotion of tourism; the first step was to divide the national territory into three areas, according to their level of tourism development (intense tourist areas, developing areas and areas requiring promotion). Popular resorts only received ordinary support, while special funds were allotted to the weakest areas (included the poorest regions of the Alps and central Apennines) through a five-year plan of the Cassa del Mezzogiorno (Southern Development Fund, law 717/1965) (Berrino, 2011, p. 271). Between 1966 and 1974, Italy lost 10% of the sun & sea market (passing from 42% to 32%). By 1974, Spain (and Majorca in particular) had become more popular than Italy among German tourists. Despite this first decline in tourist arrivals, Italy still maintained the record of accommodations, and in the late 1970s revenues from tourism could cover the cost of oil and gas imports.

7.1.2 Regionalization of Tourism Governance

In 1972, all the services, structures, and activities pertaining to tourism were transferred to the regions, while the State maintained functions linked to promotion abroad and the classification of tourist areas. If the single regions dealt with the ordinary actions, such as the planning, development and promotion of regional tourism and the organization of tourist events, the State was in charge of the extraordinary ones that would increase the quality of the offer at a national level. This was achieved by standardizing of the system of accommodation, developing and improving the structures, and supplying a greater array of services.

In 1983, a new framework law for tourism was approved (no. 217/1983). It established the Coordination Committee for Tourism Planning, composed by members of the Regional Council, and supported by experts of the sector (who formed the Advisory Committee). This committee was supposed to act as a go-between for central State and Regions, but limited its activity to the distribution of money (between

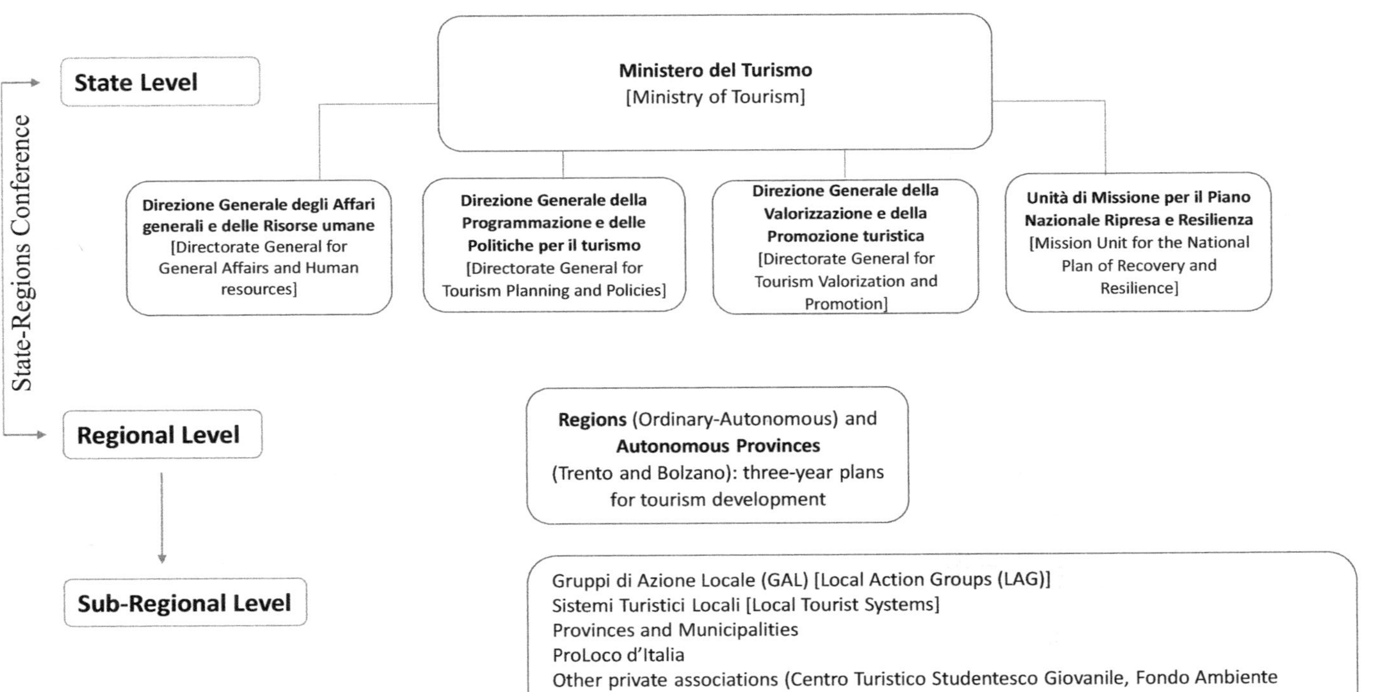

Figure 7.1: Tourism Governance in Italy (simplified version).

1983 and 1992, 1100 billion lira were handed out to the regions) and was suppressed in 1988 (Berrino, 2011, p. 291; Rotondo, 2011, p. 209). The law in question mandated the creation of APTs (Aziende di Promozione Turistica [Agencies for the Promotion of Tourism]) and IAT (Informazioni Accoglienza Turistica [Tourist Welcome and Information]) Offices.

The Constitution states that tourism falls under the jurisdiction of the regions, but for over 30 years (1959–1992) the state intervened directly to sustain the tourist industry. The Ministry for Tourism and Performing Arts was abrogated after a referendum held in 1993, and its competencies distributed among other bodies: to the Ministry for Economic Development in 1999 and to the Department for the Development and Competitiveness of Tourism in 2006. From that moment on, tourism became increasingly more of a regional affair. The year 2001 was a turning point for the national tourism governance. Law 135/2001 reformed the national legislation on the topic, introducing the so-called Sistemi Turistici Locali (Local Tourist Systems) and the National Conference on Tourism; in October 2001, the Constitutional Reform defined tourism as a regional matter. The Italian Regions have now the exclusive competence in this field and state interventions are limited to coordination activities.

On December 1st, 2009, the new Treaty of the European Union (Lisbon Treaty) was adopted into law. New matters, such as tourism, became part of the EU's objectives, and the Union was given power to direct undertakings of the Member States in this sector, but always however with the assent of the single states. After the Lisbon Treaty came into force, Italy once again required centralized governance for tourism to meet the request for a single reference institution operating at a national level.

In 2011, the Government approved Legislative Decree 79/2011, issuing the Code on Tourism, with the intent of promoting the tourist market and providing more substantial consumer protection. The Code should have ensured better coordination between the State and Regions, fixing their respective competences. It should also have played an important role in revising and rationalizing of all the regulations concerning the matter, by becoming the comprehensive text of tourism law in Italy. The Constitutional Court however substantially limited its scope, and now only the parts pertaining to private law are still in force.

In 2013, MiBAC (Ministero per i Beni e le Attività Culturali [Ministry of Cultural Heritage and Activities]) absorbed the Office for Tourism Policies (hence becoming MiBACT, where the final T stands for Tourism), thus strengthening the bond between Italy's cultural heritage (the country's greatest asset) and tourism. Its objectives included the definition of the national agenda for tourism and the elaboration of a national strategic plan. But a recent change has (yet again) given rise to a radical transformation. In March 2021, after nearly 30 years from its abolition, the Ministry of Tourism has been (re-)established (Legislative Decree 22/2021). The Ministry sets the strategy and objectives for tourism, coordinates and promotes the National tourism policy and is responsible for the relations with the Italian regions and (together with the Ministry of Foreign Affairs and International Cooperation) the EU. This political decision has been welcomed by the associations that deal with trade and tourism.

The following section briefly traces the development of tourism in Italy, from the modern era to the present day. The tourism governance in Italy is then discussed.

7.2 History of Tourism in Italy

The origins of modern tourism in Italy can be traced back to the experience of the Grand Tour, an 18th century custom among young European aristocrats, which consisted of a long journey across the continent and was considered an important part of their education. Nowadays it can be seen as a precursor of what is currently defined as cultural tourism. It was only at the beginning of 20th century that Italy discovered its tourist potential: its cultural and natural heritage. After World War II, tourist mobility increased and Italy, with its beaches and mountain resorts but also its countless art cities, endowed with a rich historical, artistic and architectural heritage, became a major tourist destination. Mass tourism started to develop in the mid-1960s, with the introduction of innovations such as the all-inclusive holiday packages and charter flights, but it was in the 1990s that tourism became an indispensable social element and a global mass phenomenon.

The reasons for traveling to the country were, and still are, manifold. Italy has more UNESCO World Heritage Sites than any other country in the world: some of these sites are whole cities, such as Verona, Siena, San Gimignano, Urbino, Matera, Pompei, Noto and Siracusa. Today Italy ranks fifth in number of international tourist arrivals, which in 2017 amounted to 58.3 million (+11%) (WTO Tourism Highlights, 2018). As far as internal tourism is concerned, according to ISTAT (Istituto Nazionale di Statistica [National Bureau of Statistics]), Italian tourist arrivals total 56.3 million. According to the World Travel and Tourism Council (2017), the value of the Italian tourism industry amounts to €77.3 billion equal to 4.6% of the GDP; this figure rises to €186.1 billion (11.1% of the GDP) when satellite activities are taken into account. As far as employment is concerned, workers in the tourism sector total roughly 1.246 million (5.5% of total employees), who become 2.867 million (12.6%) with satellite activities (i.e. jobs that are indirectly linked to the tourism industry) (WTTC, 2017).

These data have been analyzed and commented by various national institutions, whose annual reports underline the importance of tourism for the Italian economy. The World Economic Forum's (2019) analysis of competitiveness is also positive: Italy is eighth in the ranking and has the potential for further growth, if an appropriate infrastructure and commercial investment policy is adopted. Even over the long term (2004–2015), tourist arrivals and stays have grown steadily, with the only exception of the years 2008–2009 when, due to the economic crisis, there was a decrease in tourist flows. This could be however mainly ascribed to the domestic demand, which suffered more than the international one. The COVID-19 pandemic has caused a severe crisis in the tourism sector, with a 84.2% fall in flight arrivals for the first nine months of 2020, compared to the same period in 2019. Moreover, the latest bulletin published by ENIT (Ente Nazionale Turismo Italiano [Italian government tourist board]) reports a decrease by 64% of international visitors, while the domestic market dropped by 31% (ENIT, 2020)

Strong seasonality is the one of the main characteristics of Italian tourism. Tourist flows mainly occur in the summer months with the peak in August recording a rate of 15.2% of the total arrivals and the 21.1% of the total overnight stays. In the summer months (June, July, and August), tourist overnight stays are more than half (50.2%) of those throughout the entire year (Cavallo, 2016, p. 58).

The most popular resorts can be found in specific regions, which are located in central and northern Italy: Veneto, Trentino-Alto Adige, Tuscany, and Lazio. The eight regions of the Mezzogiorno (i.e., Southern Italy) are only chosen by 13.9% of the total number of foreign tourists arriving to Italy. If we consider towns instead of regions, Rome is the destination with the largest number of overnight stays (almost 29 million), followed by Milan and Venice (almost 12 million each), and Florence (9 million) (ISTAT, 2019a). These numbers confirm the positive trend of cultural tourism, which has grown both in the mid- and short term. An analysis of tourist flows reveals that Germany is the home country of the largest number of foreign tourists – 59 million according to ISTAT (2019b). Most of the others come from France, the United States, and the United Kingdom (14 million) (ISTAT 2019b). Russia is the country that has registered with the most significant increase in number of tourists travelling to Italy: from 1.2% in 2004 to 3.7% in 2014 (Becheri & Maggiore, 2016, p. 32). Likewise, China, in the years 2014–2016 the overnight stays of Chinese tourists increased by nearly 2 million (5.4 million in 2016) (TCI-Unicredit, 2017).

7.3 Tourism Governance Today

With respect to functions, competences can be either direct or indirect. A competence is considered direct when the administrative bodies are responsible for implementing the policies defined by the legislative authorities (Regional Tourism Departments or Promotion Agencies), whereas an indirect competence refers to sectors and fields that are relevant to tourism-related activities (transport, cultural and natural heritage). We will focus on the institutions and organizations that, at different levels, directly influence tourism governance (Fig. 7.1).

The governance system is extremely complex and has undergone deep and significant changes over the past thirty years. The most recent reform of the tourism sector allocates territorial competences both at central (the State and ENIT) and local (regions, APTs, provinces, municipalities, mountain communities) levels. The legislative framework regulating this sector is composite, with competences at a number of levels: EU laws, State laws, regional laws, and local regulations on tourist activities (Franceschelli & Morandi, 2013, pp. 17–18).

7.3.1 National Level

The Italian Constitution does not supply any useful indications on the governance of tourism; nonetheless, it is possible to find therein a reference to the principles that are an expression of the constitutional values. Different and conflicting interests are involved, such as those of tourists and entrepreneurs, but mostly, the public interest pertaining to the protection of the country's environmental and cultural heritage. All these interests are recognized, but with different degrees of importance, the protection of environmental and cultural assets being the most relevant one. Tourism is also involved in the sphere of social rights protected by the Constitution, such as the right to health, education, and the full development of the human person.

In late February 2014, the Minister of Cultural Heritage and Tourism, Mr. Dario Franceschini, declared, to the surprise of many, that from an economic viewpoint, the Ministry of Cultural Heritage, Activities and Tourism was the most important in the country, because it dealt with the country's greatest wealth: art and culture (Becheri, 2016, p. 689). This is by all means a public recognition of the importance of a sector that only few years beforehand was defined as unproductive by the former Minister of Economy (Becheri, 2016). Until recent times, tourism was associated with industry and production, but the liaison with the country's cultural heritage highlights both the fact that it can be considered a tool to protect and promote these assets and the pivotal role of cultural tourism in this sector.

As mentioned previously, in 2013, MiBAC (Ministero per i Beni e le Attività Culturali [Ministry of Cultural Heritage and Activities]) absorbed the Office for Tourism Policies (and consequently changing its name to MiBACT) and set off to define the national agenda for tourism and a national strategic plan. Tourism had been merged with the cultural sector and this decision represented a deliberate policy aimed at strengthening the bond between culture and tourism, as well as reinforcing the market orientation and the collaboration between public and private sectors. A decree issued in 2014 (Culture and Tourism) traced the guidelines for the intervention policies: integration (between culture and tourism, public and private, conservation and promotion) was the driving force behind all actions, which had to be aimed at modernizing the sector (Becheri, 2016, p. 700). Moreover, a collaboration between MiBACT, the Ministry of Infrastructures and

Transport, and the State-Regions Conference was supposed to play a pivotal role in the development of a touristic mobility plan aimed at promoting tourism in lesser-known destinations and the South of Italy (Becheri, 2016, p. 702).

The recent (2018) attempt to allocate the competences for tourism to the Ministry of Agriculture implied a relevant shift from the cultural assets and heritage sphere (considered the main pull factor for the tourist sector) to the domain of agriculture and food production. Fortunately, this change has been rejected by the Council of State (2019). In 2021 the Ministry of Tourism has been re-established.

7.3.1.1 ENIT

Although ENIT (Italian Government Tourist Board) has been the target of a number of reforms over the last thirty years, it has never really changed, and likewise, has never been able to properly identify the trends of the tourist market (Becheri, 2016, p.705). Since 2015, ENIT is no longer a non-economic public authority, but has become an economic one, under the supervision of the Ministry. This transformation ensures expenditure savings and better promotion and marketing of the tourist offer. Its mission is the promotion (especially through investment in digital media) of tourist resorts and products, firstly food and wine, and the development of services, such as the Tourist Card.

7.3.1.2 The Strategic Plan for Tourism (MiBACT)

The strategic plan for tourism development, issued in 2016 (for the years 2017–2022), acknowledges that Italy's competitiveness on the international tourist market is much beneath its potential, due to a variety of factors, such as the low levels of innovation displayed (both in terms of technology and organization), a difficulty to respond to market changes, skills obsolescence, and unfavorable conditions for businesses.

In 2011, the OECD Tourism Committee published a research on the state of tourism governance in Italy. The text reported the criticalities and policies that emerged from analyzing the institutional governance system, as well as the State-Regions Report, statistical information, and the education system. It revealed little or no integration of tourism policies: there were – and still are – countless (private and public) subjects involved in the development and promotion of tourism and this is a major challenge in terms of organization and governance. The report highlights the need to elaborate an integrated, long-term national strategy, which should be developed in collaboration with all the stakeholders in both the public and private sectors. According to the report, such a strategy would overcome the extreme fragmentation of the policies, contribute to optimizing the use of resources and ensure a coherent and coordinated development of tourism across the country and its regions (especially the Southern ones). According to the report:

> Long-term policy should: a) adopt an integrated administrative approach involving all the subjects involved in tourism development (e.g. the tourism industry, ministries, regional and local authorities); b) clearly define strategic priorities and themes for Italy's commitment to tourism; c) state the plans to support tourism development in the South and to reinforce cooperation among the regions. (OSCE, 2011, p. 78)

A national strategic policy seems to be essential for developing tourism, as well as measuring the impact of current and future tourism programs. In this respect, evaluation has a fundamental role, as it is the only available tool to appraise the effectiveness of public policies and investments (OSCE, 2011, p. 79).

The strategic plan elaborated in 2016 appears to respond to the recommendations, making reference to the principles of public governance. The paradigm of public governance is a substantial change in the debate about new forms of government, as it is based on the consensus and participation of public and private actors from both profit and non-profit sectors, who are called to decide about shared issues. The focus is not on the quality of the management or its efficacy (as in the New Public Management paradigm), but in the ability to activate and govern partnerships and networks, creating trust and "a sense of future among the participants" (Rotondo, 2011, p. 11). A strong network of public and private organizations and civil society allows the actors to achieve a common goal (Rotondo, 2011, p. 63) if certain conditions are respected: firstly, the actors should share the same greater purposes; secondly, information and knowledge should be shared among all the actors; finally, assumption of responsibility towards the other actors should always be acknowledged (Rotondo, 2011, p. 67).

The principles of public governance consist in cooperation between the actors, active participation, and a managerial strategy based on negotiation, communication, and interactivity. The logics behind the concept of public governance are summarized by the concept of subsidiarity, which ensures a balance among the various levels of public administration and between the latter and the economic or social institutions involved (Rotondo, 2011, pp. 12–13). The Comitato permanente per lo sviluppo del turismo (Permanent Committee for the Development of Tourism), established in 2014, brings together all the stakeholders who, at national and local levels, deal with tourism. The Committee is composed of 39 members and is chaired by the Presidente del Consiglio dei Ministri (President of the Council of Ministers), or a delegate. The Decree providing for the creation of this Committee also promotes actions in various fields, with the aim, amongst others, of establishing agreements with the regions to develop tourism-promoting projects, as well as endorsing cooperation between the various authorities involved (regions, provinces, and municipalities). The goal is to define tourism policies, assist the enterprises operating in this sector, promote the image of Italy both at a domestic and an international level, and ensure equal opportunities in the communication sector (this objective reveals the awareness of the different visibility and attractiveness of tourist sites).

In September 2016, the Committee approved the first development plan for tourism, outlining the guidelines for the 2017–2022 period. The goals are rather ambitious:

> The Plan aims to promote a new method of using Italy's heritage in tourism, based on renewal and expansion of tourism supply of strategic destinations, and enhancing new destinations and new products, in order to generate economic, social and sustainable benefits and thus revive, on new foundations, Italy's position of leadership on the international tourism market. (Strategic Plan, p. 8, English version)

Beyond the rhetoric, there is an acknowledgment that tourism is an important economical factor but lags behind in certain fields: technology and digital communication need to be improved, and there is a little attention to today's consumer needs (insufficient personalization of the offer) and the markets of origin (p. 8). Sustainability is also a core problem, as several tourist sites and resorts are subject to heavy pressure, hence risking a loss of territorial identity (p. 9). The three key words on which the plan is based are sustainability, that is promoting both the cultural and environmental heritage of the area and endorsing a greater involvement of the locals, innovation, and competitiveness (p. 10).

The Plan sets out a series of strategic objectives that address these weaknesses: an integration of the currently fragmented local supply, improvement of competitiveness, by means of a more effective use of marketing policies, and of the governance, with the goal of promoting participation at various levels. These general objectives are, in turn, articulated at different levels (specific objectives, strategic intervention lines, actions) to achieve, through concrete actions, maximum operational efficiency. In promoting the plan, the Committee has placed great emphasis on the integration of different areas (resources, territories, productive processes) and on participation of the national, regional, and sub-regional actors.

The ambition is to define a strategy that takes into account the advice and proposals made by local stakeholders and the civil society. There is the will to maintain a high level of participation even after the approval of the plan, during the implementation phase, although it should be noticed that the official website has no information or updates on how work is progressing. The implementation strategy provides for a continuous involvement of the stakeholders:

> The participation and consultation method has had a broad territorial aspect, in line with the overall approach of the Plan, which views territorial differentiation and enhancement of supply as a means of overall growth of competitiveness of the Italian tourism system. (p. 21)

The Plan aims at reinforcing innovation at different levels and its efficacy will be primarily assessed by analyzing this element. Great importance is given to cooperation among all the actors involved in tourism development, with a particular eye to the participation of citizens and local communities at different territorial levels. The Plan reveals an awareness of all the weak points of the sector and clearly states that "the tourism sector and its related industries are heavily divided in terms of issues and organization and this frequently leads to fragmentation and lack of consistency" (p. 17).

For the first time there is the will to elaborate a strategy, starting from a clear vision of the future of Italian tourism, which includes long-term management and a sustainable use of the cultural and territorial heritage, an increase in the competitiveness of the national tourism system, with an eye to the growth of employment rates, attention to the quality of the experience perceived by tourists, and a better integration of the tourism system (highlighting once again the need for participatory governance) (p. 21). The Plan was elaborated using a "general to specific" approach, but always following the three principles mentioned previously: sustainability, which is not only restricted to the environment; accessibility, hence making all resorts easy to reach for all potential tourists, whatever their age or health conditions; innovation, applied to technology, management, communication and marketing (p. 24; pp. 44–48).

The Strategic Plan has been translated into implementation measures (for the improvement of the cycling mobility at national and transnational level, the development of a digital platform, the promotion of initiatives in tourism management schools, etc . . .), but its effectiveness can only be appraised after a certain length of time. The national plan must also take into account the regional thirst for autonomy. In this respect, the State can play an important role in addressing the international promotion activities, as well as coordinating the actions, thus ensuring adequate support for the initiatives of development and promotion of tourism devised by the regions and local authorities.

7.3.2 Regional Level

The regulations of the Ordinary Regions are moving along the directives outlined by the state legislation, which has greatly evolved from the first law on tourism (Law 217/1983, then repealed by Law 135/2001) to the most recent intervention implementing the so-called Code on Tourism (Legislative Decree 79 dated May 23rd, 2011), which in turn repealed framework Law 135/2001.

The constitutional reform (Constitutional Law no. 3/2001) has made tourism the exclusive competence of the Regions; hence the latter are no longer subject to the limits of the fundamental principles established by state laws. This in theory would appear to be an ideal solution, since it promotes directly responsible management, close to the interests of the locals, and would be able to create the best opportunities, unlike a central management system that would presumably be unaware of the real problems and therefore less concerned. The coherence with the national plan is guaranteed by the State-Regions Conference that, as mentioned in the previous paragraph, ensures an added value in terms of consistency and common guidelines.

However, the governance model actually followed for tourism, with a limited central intervention, has certainly assured the effective implementation of regional policies, but has on the other hand provided sub-optimal solutions for all those problems that are not strictly local. Many of the issues involving groups of regions doubtlessly contribute to the difficulties that Italy currently meets in terms of international competitiveness. Among these:

A. The lack of infrastructures in the South and larger islands. The transport sector for example, has privileged the connection between large cities, but the smaller centers, which have a great tourist appeal, are extremely difficult to reach (Becheri & Maggiore, 2016; CNEL Report 2007).
B. The lack of uniform quality standards across the nation.
C. A non-uniformity of norms and rules at a regional level
D. The lack of a coherent promotional policy at a national level, resulting in an unclear, badly-perceived image of Italy as a destination (excluding its three most visited cities: Rome, Venice, Florence).
E. The limitations in the statistical surveys (CNEL Report, 2007).

Each Region has full and exclusive legislative competence for every aspect of tourism; therefore, the scenario is greatly fragmented and there are many differences passing from one to another. The Regions have a central role both in terms of programs and coordination. Moreover, they cover a number of functions and tasks pertaining to the following issues: tourism planning and innovation; control over the quality of services and activities linked to the regional tourist offer; promotion and creation of a unique brand for the Region; implementation and financing of specific projects; support to tourist operators, data collection, processing, and reporting regional tourism statistics; surveys and information on tourist supply and demand.

Tourism plans are made for three-year periods of time and are implemented by means of annual executive programs that outline the strategic objectives and guidelines and define the economic sources available for tourism-related projects. The three-year and annual plans are not solely defined by the regional authorities. The principle of subsidiarity implies that local authorities, private organizations, enterprises, and trade unions must also be involved. The regional administrative functions are carried out by both the Regional Department of Economic Development and Tourism and other regional agencies. Every region has its own agencies: in general, most of them have abolished the APTs (Azienda Promozione Turistica [Tourist Promotion Agencies]) established by the first framework law on tourism (Law 217/1983) and replaced them with Regional Agencies for the Promotion of Tourism having great organizational, administrative, managerial, and economic autonomy.

In brief, the activities carried out by the Regions include:

A. Programming, through the adoption of three-year plans for tourism development.
B. Promoting the regional brand in Italy and abroad.
C. Financing local development projects and providing incentives to the industry.

D. Coordinating the collection, processing, and dissemination of data on regional tourist demand and offer.

An interesting example of a regional governance tool is the recent Piano Strategico Turismo Veneto (Strategic Plan Veneto Tourism), issued in January 2019. The plan adopted a participatory method, which includes both public and private subjects involved in the tourism industry, and provides for:

A. A Regional Control Room (Cabina di Regia) with coordination functions, which is headed by the Tourism Department, with the support of CISET (International Centre of Studies on the Tourism Economy) and the Department of Economics (both from the Ca' Foscari University of Venice), as well as Veneto Innovazione S.p.A. and local trade associations.
B. Five Thematic Tables attended by Veneto Region, public (local administrations, Universities, and research institutes, etc.) and private stakeholders (trade associations, consortia, Club di Prodotto, etc.). Strategic issues are discussed, such as tourism products, infrastructures, digital tourism, welcoming actions, and communication strategies.

7.3.3 Sub-Regional Level

7.3.3.1 Gruppi di Azione Locale (GAL) – Local Action Groups (LAG)

LAGs are innovative tools within the governance of rural territorial systems. They are an example of participated planning and management for territorial development, characterized by a participative methodology that adopts a bottom-up approach. These initiatives are not directly focused on tourism; nonetheless, tourism is often included in the actions promoted by this network of actors (Tafuro, 2013; Messina, 2018). An example is the development of slow tourism initiatives in the delta of the river Po or in Alta Marca Trevigiana (http://www.slow-tourism.net/).

7.3.3.2 Sistemi Turistici Locali (Local Tourist Systems)

Law 135 /2001, mentioned previously, also provides for the establishment of Local Tourist Systems. STLs are defined as homogeneous or integrated contexts, characterized by an integrated supply of cultural, environmental, and tourist attractions, including typical products of agriculture and local craftsmanship, or by the presence of single or associated tourism enterprises. The STLs may include areas belonging to different regions and are called to intervene in different sectors, such as technological innovation, urban and environmental requalification, online marketing, promotion of typical products, ecological certification, infrastructures, safety, and service standards.

In practice, they are no more than clusters of public and private bodies that are recognized officially by the Region, cooperating for the development of the tourist chain in a specific and homogeneous territory. They endorse a bottom-up approach for the development of the area and have various operational purposes, such as organizing, managing, and promoting tourism-related assets and services, creating a network both at local supply level and on the global market, and making the territory visible and distinguishable by "branding" it. The idea behind STLs is to create innovation and development processes, without separating all the various economic sectors. In brief, the STL-based model is one that conceives and manages the territory and its resources with an integrated, networked approach (Giannone, 2002). The decision to establish STLs is made by either local authorities or private stakeholders together with trade associations. Regional accreditation is required to be eligible to receive the funds allocated for the implementation of development projects.

The creation of a network of food and wine itineraries, cultural and tourist facilities, and local means of transport requires an active role of local authorities, who thus become the promoters of a tourism development model based on the capabilities of the local economic and cultural actors. STL is not just an entrepreneurial activity, but as Costa observes (2002, pp. 255–268), a system based on social capital, where interpersonal and business relationships with the local authorities are based on trust and inter-organizational collaboration.

Dallari (2007, pp. 194–218) describes the different approaches of the regional administrations towards STLs: some regions openly decided not to refer to Law 135/01, because the existing organization was considered consistent with the national law; other approved STLs with some slight modifications or are in the midst of doing so (see also Rotondo, 2011, pp. 118–122). In brief, the regional laws define a variety of ways through which STLs are created, but all of them require a formal collaboration between public authorities and the private sector (Malo & Perini, 2013, p. 84).

This governance model, conceived as an innovative tool encouraging integrated tourism development, hides some risks, and mainly that of simply creating a standard promotional agency. There is also the risk of failing in the negotiation processes needed to find a power-balance between resorts, stakeholders, associations; limiting the actions to the usual itineraries and resorts; increasing the number of institutions, thus duplicating the competencies (Dall'Ara & Morandi, 2004, p. 95). The latter risk has been clearly highlighted by Becheri: the rush to create STLs ended soon, as the regions realized that they were often useless (Becheri, 2014, p. 17; 2016, p. 768).

An interesting study by Cerquetti et al. (2007) analyses the structural characteristics, operating modes and touristic performance of some STLs in the Marche region, and more specifically, the provinces of Pesaro and Urbino (Urbino; Montefeltro and Altamarina) and Macerata (Terre dell'Infinito). The conclusions are far from positive. According to the authors, instead of simplifying the tourist offer, the STLs gave rise to a proliferation of tourism-related institutions operating at a local level and a duplication of promotional initiatives and trademarks. Their limited entrepreneurial vocation and poor use of management and marketing techniques are the main weakness of the STLs, which therefore fail to propose competitive tourism experiences and attract new markets or customer niches. Moreover, the study highlights the political use of such a system; there is the concrete risk that STLs simply become "operational tools" of the political stakeholders (provinces and municipalities), a simple set of promotional activities created through a top-down approach. This way they would lose their innovative trait given by the involvement of the network of stakeholders, the bottom-up approach, and the attention to the needs of the territories (Cerquetti et al., 2007, pp. 347–351). Costa notes that they can become a tool used by lobbies to intercept public resources, and as such the author denounces the lack of precise local development indicators, which can serve as verifiable outputs (2002, pp. 255–268).

7.3.3.3 Provinces and Municipalities

The Regions imbue the Provinces with some administrative functions, which the Provinces can in turn delegate to the municipalities. Provinces play a key role in harmonizing the activities of the municipalities and coordinating the cross-sector activities and the territories. The Provinces are in general responsible for the management of the IATs (Informazioni Accoglienza Turistica [Tourist Information and Welcome Offices]), although these tasks may be carried out by the municipalities (as in the region Marche). The organization of all aspects pertaining to promotion, information, tourist welcoming varies according to the strategic choices of each local authority (Malo & Perini, 2013, pp 78–79).

The municipalities carry out administrative activities and actions aimed at supporting and endorsing local initiatives. They play a central role in the promotion of an integrated tourist offer and the creation of public-private cooperation networks, such as those of the Pro Loco (Tubertini, 2007, p. 30; Malo & Perini, 2013, p. 86). The first Pro Loco was established in 1881 in Pieve Tesino (Province of Trento), which at the time was under the Austro-Hungarian Empire. The name Pro Loco is a Latin expression that means "in support of the district" and refers to the goal that these private, non-profit societies have: to promote awareness and understanding of the local history and traditions and protect the cultural heritage of the area. The Pro Loco societies are organized at a national level in the Unione Nazionale Pro Loco d'Italia (UNPLI), which was established in 1962 (Malo & Perini, 2013, p. 86).

Municipal competences pertain to the promotion of the territory and the development of projects aimed at qualifying the local supply system: information, hospitality, promotional initiatives, and events (Tubertini, 2007, p. 31). An example of the role played by the municipalities is the Project of territorial governance of tourism in Venice (2017), addressing the problems linked to tourist seasonality and the over-tourism that put this World Heritage Site at risk. Once again, the tool is a document, whose usefulness is debatable.

Other private associations have played and still play an important role in the development and promotion of tourism: Touring Club Italiano (TCI, established in 1894), Club Alpino Italiano (CAI, established in 1893), Centro Turistico Studentesco Giovanile (CTS, established in 1974), and Fondo Ambiente Italiano (FAI, established in 1975), professional associations (Confturismo, Fiavet, Federalberghi, Assotravel etc.).

7.4 Conclusion

The weaknesses of the Italian tourist system are listed in the National Strategic Plan published in 2013: problems in tourism governance, highly fragmented promotional actions, especially abroad, large number of small-sized companies, scarce ability to create competitive tourist products, inadequate infrastructures, poor training of the human resources operating in this field, difficulty in attracting international investments, and so forth. Although it makes no explicit reference to these problems, the new strategic plan seems aware of these issues and suggests a solid framework, with a focus on integration of all the stakeholders involved and defining precise plans that will translate the strategic goals into actions: multilevel governance is considered pivotal for the coordination of the different actors. At the same time, there is an urgent need to develop evaluation tools to monitor the implementation of the plan and the effects on the communities involved in tourism development, and therefore intervene whenever the results fail to meet the expectations.

References

Alvisi, C. (2015). *Il diritto del turismo nell'ordine giuridico del mercato*. Torino: Giappichelli.

Bardelli, D. (2004). *L'Italia viaggia: il Touring Club, la nazione e la modernità*. Roma: Bulzoni.

Banca d'Italia. (2016). Indagine sul Turismo internazionale. Retrived from: http://www.bancaditalia.it/pubblicazioni/indagine-turismo-internazionale/2017-indagine-turismo-internazionale/statistiche_ITI_17072017.pdf Date accessed: 01/ 10/2019.

Battilani, P. (2011). *Storia del turismo: le imprese*. Bologna: il Mulino.

Battilani, P. (2001). *Vacanze di pochi, vacanze di tutti*. Bologna: il Mulino.

Becheri, E. et. al. (2010). *Il turismo in Italia e in Emilia Romagna. Dall'ordine sparso alla geometria variabile*. Milano: Franco Angeli.

Becheri, E. (2014). Editoriale. I Mutamenti del turismo. Turistica. *Italian Journal of Tourism*, XXIII (1/2), 5–19.

Becheri, E., Maggiore, G. (Eds) (2016). *Rapporto sul turismo italiano 2015/2016. XX Edizione*. Napoli: Rogiosi.

Becheri, E. (2016). Governo del turismo. In E., Becheri, G., Maggiore (Eds). *Rapporto sul turismo italiano 2015/2016. XX Edizione* (pp. 689–710). Napoli: Rogiosi.

Berrino, A. (2011). *Storia del turismo in Italia*. Bologna: il Mulino.
Cavallo, L. (2016). I flussi turistici in Italia nell'ultimo decennio. In E., Becheri, G., Maggiore (Eds). *Rapporto sul turismo italiano 2015/2016. XX Edizione* (pp. 55–68). Napoli: Rogiosi.
Cerquetti, M., Forlani, F., Montella, M. & Pencarelli, T. (2007). I Sistemi Turistici Locali nelle Province di Pesaro e Urbino e Macerata. In S. Sciarelli (Ed), *Management dei Sistemi Turistici Locali* (pp. 315–351). Torino: Giappichelli.
CNEL & Casadio G. (2007). Il turismo come settore produttivo Retrived from: https://www.cnel.it/Documenti/Studi-e-indagini Date accessed: 06/ 06/2019.
Coletti, R. & Sanna, V.S. (2012). *La governance pubblica del turismo in Italia: elementi chiave e nodi critici. Annali del Dipartimento di metodi e modelli per l'economia, il territorio e la finanza 2010–2011*. Bologna: Patron Editore.
Comune di Venezia. (2017). Project of Territorial Governance of Tourism in Venice. Retrived from: https://www.comune.venezia.it/sites/comune.venezia.it/files/documenti/documenti/territorial%20governance%202017.pdf Date accessed: 06/ 06/2019.
Costa, N. (2002). Verso la progettazione e gestione dei sistemi turistici locali. In F. Sangalli (Ed). *Le organizzazioni del turismo* (pp. 255–268). Milano: Apogeo.
Costa, P., Manente, M. & Furlan, M.C. (2001). *Politica economica del turismo: lezioni, modelli di gestione e casi di studio italiani e stranieri*. Milano: Touring Club Italiano.
Dall'Ara, G., Morandi, F. (2004). *I Sistemi Turistici Locali. Normativa, progetti e opportunità*. Matelica: Halley.
Dallari, F. (2007). Distretti turistici tra sviluppo locale e cooperazione interregionale. In F. Bencardino & M. Prezioso (Eds). *Geografia del turismo* (pp. 194–218). Milano: McGraw-Hill.
Degrassi, L., Franceschelli V. (Eds) (2010). *Turismo. Diritto e diritti*. Milano: Giuffré.
ENIT (2020). Bollettino n.9. Retrieved from: https://www.enit.it/wwwenit/images/multimedia/Bollettino_Ufficio_Studi/Bollettino_9/BOLLETTINO-ENIT-N9.pdf Date access: 25/ 11/2020.
Franceschelli, V. & Morandi, F. (2013). *Manuale di diritto del turismo*. Torino: Giappichelli.
Giannone, M. (2002). I sistemi turistici locali: un approccio alla dimensione territoriale dello sviluppo turistico. In E. Becheri (Ed.). *XI Rapporto sul turismo italiano* (pp. 601–608). Firenze: Mercury.
Gola, M., Zito, A. & Cicchetti, A. (Eds) (2012). *Amministrazione pubblica e mercato del turismo*. Rimini: Maggioli.
Grasselli, P. (2001). *Economia e politica del turismo*. Milano: Franco Angeli.
Guidicini, P. & Savelli, A. (1999). *Strategie di comunità nel turismo mediterraneo*. Milano: Franco Angeli.
ISTAT. (2019). Movimento Turistico in Italia. Anno 2018 Retrived from: https://www.istat.it/it/files//2019/07/infograficaIT.pdf Date accessed: 02/ 12/2019.
ISTAT. (2019). Turismo. Retrived from: https://www.istat.it/it/files//2019/12/C19.pdf Date accessed: 02/ 12/2019.
Malo, M. & Perini, A. (2013). Promozione, informazione, accoglienza turistica. In V., Franceschelli & F., Morandi (Eds). *Manuale di diritto del turismo* (pp. 69–102). Torino: Giappichelli.
Manente, M. & Mingotto, E. (2016). *Gestire il turismo: come valutare strategie e azioni*. Milano: Franco Angeli.
Mariotti, G. (1958). *Storia del turismo*. Roma: Ed. Saturnia.
Messina, G. (2018). La governance delle aree rurali: l'esperienza del GAL Elimos. *Geotema*, 57, 239–246.
MiBACT. (2015). Stati Generali del Turismo Sostenibile, Pietrarsa. Retrived from: http://www.beniculturali.it/mibac/multimedia/MiBAC/documents/1450261619430_Documento_Finale_Pietrarsa-REV.pdf Date accessed: 26/ 08/2019.

MiBACT. (2017). The Strategic Plan for Tourism 2017–2022. Retrived from: http://www.pst.beniculturali.it/wp-content/uploads/2017/05/PST_2017_ENG_21apr17.pdf Date accessed: 07/ 06/2019.

Moro, E. (2007). *Nuovi orizzonti della politica locale per lo sviluppo del turismo sostenibile.* Empoli: Barbieri, Noccioli & C.

Morvillo, A. & Becheri, E. (Eds) (2020). *Dalla crisi alle opportunità per il futuro del turismo in Italia.* Napoli: Rogiosi.

Notarstefano, C. (2010). *Genesi, evoluzione giuridica e orientamenti comunitari del turismo sostenibile.* Bari: Cacucci.

OECD. (2011). OECD Studies on Tourism: Italy Retrived from: https://read.oecd-ilibrary.org/industry-and-services/oecd-studies-on-tourism-italy_9789264114258 Date accessed: 14.11.2019.

Petti, C. (2009). *La gestione innovativa dei sistemi turistici.* Milano: Franco Angeli.

Presidenza del Consiglio dei Ministri- Ufficio per le Politiche del Turismo, ONT. (2012). Rapporto sul turismo 2012. Retrived from: http://www.ontit.it/opencms/export/sites/default/ont/it/documenti/files/ONT_2013-08-09_02992.pdf Date accessed: 06/ 06/2019.

Regione Veneto. (2019). Programma Strategico Turismo Veneto (PSTV). Retrived from: http://www.regione.veneto.it/c/document_library/get_file?uuid=fe8e70ab-ac5a-4ed9-8d1e-703ca72d87af&groupId=10813 Date accessed: 07/ 06/2019.

Rotondo, F. (2011). *Principi di public governance nei sistemi integrati di offerta turistica.* Torino: Giappichelli.

Sainaghi, R. (2004). *La gestione strategica dei distretti turistici.* Milano: EGEA.

Sangalli, F. (2002). *Le organizzazioni del turismo.* Milano: Apogeo.

Tafuro, A. (2013). *Gruppi di azione locale: governance e sviluppo del territorio.* Bari: Cacucci.

Tomati, M. (1973). *I fabbricanti di vacanze. La rivoluzione nella industria turistica.* Milano: Franco Angeli.

Tubertini, C. (2007), Il turismo tra Stato, Regioni ed enti locali: alla ricerca di un difficile equilibrio delle competenze. *Le istituzioni del federalismo suppl.* 1, 21–40.

Unicredit, TCI. (2016). Rapporto sul turismo 2016. Retrived from: https://www.unicredit.it/content/dam/ucpublic/it/chi-siamo/documents/noieleimprese/UCI–TCI-2016_pagina-doppia.pdf Date accessed: 27/ 03/2019.

Unicredit, TCI. (2017). Rapporto sul turismo 2017. Retrived from: https://www.unicredit.it/content/dam/ucpublic/it/chi-siamo/documents/noieleimprese/UC–TCI-2017-low.pdf Date accessed 27/ 03/2019.

UN-WTO Tourism Highlights. (2018) Retrieved from: https://www.e-unwto.org/doi/pdf/10.18111/9789284419876 Date accessed: 07/ 06/2019.

World Economic Forum. (2019). The Travel & Tourism Competitiveness Report. Retrieved from: http://www3.weforum.org/docs/WEF_TTCR_2019.pdf Date accessed: 23/ 09/2020.

World Travel Tourism Council. (2017). Travel &Tourism Economic Impact 2017 Italy. Retrived from: https://www.wttc.org/-/media/files/reports/economic-impact-research/countries-2017/italy2017.pdf. Date accessed: 27/ 03/2019.

Kamil Hamati

8 Lebanon

8.1 Introduction

Tourism is becoming the sector of activity where global dynamics can create multiple local externalities as well as new pressures on local communities and natural ecosystems. Nevertheless, tourism development can produce positive outcomes if the sector and the development strategies of stakeholders involved is connected to Global agendas such as Agenda 2030 in order to mitigate, adapt, and integrate elements of Global change while offering innovative thinking and processes connecting global tourism to the local priorities.

The case of tourism in Lebanon illustrates the multiple sets of challenges and opportunities towards timely and positive localization of global tourism's trends and dynamics. Indeed, during its "golden era" Lebanon was often identified as the emblem of successful tourism development policy until 1970s. Often referred to as the "Switzerland of the Middle East" due to the high prosperity experienced by the country, Lebanon experienced a surge of tourists flowing into its capital, Beirut. The economic flourishing was also attributed to a booming in its agriculture, commerce, and banking sectors (Ladki & Sadik, 2004; Lebanese Republic, Ministry of Foreign Affairs, Embassy of Lebanon to the Arab Republic of Egypt – Cairo, 2018; Rowbotham, 2010). The country gained much of its fame during the "golden era" thanks to its connection to the global cultural scene showcasing the richness of its natural capital and the hospitality of its rich and diverse communities (Bkhairia, 2017; Mortimer Winterthorpe, 2018). The country's tourism development also benefited from its connection with its own elite diaspora reminiscing and connecting a global audience to the depth of Lebanon's natural, cultural, and human capital (Gibran Khalil Gibran, Amin Maalouf, and others) (Bkhairia, 2017; Mortimer Winterthorpe, 2018). Thus, the country's reputation gave it a historical appeal to tourists from all across the world; however, its geographical location at the heart of the Middle East in an era of political turmoil slowly took its toll on its fame, accomplishments, and success story (Bassil et al., 2015; Ladki & Sadik, 2004).

8.1.1 Historical and Geographical Background

Lebanon's geographical location in Western Asia, at the eastern shores of the Mediterranean, made it a natural passageway between the East and the West as the intersection of migratory and empire-building routes. Thus, a host of foreign conquerors as well as ordinary, traders and settlers made Lebanon their temporary and permanent home. The Phoenicians (around 3000 B.C.), renowned for their interregional

https://doi.org/10.1515/9783110638141-008

trade, were the first known civilization to have lived and prospered collectively on this land. Other nations and communities followed, the most noticeable were the Assyrians, the Greeks, the Romans, the Persians, the Arabs, the Crusaders, and lastly the Ottomans then the French (Collelo et al., 1989; Rowbotham, 2010; Salibi, 1971).

Lebanon as an autonomous political entity was first acknowledged during the Ottoman rule as the "Emirate of Mount Lebanon," then as the Ottoman administrative unit, "Mutasarrafiya of Mount Lebanon" which had a relative degree of autonomy within the Ottoman Empire, guaranteed by European powers of France, Britain, Austria, Prussia, Russia, and Italy (Traboulsi, 2012a). However, the country's size and borders were not clearly defined until the declaration of "Greater Lebanon" in 1920 under the French mandate that was a result of the 1916 Sykes-Picot Accord (Collelo et al., 1989; Traboulsi, 2012a).

The confessional system at the heart of Lebanon's current political system laid the ground for what was foreseen to be a transitional phase towards a more secular political system (Library of the Congress, 2018; Maila, 1992). However, recent history proved that the Lebanese religious groups and communities were unable to reach a mutually agreed consensus towards collective wealth distribution and development vision. Several civil unrests broke out since, the most infamous of which is the Lebanese Civil War of 1975–1990 that caused an exodus of millions of Lebanese people, claimed hundreds of thousands of civilian lives, resulted in tens of thousands of permanent disabilities and internally displaced people, in addition to millions of dollars in economic losses (Norkonmaa, 1995). In addition, the country's southern neighbor's attack spread additional havoc across the country in 1982, 1996, and 2006 (BBC, 2018; P. Salem, 1998; Security Council Report, 2018; Traboulsi, 2012b).

Finally, after the civil war and as a result of regional powers competitive agendas, Lebanon witnessed an intermittent number of both political and military shocks. Consequently, localized tensions adversely affected its security, stability, and capacity to trigger innovation needed to promote many sectors of activities including tourism, the latest of which is the Syrian civil war that has over-spilled on the Lebanese economy, security, infrastructure, and politics (Ladki & Dah, 1997; G. Salem & Azoury, 2017).

8.1.2 Ethno-Geographic and Economic Landscape

The country is home to approximately 6 million people. Around 3% are Palestinian refugees, 16% are Syrian refugees, and 2.5% are Armenians, with a minority of other nationalities that include Ethiopian, Filipino, and Sri Lankan (Central Intelligence Agency, 2020; Embassy of Armenia to Lebanon, 2017; United Nations, Department of Economic and Social Affairs, Population Division, 2019). The country's diverse topography offers coastal zones on the Mediterranean with a moderate climate across the seasons, and a mountainous hinterland providing privileged connections with a different natural ecosystem to urban dwellers. However, the highest agglomeration of residents (87%) is settled in urban areas putting exorbitant pressure on infrastructure (World Bank, 2019).

This mixture of cultures and identities creates the foundation for multicultural sets of innovation models enabling entrepreneurship and regional leadership in various sectors of activity, and above all, tourism. In total, 18 religious sects are recognized in Lebanon, with Muslim sects representing 58%, 12 Christian sects (36%), Druze (5%), and Jewish (<1%) (Central Intelligence Agency, 2020; United States Department of State. Bureau of Democracy, Human Rights and Labor, 2011). This is a matter of great pride for the Lebanese who claim to set an example of coexistence in the Arab world, despite their history failing to substantiate the case (Raymond, 2013; Spaan, 2013; van Ommering, 2015). Albeit Arabic being the country's official language, French and English are commonly used by the Lebanese, thus supporting its connectivity to the region and the world (Central Intelligence Agency, 2020).

Having been a hub for interregional trade along the Silk Road, the country's economy relied historically on trade and services, which resulted in turning Lebanon into a multi-cultural metropolitan center. To cater for their visitors, the Lebanese soon evolved into first-rate service providers, and the services sector was molded as a pillar of their economy (Bassil et al., 2015). In 2017, 76% of the country's US$ 43.78 billion GDP, was contributed by the services sector, its main components being real-estate and construction, finance, and tourism (Harake et al., 2018; Lebanese Republic, Presidency of the Council of Ministers, The Investment Development Authority of Lebanon, 2018b; World Bank, 2019). It is noteworthy to highlight however, that in 2017 Lebanon was still suffering from the overspill of neighboring Syria's war, negatively impacting its geostrategic role on the regional trade routes (G. Salem & Azoury, 2017).

Despite these negative developments, the Lebanese diaspora, estimated to outnumber the country's resident population, including many entrepreneurs have been investing in the country's tourism infrastructures and the rest adopted it as their holiday destination. This connectivity has created a vital economic line between Lebanon and the rest of the world. It has boosted tourism both directly and indirectly and attracted different forms of foreign investments which ended up reshaping the real-estate business in the country and fueling the banking sector (Lebanese Republic, Central Administration of Statistics, 2018; Lebanese Republic, Presidency of the Council of Ministers, The Investment Development Authority of Lebanon, 2018b).

Furthermore, Lebanese authorities created a welcoming environment for foreign investors and the economic elite notably through a policy of banking secrecy (Banking Secrecy Law of September 3rd, 1956) (قانون سرّية المصارف الصادر بتاريخ 09/03/1956 – Banking Secrecy Law of 03/09/1956, 1956), a free banking zone under a free exchange system, and a free movement of capital and earnings. Additionally, in order to reinforce trust in the sector, the Central Bank of Lebanon (BDL) pegged the country's currency at 1507

Lebanese Lira (L.L.), sometimes referred to as the Lebanese Pound, to the dollar. Both currencies are interchangeably used across the country (Banque du Liban, 2018; Schimmelpfennig & Gardner, 2008; Tax Justice Network, 2018; The Economist, 2018).

However, Lebanese economy remains highly affected by the country's political stability, which in turn controls its security situation (Bassil, 2013; Harake et al., 2018). And, as historically proven, Lebanon's political stability hinges around both regional and international politics.

Tourism has always been one of the major pillars of Lebanon's economy. The country capitalizes on its rich natural and cultural capital along with its long heritage in the services sector, to develop a tourism industry that stays vulnerable vis-à-vis a constantly volatile political situation. This chapter aims to present the complex dynamics that govern the Lebanese tourism sector and to draw an overall picture of the sector's prospects and challenges, and to devise policy recommendations to ameliorate the equity of the sector's development.

8.2 National Governance Structure and Local Politics of Tourism

Tourism development in Lebanon and its evolution has been tightly linked to the country's political systems. Multiple divergences between the legal and constitutional framework and the political practices and their territorialization in reality have shaped the tourism industry in the country.

Indeed, "Lebanon is a parliamentary democratic republic (. . .)" according to the preamble of its constitution. The state power is thus represented by three main actors: the President of the Republic, the Prime Minister, and the Speaker of the House (Lebanese Constitution, 1995). It is noteworthy to highlight, however, that despite Lebanon not being a centralized autocratic political regime, such as monarchy, in practice it is ruled within a general practice of socio-cultural hereditary rule that has been governing its political scene since the Ta'if Agreement (1989). The "National Pact" of 1943 instated customs of representation in political power on confessional basis. The president of the republic was thus to be Christian Maronite, the prime minister Muslim Sunni, and the speaker of the house Muslim Shia (Library of the Congress, 2018; Maila, 1992; P. Salem, 1998; Saliba, 2012).

The constitution divides state power in Lebanon to three branches: the judiciary, the executive power, and the legislative power. The legislative power is vested in the Chamber of Deputies, comprising 128 deputies elected by Lebanese citizens for 4-year terms, according to three main principles (Lebanese Constitution, 1995):

A. Equal religious representation (Christians and Muslims);
B. Proportional representation within religious sects, and;
C. Proportional representation among geographic regions.

Lebanon is administratively divided into 8 governorates, also referred to as "Muhafazat," that are responsible for 25 districts and the group of municipalities that fall under their geographical jurisdiction. All of which report to the Ministry of Interior and Municipalities (Localiban, 2014). This administrative division has given a spatial dimension to the socio-hereditary rule and it further reinforced the practice of clientelism on the basis of geo-cultural coalitions and enshrined territoriality of the political system (Saliba, 2012). As a result, Lebanon's lack of a standardized governance system led to sporadic development in tourism and in urbanization. Consequently, local authorities have a large space to nationally reflect the global tourism trends and dynamics within their local specificities.

Additionally, a plethora of scholars describe the political system in Lebanon as clientelist, relying heavily on one's confessional affiliation, especially for high-level positions. This has given each community its own territorial encroachment with responsibility towards the wellbeing, prosperity, and security of citizens living under its geographical custody (Maila, 1992; P. Salem, 1998; Saliba, 2012). However, with an effective governance scheme, the Lebanese context could present an opportunity for living-lab based innovation where stakeholders' engagement and connection are at play and practiced in different ways and insulated from heavy-handed interventions of central government. Thus, ideally, the integrative public-private-people dynamics of consultation and monitoring support the wellbeing and the betterment of the community would suggest the sustainability of an innovative development model.

Despite the government being responsible for producing long term vision and policy guidance for development ventures, the financial strength of the private sector gave it influential power over the political system, and thus contributed to the development of short-term policies in favor of quick profit rather than developmental longer-term policies. Consequently, this power enabled the private sector to either entirely bypass legislation or to maneuver its way around it (Abir Saksouk, 2015; Kisirwani, 1997; Preston, 2018). Additionally, legislation is vague on leveling standards and rating systems, especially in the geographic context (Law nb 646 of 11/12/2004, and decree nb 69 of 9/9/1983) (قانون التنظيم المدني – The law of urban planning, 1983; (قانون 16/09/1983 تعديل المرسوم الاشتراعي رقم 148 تاريخ) – Amendment of the legislation nb 148 of 16/09/1983 (the building law), 2004), giving considerable room for local authorities and developers to reshape the building envelope and skyline of different urban centers across the country (Abir Saksouk, 2015; Ghandour, 2011; Preston, 2018).

In theory, the public institutional governance system is comprised of specialized ministries with specific mandates insinuating sectoral centralization. However, in practice, most ministries' work is linked to that of others (Figure 8.1). Tourism, theoretically vested in the Ministry of Tourism (MoT), is in reality governed through a multi-institutional scheme. The MoT itself is only mandated with promoting and monitoring the industry (Lebanese Republic, Ministry of Tourism, 2010). To do so, the ministry receives its financial allocations – which merely cover its running

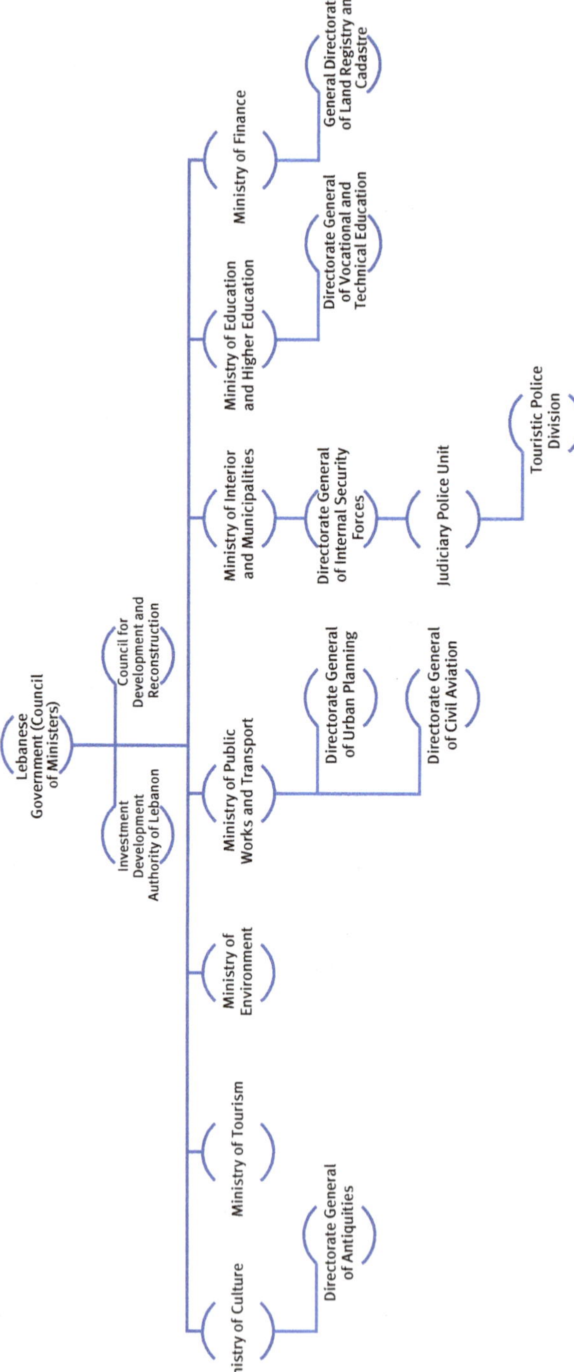

Figure 8.1: Governance structure of the tourism sector in Lebanon.

costs – through a budget presented by the Ministry of Finance (AlBustani, 2018; Ministry of Finance, 2018), whereas tourism law enforcement and monitoring are carried out by tourism police, a branch of the Ministry of Interior and Municipalities (Lebanese Republic, Internal Security Forces, 2018). The wealth of Lebanese tourism industry, the historical sites and museums, are under the mandate of the Ministry of Culture (MoC) (Lebanese Republic, Ministry of Tourism, 2010). Additionally, hospitality management education falls under the mandate of the Ministry of Education and Higher Education (MEHE), the custodian of public education institutes. In addition, this has not prevented the development by private universities of curricula on hospitality management (Lebanese Hotel Association, 2018; Lebanese Republic, Ministry of Education and Higher Education, 2018).

As Lebanese tourism has slowly converged into a leisure industry, major tourism development projects form the bulk of the sector's expansion (Rowbotham, 2010). Depending on the size and nature of the project, governance is either practiced by: the Investment Development Authority of Lebanon (IDAL) (reporting directly to the council of ministers (CoM)), or the Directorate General for Urban Planning (DGUP) (which falls under the Ministry of Transport and Public Works), in the case of private projects; or by the Council for Development and Reconstruction (CDR), for the public projects. The CDR is also affiliated with the CoM (Lebanese Republic, Ministry of Environment, 2003; United Nations Human Settlement Programme, 2013).

The following chart illustrates all stakeholders who share a responsibility in tourism development, across the political system. This type of extremely bureaucratic governance scheme proved to be cumbersome to all stakeholders involved. The decentralization of decision-making and the large spectrum of stakeholders gives homage to the power-sharing political system that governs the entire country and complicates the administration of reforms.

8.3 Integrating Tourism in Lebanon's National Development Plans

Since 1995, the volume of tourism in Lebanon has been oscillating between profitable highs and costly lows. Post-war tourism suffered from several shocks relating one way or another to political turmoil, the sector's golden era spanning from the 1950s until the 1970s is yet to be matched. The spillover of the Syrian war extended to a deteriorating security situation in Lebanon, further hindering the tourism industry (G. Salem & Azoury, 2017). Shortly after the debut of the war, over 1 million Syrians took refuge in Lebanon, and their number soon reached an approximate quarter of the Lebanese population (United Nations High Commissioner for Refugees, 2018). As a result, population density swelled from 450 to around 600 persons/km2 between 2011 2018 (World Bank, 2019), stretching the limits of the country's already suffering infrastructure (Mariusz Jarmuzek et al., 2014). Living conditions deteriorated significantly for both, guests and hosts, and social friction ensued.

Visitors from Arab countries make the biggest share of tourists visiting Lebanon (almost 50%, i.e. around 4 million entries annually) (Lebanese Republic, Central Administration of Statistics, 2018). Its location, nature, climate, and multi-cultural façade merging Western culture with Eastern traditions appealed to them, as this combination was unique in the mostly homogenous region. Prior to the war in neighboring Syria, tourists originating from Iraq, Syria, Jordan, and the Gulf countries, entered Lebanon through its Eastern and Northern borders. Many international tourists also used the opportunity to make a Middle Eastern tour, spending a day or two in Lebanon and visiting neighboring countries by land (G. Salem & Azoury, 2017). The second major group of inbound travelers are Lebanese diaspora whose numbers reach up to 46% of the entirety of incoming travelers (i.e. around 2.5 Million entries, annually) (Lebanese Republic, Central Administration of Statistics, 2018).

The multitude of cultures that have shaped Lebanese heritage, provide the country with a richness in archaeological spots and historical monuments that placed several of its cities on the UNESCO list of World Heritage Sites. Heritage tourists typically admire the cities that boast of history and culture: from Beirut, to Sidon and Tyre in the south, passing by Baalbek and Anjar in the east, and all the way north, to Byblos (also referred to as "Jubail") – world's oldest continuously inhabited city (United Nations Educational, Scientific and Cultural Organization, World Heritage Centre, 1984).

Some tourists from the Gulf appreciate the moderate climate of Mount Lebanon and have adopted it as their summer escape. Another array of Gulf and other Arab tourists favor Beirut as a destination; they admired the lush services that were provided to them in a posh setting (Ladki & Sadik, 2004; Rowbotham, 2010). Other tourists are mostly attracted to the Mediterranean charm and resort to beach retreats. In short, tourists across an array of interests, found in Lebanon exactly what they were looking for.

Tourism in Lebanon has been consistently an integral part of its modern economy. It directly contributes to around 6.85% of Lebanon's GDP, and indirectly to around 12.4%. Additionally, it creates around 338,500 direct and indirect jobs. International tourism receipts reached their highest at US$8 billion in 2010 and are slowly relapsing after their significant drop in 2011 (Lebanese Republic, Central Administration of Statistics, 2018; Lebanese Republic, Presidency of the Council of Ministers, The Investment Development Authority of Lebanon, 2018b; World Bank, 2018).

Realizing the importance of the sector, and in an effort to compensate for the MoT's lack of funding, Lebanese authorities created an inviting environment for foreign investment in the tourism sector (such as Resolution 51/2002 by the Ministry of

Tourism in Lebanon) (Lebanese Republic, Presidency of the Council of Ministers, The Investment Development Authority of Lebanon, 2018b). Several institutions, notably the Ministry of Tourism, BDL through support to commercial banks, a loan guarantee agency called Kafalat, and IDAL) provide credit facilities through long term loans with low interest rates, at times reaching 50–100% exemptions from corporate tax. Additional credit facilities included reductions on, or exemptions from administrative fees, and some exemptions from customs duties (Lebanese Hotel Association, 2018; Lebanese Republic, Presidency of the Council of Ministers, The Investment Development Authority of Lebanon, 2018a). They managed to attract many international stakeholders, mainly from the UK, Saudi Arabia, UAE, Kuwait, Libya, and Iraq. Between 2003 and 2011, the "Investment Development Authority of Lebanon" (IDAL), single handedly mediated 12 tourism related projects ranging in size from US$18 million to US$155 million (Lebanese Republic, Presidency of the Council of Ministers, The Investment Development Authority of Lebanon, 2014, 2018b; Tax Justice Network, 2018).

On the other hand, the Arab countries' funding injected in Mount Lebanon projects caused a patchy blooming of apartment hotels and restaurants. Visitors rapidly realized the financial advantage of buying apartments rather than renting and demand for property steeply increased. A similar phenomenon dotted Beirut's skyline, and the high demand for luxury condominiums and apartments resulted in a ballooning of the real-estate market. Consequently, the Syrian refugee crisis that impeded Lebanese economy inversely hiked the prices of the real-estate sector to a point that it adversely affected locals' ability to invest in real estate (Mariusz Jarmuzek et al., 2014; Schimmelpfennig & Gardner, 2008; The Economist, 2018). This exemplifies the multiple connections between national development strategies and the country's volatile regional geopolitics affecting the capacity of central government as well as local stakeholders to implement long-term tourism promotion strategies. This context is worsened by the different politico-confessional configurations and clientelist agendas resulting in ministries coming up with silo-based individual strategies rather than working together through administrative, political, confessional boundaries (Lebanese Republic, Ministry of Tourism, 2010).

Nevertheless, having the private sector as a backbone of tourism provided the industry with a great adaptative capacity and agility in face of socio-political and economic volatility. In the occurrence of crisis those regions and stakeholders are insulated by the fragmented national landscape and continue to prosper during the crisis while improving their models based on the experience of those affected. Not surprising that despite all the shocks that it has sustained throughout history, tourism is one of the fastest recovering sectors in Lebanese economy (Bassil et al., 2015; G. Salem & Azoury, 2017).

The positive impact of the highly privatized tourism industry should not distract one from multiple pitfalls associated to the lack of oversight and scrutiny of private sector's involvement in tourism. The case of Solidere (the Lebanese Company for the

Development and Reconstruction of the Beirut Central District) illustrates those drawbacks after the civil war. Indeed, Lebanon was left with major developmental challenges across the social, economic, and environmental fronts. In an effort to reverse the detrimental effects of the war, the government inaugurated the "Horizon 2000" reconstruction plan that was designed to boost the economy and rehabilitate the country's infrastructure, including expansion and reconditioning of the airport, and an overhaul of downtown Beirut. As such, the plan was foreseen to build a network of vital structures that would promote tourism services (Ladki & Dah, 1997; Lebanon.com, 2018).

The retrofitting and renovation of downtown Beirut, dubbed "Beirut Central District" (BCD), was tasked through legislation to the company Solidere, granting it expropriation rights and authoritative control over the country's urban core. Solidere's vision of Beirut was a glamorous post-war touristic scheme (Makarem, 2018). Yet, the company ended up creating a "privatized public space" by overlooking one of the pillars of sustainable tourism development: maintaining a balance between locals, tourists, and the site (Al-hagla, 2010). On the negative side of the private sector's prominence there's a risk of having impact at a scale beyond one infrastructure of neighborhood. In this case, the sheer size of Solidere and its own size of activities, affected the whole downtown of the capital city (Barthel, 2010; Daher, 2018; Makarem, 2018). Locals, who now frequently refer to their capital's downtown as "Solidere," were excluded from the development configuration, leading to an urban bubble accessible only to the social elite and the luxurious tourist. However, the viability of this scheme was soon put at stake due to increasing inflation and changing global dynamics of tourism (Naylor, 2015).

8.4 Tourism Development in Flux

8.4.1 Private Sector-Led Tourism

8.4.1.1 Tourism Development Responses to Laws & Regulations

The government's policy for tourism development is mainly focused on both foreign and local private sector investments (Vlachos, 2018). On one hand, this has given tourism a large extent of resilience against all security and political shocks experienced by the country across the years. On the other hand, it has granted the private sector large control over the country's industry and its infrastructures (Harake et al., 2018; Ladki & Dah, 1997; Schimmelpfennig & Gardner, 2008). Consequently, Lebanon has become a leisure destination with little connection with its cultural wealth.

Tourism industry also draws extensively from the involvement of NGOs, activists, political parties, and the government that all take part in developing tourism destinations in Lebanon (Vlachos, 2018). The relation between various stakeholders however is not always symbiotic due to lack of coordination and more generally development strategies based on co-benefits and cross-sector value creation. Finally, the geographical distribution of the cultural sites throughout the country sometimes hampers development efforts (Vlachos, 2018). Different geographical areas in Lebanon are under different politico-religious influences, and the prominent political party in any area usually has the lion's share of decision-making on the ground.

8.4.1.2 The Role of Central Government

As Lebanon's cultural heritage is not restricted to Beirut, the government implemented through the Council for Development and Reconstruction, a World Bank funded "Cultural Heritage and Urban Development" project. The project's aim was to improve the quality of life in the historical centers of the cities of Baalbek, Byblos, Saida, Tripoli, and Tyre, while safeguarding their cultural heritage. This initiative started 10 years after the end of the civil war as a mitigative response to the chaotic urban "development" and reconstruction that were happening across the country. The project was initially conceived to be completed between 2004 and 2010; however, it took 13 years to conclude mainly because of protracted conflict in some of the cities of implementation (the poorest cities) (Al-hagla, 2010; The World Bank, 2017).

The project depicts massive fragmentation in relation to tourism in Lebanon as despite having a tourism-related facet, only the Ministry of Culture and the Ministry of Transport and Public Works were identified as partners of the project and counterparts for policy reform, and the MoT was left out of the equation (The World Bank, 2017).

8.4.1.3 The Role of the Private Sector

Only few restrictions are present on the private sector in terms of tourism development. In light of little regulation and minimal control that thwarts the country's advancement, and the hastened widening of wealth gap, financial power can easily influence the decision-making process (Abir Saksouk, 2015; Harake et al., 2018; United Nations Human Settlement Programme, 2013). Coupled with the absence of monitoring, and a sizable shadow economy, tourism in Beirut and by extension in Lebanon provides an excellent visualization of economic disparity and uneven distribution of wealth. Its skyline crowded with high-rises and tower-cranes, hides a large chaotic agglomeration of old and unregulated buildings that stand out in the landscape's context as a result of incapacitated decision-making process (The Economist, 2018).

A recent tourism development project, Lancaster Eden Bay Hotel, in Beirut by a prominent developer, illustrates the controversy. The project stretched and committed several violations of building code and bent several public urban regulations and limits such as its set-back from the sea, parcel classification, and height and size declarations (Preston, 2018). This project has contributed to the gentrification of whole neighborhoods pushing communities outside the perimeter of the new touristic neighborhood that has replaced what was once public spaces by the sea, favored by the locals.

8.4.1.4 The Role of NGOs

The Shouf Biosphere Reserve is the largest nature reserve in Lebanon, it crosses several villages in Mount Lebanon area, and was declared a nature reserve through government legislation (Law No. 532 of July 24th, 1996). Today the project can be labeled as an ecotourism initiative launched by a NGO and environmental activists who reached out to political authorities for their security umbrella. It is operated and managed by a NGO, Al-Shouf Cedar Society (ACS), that works under the umbrella of the Ministry of Environment, and in cooperation with a committee whose members are the mayors of major neighboring villages and a number of potent political figures from the dominant political party in the Shouf area; the committee is headed by the leader of that party (Shouf Cedar Reserve, 2016; United Nations Educational, Scientific and Cultural Organization, 2018).

In a highly fragmented political system and a weak central government setting, the project has successfully managed to sustain and promote efforts of reforestation and community engagement while promoting ecotourism in the Shouf area thanks to the involvement and patronage of Walid Jumblat, a prominent political figure in the Druze community. Through the launching of nature exploration and biodiversity monitoring activities, and the creation of traditional and community-operated guesthouses, Shouf Biosphere has been successful in connecting local formulation and implementation of development strategies and policies with global dynamics and trends in ecotourism.

8.4.1.5 Others

Other sporadic initiatives exist throughout the country; they are usually vested in non-politicized NGOs and small groups of civil society activists (in the forms of small-scale tour operators and individual adventurers or nature lovers) (Vlachos, 2018), but they all lack financial support. Additionally, their vast majority is not registered nor regulated by any government body.

8.4.2 Public Sector-Led Tourism: National and Local Contexts

Unlike the traditional role of central governments elsewhere in the Middle East, in Lebanon the central government has been playing the role of a mediator for the different involved stakeholders without having a substantive input to the sector (Vlachos, 2018). The biggest share of government's revenues from tourism either comes indirectly from taxes, or directly through income generated through shadow economy (Mariusz Jarmuzek et al., 2014; Vlachos, 2018). Additionally, despite of its role localizing the global context, the Lebanese central government also seems to be lagging behind.

Local governments are subsidiary organs of the central government through the Ministry of Interior and Municipalities (Localiban, 2014). However, local governments have a large extent of autonomy when it comes to decisions on built tourism projects (hotels, shopping malls, etc . . .). In the cases of Beirut and Tripoli, the developer files for a construction permit through an architect or a civil engineer (who is registered in the Order of Engineers and Architects (OEA) in Beirut or Tripoli), and the request is sent to the corresponding municipality for approval. In the case of all other municipalities, the permit application is presented to the Urban Planning Regional Office or to the Federation of Municipalities (such is the case of Byblos, Kesrouane, and Metn), and the construction permit is issued by the President of the corresponding municipality (Lebanese Republic, Ministry of Environment, 2003; United Nations Human Settlement Programme, 2013).

Theoretically, the construction permit is only approved if it complies with the building law and the master plans devised by the Directorate General for Urban Planning (DGUP) that falls under the authority of the Ministry of Public Works and Transport. In practice, deviations are frequent and often happen due to either: financial enticement and a lack of monitoring, or political/religious affiliations and exertion of power (Abir Saksouk, 2015; United Nations Human Settlement Programme, 2013).

The Ministry of Tourism is an administrative body that does not possess any tangible legislative power. Its institutional legal framework is vague and mostly administrative; as such, its role has effectively been restricted to promotion of tourism (Lebanese Republic, Ministry of Tourism, 2010; Vlachos, 2018). However, in light of its serious lack of funding, its promotional efforts are insufficient and are turning it into a freeloader on private sector ventures. The ministry's role is even overshadowed when talking about ecotourism efforts, by that of the MoE, and in the case of cultural events by the MoC, and the list goes on (Lebanese Republic, Ministry of Tourism, 2010, 2018).

The municipality that receives the building permit application has a responsibility, by law, to consult with the relevant authorities before approving and issuing the permit. These authorities include the Ministry of Environment, the Ministry of Culture (through the Directorate General of Antiquities), the Ministry of Public Works and

Transport (either through the DGUP, or the Higher Council for Urban Planning (HCUP), depending on the size of the project; bigger development projects should be reviewed by the HCUP) (United Nations Human Settlement Programme, 2013).

8.4.3 Are They Integrated or Competing?

Since tourism development projects are usually sizable, such as resorts or hotels, they go through the HCUP, which is presided by the DGUP, and comprises of high level members from different authorities; namely the Director General of Urban Planning, Justice, Interior and Municipalities (Local Councils and Administration), Public Works and Transport (Roads and Buildings), Housing, and Environment, in addition to the Director of Programs at CDR, the President of the OEA in Beirut and Tripoli, and three experts (sociology, urban planning and environment, and architecture) (Lebanese Republic, Ministry of Environment, 2003; United Nations Human Settlement Programme, 2013). In general, members of the HCUP are in accord, but as their political affiliations are diverse and usually serve the higher system of clientelism, this situation is under constant assessment.

8.5 Cross-Sectoral Impacts and Models of Tourism

8.5.1 Ecological Impact

Within Lebanon's specific setting, especially in light of the high levels of pollution across the country and across resources, it is difficult to determine exactly how much tourism had an ecological impact (Vlachos, 2018). However, based on the authors' personal observations, ecotourism played a major role in sensitizing the people on the importance of maintaining a clean environment and being more responsible.

Activists and NGOs are among the first entities to launch eco-friendly initiatives; a recent one that has been gaining popularity in Lebanon is a yearly participation in the World Cleanup Day (September 15) where NGOs and scuba diving clubs across the country join efforts to clean the country's streets and beaches and the waterbed underneath the famous maritime landmark Pigeon Rock (Rahayel, 2018; United Nations Development Programme, 2018). On the other side, and despite the obligation of presenting an "Environmental Impact Assessment" study prior to construction, large scale development projects such as resorts and hotels impose strains on the country's environment beyond its coping capacities (Abir Saksouk, 2015; Preston, 2018; United Nations Human Settlement Programme, 2013).

8.5.2 Social Impact

Aside from widening the social divide in certain areas and around specific services (especially higher scaled resorts, restaurants, and bars), big tourism projects are putting extreme pressures on the country's already poor infrastructure. Catastrophe strikes the country with every rainfall as streets, highways, and tunnels flood due to clogged sewage systems that are operating without maintenance, beyond their capacities. On the other hand, tourism contributes to creating better livelihoods for people who were living in dire conditions. Thanks to the "Cultural Heritage and Urban Development" project, local communities are thriving in rehabilitated housing units after having lived in makeshift settlements prior to the project's implementation (Al-hagla, 2010; The World Bank, 2017).

8.5.3 Economic Impact

The dominant "laisser-faire" that distinguishes the business sector in Lebanon has created large room for competition as well as innovation (Tax Justice Network, 2018). The hospitality sector thus thrives on its reputation as one of the best in the world (Bassil et al., 2015; Ladki & Sadik, 2004; Rowbotham, 2010). Even though this accelerated the economic wheel in the country, it also contributed to a fast and disproportionate distribution of wealth (Harake et al., 2018). Lebanon, despite of its relatively small size as an economy as well as a polity, is matching some of the most expensive developed countries in the world. The rapid economic gains lured developers and landlords to impose hefty rental prices that investors struggle to meet (Mariusz Jarmuzek et al., 2014). For instance, the Lebanese coast has become infested with large and luxurious beach resorts that left no place for a public beach. Thus, a family beach outing for an average Lebanese family would cost well over US$100 per day for a family of four (with a minimum wage of around US$400). Another example is the rampant prices of cafés and restaurants in posh Downtown Beirut, where a cup of coffee could set you back US$10 (noting that in small popular places its cost could average around US$4) (Dan Azzi, 2018).

8.6 Conclusion

The lack of a solid regulatory framework and the "laisser-faire" attitude even in governance is having adverse implications on tourism. The lack of an effective monitoring and regulatory framework is making developers quickly increase their profits, while foreign tourists are being repulsed by the excessively costly services and commodities that they are receiving.

Political considerations that cement certain institutions under the current political system, are in no way sustainable and do not promote development. This is evident from the constantly changing strategies of different ministries under different ministers, where each new minister comes with a different agenda and restarts from zero.

The existing government system is too complex to be viable. Additionally, the lack of digitalization in the age of digitalization will soon catch up with the government and hinder much of its activities. The public sector is not the only one to be blamed in this case. The greed of the private sector and its drive towards less social responsibility is contributing to the status quo in Lebanon. Privatizing public spaces, such as the case of Solidere, will surely backfire one way or another especially in the Lebanese context; lack of public spaces, rapidly increasing poverty rates, exorbitant prices, poor infrastructure, dangerous levels of pollution, and extremely high levels of corruption, will eventually lead the people to overthrow their governing system and demand reforms.

A restructuring of the institutional framework to centralize all tourism-related decisions is necessary for the effective and smooth operation of the sector. Bureaucratic processes need to be revised as the current state of affairs is extremely lengthy and inefficient. Under the current circumstances, resources need to be better allocated for promotion and advertisement of tourism destinations in Lebanon, through the engagement of professional and experienced agencies who adopt an innovative and appealing approach. Additionally, a better control of corruption and promotion of the rule of law is necessary to put a limit to the resources drain, and to reinstate security and peace throughout the country.

Stakeholders should take further initiative and coordinate with the government to identify areas of possible remediation. Stakeholders should not rely on their reputation and previous achievements. They should always look to improve their services. Stakeholders should also assess the social impact that their development schemes are entailing. Creating incentives for the locals to open small shops and businesses in close proximity to the touristic sites, to promote local produce and traditional practices. Shops and businesses should be regulated by an authoritative body for a control over service/commodity prices. Decision-makers should work on promoting the culture of communal spaces for the population, they should also include alternative freely accessible public spaces in their planning, to allow for a cultural exchange between different factions of the local population and between the local population and tourists. Decision-makers should also impose better control over new projects by introducing efficient and consistent schemes of monitoring and evaluation that also promote the rule of law and minimize illegal practices; they should expand the monitoring scheme in a second phase to include older and already-approved projects, to homogenize practices across the sector. Decision-makers should also familiarize themselves with best practices in policy in the field of tourism development in order to fully exploit the potential of the sector in Lebanon.

It is noteworthy to mention that the failed Solidere experience in Beirut should be one of the lessons learned for the restructuring of neighboring Aleppo. There is a dire need for innovation and youth to promote relatable content in terms of promotion of tourism and maximization of the country's tourism potential. Introduction of technology to cultural sites in order to increase user interaction and promote user engagement.

References

Abir Saksouk. (2015). Where is Law? Investigations from Beirut (Neighborhood Initiative, p. 26) [Working Paper]. American University of Beirut. https://www.aub.edu.lb/Neighborhood/Documents/aubAbirNIReport.pdf

AlBustani, I. (2018). Citizen Budget 2018. Bassel Fleihan Institut des Finances. http://www.finance.gov.lb/en-us/Finance/BI/ABDP/Annual%20Budget%20Documents%20and%20Process/Citizen%20Budget%202018.pdf

Al-hagla, K. S. (2010). Sustainable urban development in historical areas using the tourist trail approach: A case study of the Cultural Heritage and Urban Development (CHUD) project in Saida, Lebanon. Cities, 27(4),234–248. https://doi.org/10.1016/j.cities.2010.02.001

Banque du Liban. (2018, December 14). Banking and Finance | Main Characteristics. http://www.bdl.gov.lb/pages/index/4/25/Main-Characteristics.html

Barthel, P.-A. (2010). Yasser Elsheshtawy (ed), *The Evolving Arab City: Tradition, Modernity and Urban Development. Géocarrefour*, Vol. 85/2, 164–166.

Bassil, C. (2013). Intervention Model for Analyzing the Lebanese Tourism Sector. *Review of Middle East Economics and Finance*, 8(3),1–15. https://doi.org/10.1515/rmeef-2012-0022

Bassil, C., Hamadeh, M., & Samara, N. (2015). The tourism led growth hypothesis: The Lebanese case. *Tourism Review*, 70(1),43–55. https://doi.org/10.1108/TR-05-2014-0022

BBC. (2018, April 25). Lebanon profile – Timeline. https://www.bbc.com/news/world-middle-east-14649284

Bkhairia, H. (2017). La Méditerranée dans le Voyage en Orient de Lamartine. Babel. Littératures plurielles, 36, 87–98. https://doi.org/10.4000/babel.4968

Central Intelligence Agency. (2020). Middle East: Lebanon – The World Factbook. The World Factbook. https://www.cia.gov/library/publications/the-world-factbook/geos/le.html

Collelo, T., H. Smith, & Library of Congress. Federal Research Division. (1989). Lebanon: A country study (Third Edition). Federal Research Division, Library of Congress. https://www.loc.gov/item/88600488/

Daher, R. F. (2018). Tourism, heritage, and urban transformations in Jordan and Lebanon: Emerging actors and global-local juxtapositions. Channel View Publications. http://www.academia.edu/2263131/Tourism_heritage_and_urban_transformations_in_Jordan_and_Lebanon_Emerging_actors_and_global-local_juxtapositions

Dan Azzi. (2018, October 28). Why is Beirut the 7th most expensive city in the world. Annahar.Com. https://www.annahar.com/english/article/888418-why-is-beirut-the-7th-most-expensive-city-in-the-world

Embassy of Armenia to Lebanon. (2017). Armenian community in Lebanon. Embassy of Armenia to Lebanon. http://lebanon.mfa.am/en/community-overview/

Ghandour, M. (2011). Building Law: A Critical Reading of the Lebanese Case. In T. Fisher & C. Macy (Eds.), Paradoxes of Progress: Vol. 89th ACSA Annual Meeting Proceedings, Paradoxes of

Progress (p. 571). Association of Collegiate Schools of Architecture. http://apps.acsa-arch.org/resources/proceedings/indexsearch.aspx?txtKeyword1=74&ddField1=4

Harake, W., Hamadeh, N., Kostopoulos, C., Carey, K., Mobarek, S. I., & Ziade, M. (2018). Lebanon Economic Monitor – De-Risking Lebanon (pp. 1–41). The World Bank. http://documents.worldbank.org/curated/en/615661540832875043/Lebanon-Economic-Monitor-De-Risking-Lebanon

Kisirwani, M. (1997). The Rehabilitation and Reconstruction of Lebanon. In P. J. White & W. S. Logan (Eds.), *Remaking the Middle East* (English edition, p. 340). BERG. http://ddc.aub.edu.lb/projects/pspa/kisirwani.html

Ladki, S. M., & Dah, A. (1997). Challenges Facing Post-War Tourism Development: The Case of Lebanon. *Journal of International Hospitality, Leisure & Tourism Management*, 1(2),35–43. https://doi.org/10.1300/J268v01n02_04

Ladki, S. M., & Sadik, M. W. (2004). Factors Affecting the Advancement of the Lebanese Tourism Industry. *Journal of Transnational Management Development*, 9(2–3), 171–185. https://doi.org/10.1300/J130v09n02_09

Lebanese Hotel Association. (2018). Why invest in Lebanon? Investor – Lebanese Hotel Association. https://lebhoa.com/lebanese-hotel-association-investor/

Lebanese Republic, Central Administration of Statistics. (2018, December 13). Central Administration of Statistics – Thematic Time Series. http://www.cas.gov.lb/index.php/thematic-time-series

Lebanese Republic, Internal Security Forces. (2018). Hierarchy of units. ISF Official Website. http://www.isf.gov.lb/ar/articles/149/Judicial-Police

Lebanese Republic, Ministry of Education and Higher Education. (2018, December 19). دليل المعاهد الرسمية. Guide of Public Education Institutes. http://edu-lb.net/schoolrasmelist.php

Lebanese Republic, Ministry of Environment. (2003). Key players in urban planning. http://www.databank.com.lb/docs/Urban%20Planning%20Heirarchy.doc

Lebanese Republic, Ministry of Foreign Affairs, Embassy of Lebanon to the Arab Republic of Egypt – Cairo. (2018, December 11). Embassy of Lebanon to the Arab Republic of Egypt – Cairo. http://mfa.gov.lb/egypt/english/lebanon

Lebanese Republic, Ministry of Tourism. (2010). Ministry of Tourism Business Plan, 2010–2014. http://www.databank.com.lb/docs/Ministry%20of%20Tourism%20Business%20Plan%2C%202010-2014.pdf

Lebanese Republic, Ministry of Tourism. (2018). Rural Tourism Strategy. http://www.mot.gov.lb/Publications/Miscellaneous

Lebanese Republic, Presidency of the Council of Ministers, The Investment Development Authority of Lebanon. (2014). Tourism Fact Book (p. 17) [Fact Book]. The Investment Development Authority of Lebanon. https://investinlebanon.gov.lb/Content/uploads/Publication/140617020022273~Tourism%20Fact%20Book.pdf

Lebanese Republic, Presidency of the Council of Ministers, The Investment Development Authority of Lebanon. (2018a). Tourism Sector Investment Incentives (p. 5) [Guide Book]. The Investment Development Authority of Lebanon. https://investinlebanon.gov.lb/Content/uploads/SideBlock/171024104452052~Tourism%20Incentives.pdf

Lebanese Republic, Presidency of the Council of Ministers, The Investment Development Authority of Lebanon. (2018b, December 18). IDAL – Sectors in Focus – Tourism. http://investinlebanon.gov.lb/en/sectors_in_focus/tourism

Lebanon.com. (2018, December 17). Beirut Central District [News website]. Lebanon.Com. http://www.lebanon.com/construction/beirut/airport.htm

Library of the Congress. (2018, December 20). The National Pact (Lebanon). https://country-studies.com/lebanon/the-national-pact.html

Localiban. (2014, July 25). الإدارة الإقليمية في لبنان—لوكاليبان. Localiban.Org. http://www.localiban.org/%D8%A7%D9%84%D8%A5%D8%AF%D8%A7%D8%B1%D8%A9-%D8%A7%D9%84%D8%A5%D9%82%D9%84%D9%8A%D9%85%D9%8A%D8%A9-%D9%81%D9%8A-%D9%84%D8%A8%D9%86%D8%A7%D9%86-388-388-388-388

Maila, J. (1992). The Document of National Understanding: A Commentary. Centre for Lebanese Studies. https://www.lebanesestudies.com/portfolio-item/the-document-of-national-understanding-a-commentary/

Makarem, H. (2018). The Bottom-Up Mobilization of Lebanese Society against Neoliberal Institutions: The Case of Opposition against Solidere's Reconstruction of Downtown Beirut. http://www.academia.edu/16641731/The_Bottom-Up_Mobilization_of_Lebanese_Society_against_Neoliberal_Institutions_The_Case_of_Opposition_against_Solidere_s_Reconstruction_of_Downtown_Beirut

Mariusz Jarmuzek, Francisco Parodi, & Najla Nakhle. (2014). IMF Country Report No. 14/238 (Selected Issues, p. 48). International Monetary Fund. https://www.imf.org/external/pubs/ft/scr/2014/cr14238.pdf

Ministry of Finance. (2018, December 19). Budget Law 2018 [Government website]. Ministry of Finance. http://www.finance.gov.lb/en-us/Finance/BI/ABDP/Documents/Budget%20Law%202018.pdf

Mortimer Winterthorpe. (2018, December 17). Beyrouth, ton horizon m'appelle – Le Pays du cèdre en musique | Le Net plus ultra de la chanson française. https://lachansonfrancaise.net/2016/08/18/beyrouth-ton-horizon-mappelle-le-pays-du-cedre-en-musique/

Naylor, H. (2015, January 1). The bling in Beirut's rebuilt historic downtown pushes Lebanese away. Washington Post. https://www.washingtonpost.com/world/middle_east/beirut-rebuilt-its-downtown-after-the-civil-war-now-its-got-everything-except-people/2014/12/31/3b72e8b5-1951-409e-8b3d-1b16275d7f3d_story.html

Norkonmaa, K. (1995, June 19). The reconstruction of Lebanon. The Third Nordic Conference on Middle Eastern Studies. Ethnic encounter and culture change, Joensuu, Finland. https://org.uib.no/smi/paj/Norkonmaa.html

قانون سرّية المصارف الصادر بتاريخ 09/03/1956–Banking Secrecy Law of 03/09/1956,(1956) (testimony of Presidency of the Republic of Lebanon). https://cyrilla.org/ar/entity/vj24q5xwaxx1hhm4h48b5u3di?page=1

قانون التنظيم المدني–The law of urban planning, no. 69, Lebanese Republic, Council of Ministers (1983). http://www.legallaw.ul.edu.lb/Law.aspx?lawId=244511

Preston, S. (2018, April 16). The untouchable hotel. Executive Magazine. http://www.executive-magazine.com/real-estate-2/the-untouchable-hotel

Rahayel, A. (2018, September 19). 3500 volunteers Clean More Than 30 Locations in Lebanon on World Cleanup Day [Blog]. NoGarlicNoOnions. http://www.nogarlicnoonions.com/3500-volunteers-clean-more-than-30-locations-in-lebanon-on-world-cleanup-day/

Raymond, C. (2013). Vie, mort et résurrection de l'histoire du Liban, ou les vicissitudes du phénix. Revue Tiers Monde, 216(4),71–87.

Rowbotham, J. (2010). 'Sand and Foam': The changing identity of Lebanese tourism. *Journal of Tourism History*, 2(1),39–53. https://doi.org/10.1080/17551821003777857

Salem, G., & Azoury, A. (2017). The Repercussions of the Syrian crisis on Tourism in Lebanon. *Arab Economic and Business Journal*, 12(2),121–127. https://doi.org/10.1016/j.aebj.2018.01.001

Salem, P. (1998). Framing post-war Lebanon: Perspectives on the constitution and the structure of power. *Mediterranean Politics*, 3(1),13–26. https://doi.org/10.1080/13629399808414638

Saliba, I. (2012). Lebanon: Constitutional Law and the Political Rights of Religious Communities | Law Library of Congress [Web page]. https://www.loc.gov/law/help/lebanon-constitutional-law.php

Salibi, K. S. (1971). The Lebanese Identity. *Journal of Contemporary History*, 6(1),76–86. JSTOR.
Schimmelpfennig, A., & Gardner, E. H. (2008). Lebanon-Weathering the Perfect Storms. IMF Working Papers, 08(17), 1. https://doi.org/10.5089/9781451868791.001
Security Council Report. (2018, December 11). Lebanon Chronology of Events: Security Council Report. https://www.securitycouncilreport.org/chronology/lebanon.php?page=all&print=tr
Shouf Cedar Reserve. (2016, December 24). Shouf Biosphere Reserve. https://www.Shoufcedar.Org.http://shoufcedar.org/profile-2/
Spaan, S. (2013, March 14). The failure of a National History Textbook: Preserving diverse Civil War narratives in Lebanon. Asfar. https://www.asfar.org.uk/the-failure-of-a-national-history-textbook-preserving-diverse-civil-war-narratives-in-lebanon/
Tax Justice Network. (2018). Narrative Report on Lebanon (Financial Secrecy Index, p. 6). http://www.financialsecrecyindex.com/PDF/Lebanon.pdf
The Economist. (2018, August 30). Lebanon's economy has long been sluggish. Now a crisis looms. The Economist. https://www.economist.com/middle-east-and-africa/2018/08/30/lebanons-economy-has-long-been-sluggish-now-a-crisis-looms
(تعديل المرسوم الاشتراعي رقم 148 تاريخ 16/09/1983 (قانون البناء)–Amendment of the legislation nb 148 of 16/09/1983(the building law), no. 646, Lebanese Parliament, 16 (2004). http://www.legallaw.ul.edu.lb/Law.aspx?lawId=210543
Lebanese Constitution, 23 (1995). https://www.lp.gov.lb/backoffice/uploads/files/Lebanese%20%20Constitution-%20En.pdf
The World Bank. (2017). Lebanon – Cultural Heritage and Urban Development Project (pp. 1–65). The World Bank. http://documents.worldbank.org/curated/en/604321498843122368/Lebanon-Cultural-Heritage-and-Urban-Development-Project
Traboulsi, F. (2012a). The Emirate of Mount Lebanon (1523–1842). In A *History of Modern Lebanon* (pp. 3–23). Pluto Press; JSTOR. https://doi.org/10.2307/j.ctt183p4f5.7
Traboulsi, F. (2012b). The War Order (1983–1990). In A *History of Modern Lebanon* (pp. 226–245). Pluto Press; JSTOR. https://doi.org/10.2307/j.ctt183p4f5.19
United Nations, Department of Economic and Social Affairs, Population Division. (2019). World Population Prospects 2019. World Population Prospects 2019. https://population.un.org/wpp/DataQuery/
United Nations Development Programme. (2018, September). UNDP Live Lebanon – World Cleanup Day. Live Lebanon. http://live-lebanon.org/activity-list/world-cleanup-day
United Nations Educational, Scientific and Cultural Organization. (2018, October 24). Shouf Biosphere Reserve, Lebanon. UNESCO. https://en.unesco.org/biosphere/arab-states/shouf
United Nations Educational, Scientific and Cultural Organization, World Heritage Centre. (1984). Lebanon – UNESCO World Heritage Centre. UNESCO World Heritage Centre. https://whc.unesco.org/en/statesparties/lb
United Nations High Commissioner for Refugees. (2018, December 21). Situation Syria Regional Refugee Response. Operational Portal – Refugee Situations. https://data2.unhcr.org/en/situations/syria/location/71
United Nations Human Settlement Programme. (2013). Reforming Urban Planning System in Lebanon: Findings of the Research Assessment (p. 50). United Nations Human Settlement Programme. https://www.alnap.org/system/files/content/resource/files/main/reforming-urban-planning-system-in-lebanon-e.pdf
United States Department of State. Bureau of Democracy, Human Rights and Labor. (2011). International Religious Freedom Report for 2011. United States Department of State. https://www.state.gov/documents/organization/193107.pdf

van Ommering, E. (2015). Formal history education in Lebanon: Crossroads of past conflicts and prospects for peace. International Journal of Educational Development, 41, 200–207. https://doi.org/10.1016/j.ijedudev.2014.06.009
Vlachos, P. (2018). Tourism Sector in Lebanon Policy Recommendations (p. 22). Lebanese Republic, Office of the Minister of State for Administrative Reform. https://www.omsar.gov.lb/Assets/Tourism-Sector-Report.pdf
World Bank. (2018, December 17). Travel and Tourism direct contribution to GDP. Open Trade and Competitiveness Data. https://tcdata360.worldbank.org/indicators/tot.direct.gdp?country=LBN&indicator=24648&viz=line_chart&years=1995,2028&compareBy=region
World Bank. (2019). World Development Indicators. World Bank Open Data. https://data.worldbank.org/

Matilde Córdoba Azcárate

9 Mexico

9.1 Introduction

This chapter offers a historical approach to tourism governance in Mexico. It creates a timeline for the articulation of tourism as a state development tool. The intention of this timeline is twofold. First, to trace the shifting roles of tourism in national development plans. And second, to offer glimpse at the historicity and centrality of public-private alliances in the design and planning of Mexico's contemporary mainstream tourism models. I also view this timeline as part of wider efforts to theorize tourism governance along a socio-cultural lens that pays attention to tourism and development as cultural categories.

This timeline is by no means exhaustive of the ways in which tourism governance occurs in Mexico. However, and as incomplete as it might be, it helps us to build a nuanced understanding of the uninterrupted yet heterogenous ways in which tourism has been articulated as a state development plan. The timeline points at five turning moments where tourism governance has significantly shifted roles within national development policies: from having a liminal role, to be a national aid, to become a national priority, and to today's ambivalent position in-between defunding and/or its strategic articulation in support of mega infrastructure projects for development. Moreover, each moment in the timeline also traces Mexico's synergetic relationship with major international development discourses: tourism as modernization, tourism as sustainable development, tourism as a tool of socio-cultural inclusion, tourism as empowerment and guarantor of social justice. The timeline also pinpoints the moments where shifts in institutionalized cultural understanding of development occurred in and beyond Mexico giving us elements to understand Mexico's global centrality in tourism matters. Each moment in the timeline also helps us to trace when, where, why, and how existing tourism infrastructure and architectonical projects came to exist- the all-inclusive beach resort, the natural enclave, archeological and cultural sites, and boutique hotels. Lastly, the moments in the timeline are representative of the continuity of a particular governance model as it relates to funding and implementation schemes that privilege public-private arrangements in the pursuit of tourism development.

In order to illustrate these different moments in the timeline, the chapter draws on empirical evidence from the Yucatán Peninsula, a region conceived by historians as a laboratory and privileged example of Mexican corporatist state practices.

https://doi.org/10.1515/9783110638141-009

9.2 Tourism Governance and Economic Development in Mexico

Tourism governance in Mexico, that is, the way in which Mexico decides how tourism is done, is largely state-led and it has a strong top-down technocratic orientation. Tourism governance is also inseparable from development discourses and it is linked to national policies of economic growth. In most cases, tourism ventures in Mexico are articulated as vectors to attract foreign capital and they work through public-private alliances where government officials and businessmen work hand in hand. For this reason, tourism development and, by extension, tourism governance in Mexico have been highlighted by critical scholars as state tools for the modernization and the civilization of Indigenous landscapes and peoples.

The use of tourism as a state tool to modernize landscapes and to discipline cultural diversity is not unique from Mexico. Studies have raised concrete concerns regarding the unevenness of using tourism as a vehicle for state led economic, socio-cultural, regional and community development around the globe (Aramberri & Butler, 2005; Chambers, 2009; Sharpley & Telfer, 2015; Wahab & Pigram, 1997). These studies highlight tourism's negative impacts in local livelihoods, environments and cultural heritage (Castellanos, 2010). They refer to tourism as a colonial project pointing to a lack of involvement of local and Indigenous peoples in shaping the tourist industry and its products (Butler, 2006; González; 2013; Weaver, 2013). Critical studies also highlight tourism's pervasive commodification of traditions and natural resources (Bunten, 2008) and tourism's indisputable role in processes of rapid acculturation (Nuñez Theron, 1963).

And yet, whilst many highlight the social and environmental risks and barriers to tourism as a vehicle of economic growth, others emphasize its revenues as an engine for job creation, infrastructure development and economic growth. Tourism for example, is widely celebrated as a sustainable economic development tool and most of the times, governments around the world embrace it in their national plans for economic growth without further inquiry into its practical modes of operation.

Tourism's inclusion into state development agendas and its embrace as a tool for socio-economic growth has followed different moments in Mexico. With important antecedents in the late 1920s–1940s, tourism was not an important topic for national governance up to the mid-1960s. It was then, when the tourism and travel industries increased exponentially at an international level. Since the mid-1960s, tourism has acquiered a more prominent position in both national and regional development plans as the favorite tool for planned economic growth and, concomitantly, to thrive socio-culturally and environmentally speaking. Mexico is not unique in these shifts, which are also observable in other economies that have adopted tourism as a main vehicle for economic growth such as the Caribbean region.

In what follows, I situate the origins of tourism as a matter of national governance in Mexico in the late 1920s, and I describe the gradually proactive role that

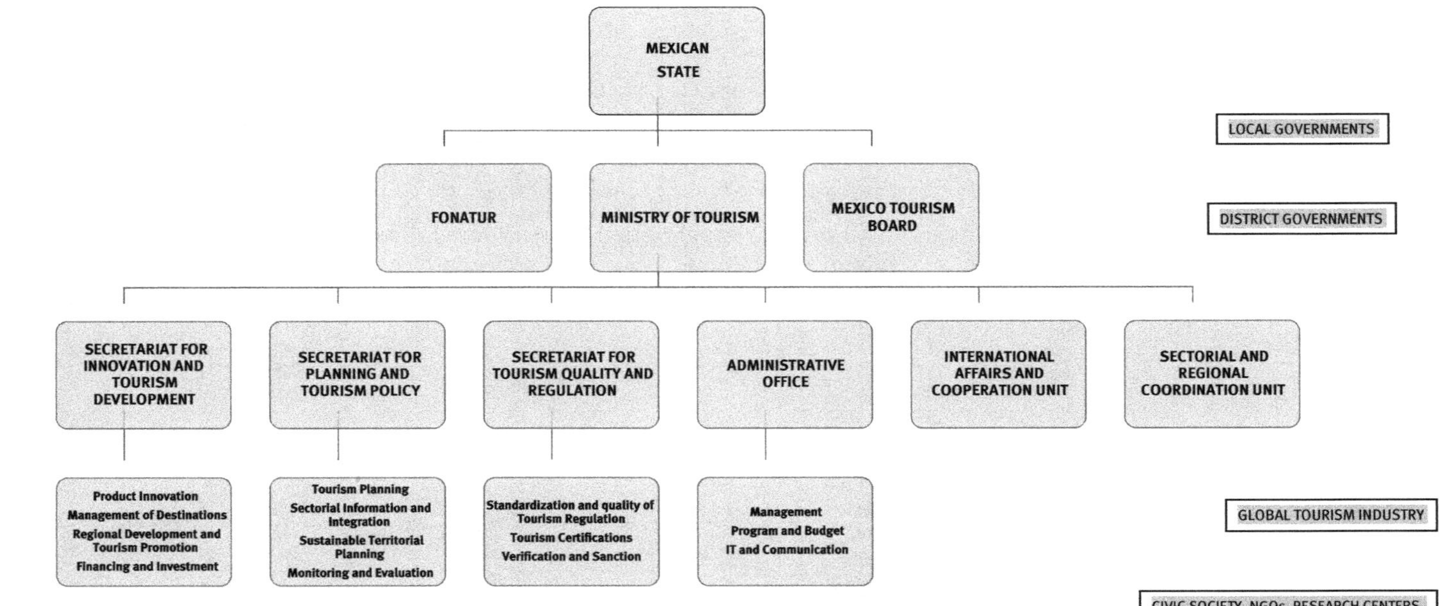

Figure 9.1: Tourism Governance Mexico. (Source: Figure by author, based on OECD Mexico, adapted from Ministry of Tourism, 2018).

tourism has had in National Development Plans first as a public policy of targeted intervention in the form of tourism development poles (1960s–1970s); then as a national aid to overcome national crisis (1980s–1990s); after, as a national priority in the pursuit of neoliberal policies (2000s); and today's ambivalent position in-between being a sector to defund or a main state-led driver for social justice (2012-present). Often times, and because Mexico is a federal republic, the regional and local governments have been assigned with the tasks to make tourism development work in the state-envisioned way. Not always do these tasks receive state support of any kind.

9.3 Tourism as a Matter of State Governance (late 1940s-mid-1960s)

Dina Berger (2006) situates the start of a governmental interest in tourism in the late 1920s, with the creation of the Mixed Pro-Tourism Commission by Mexico's revolutionary elite in 1928. According to Berger, the Commission begun a dialogue on tourism and brought together government officials and businessmen in the study of tourism development in Canada, the US, and Cuba. It is here she says, that an interest in tourism was forged for the first time as a matter of national governance. Berger's research shows that it was private organizations like the Mexican Tourism Association (MTA) and the Mexican American Automobile Association (AAMA) that helped promote tourism in Mexico through the generation of advertisements and lectures in the United States that created positive images of Mexico as a tourist destination abroad. Most of these lectures and advertisements, she says, had the intention to clean negative stereotypes of Mexico as a violent and corrupt country. With that intention in mind, the first National Tourism Committee was created in 1938. Composed of government officials and businessmen the committee intentionally promoted Mexico abroad in contraposition to Cuba's overtly dependency on foreign tourists and foreign representations of place. Not wanting to become another Cuba, the 1938 National Tourism Committee maintained a strong regional focus that would help two goals in one: promoting Mexican identity abroad as diverse, and maintaining revolutionary goals internally (Berger, 2006).

As a result of the private impulse in promoting tourism to Mexico, foreign tourism increased steadily between 1930 and 1934 showing from the very beginning not only the importance of this public-private alliance in the way tourism governance would take place in the future, but also, the kind of tourism models the state was going to prioritize. Mostly, beach and resort tourism. It is important to remember here that mass tourism, often referred to as "sea, sand, and sun" tourism, resort tourism, packaged tourism, charter tourism, or mature tourism, largely depends on the provision of standardized experiences and products and that, as its name announces, it is a form of travel rigidly packaged, often involving the movement "from the airport to the hotel, and back home." As such, resort tourism feeds on a largely undifferentiated clientele that demands to consume en masse, not only beaches but also through increased commercialism and sightseeing (Patullo, 2011).

Resort tourism has depended since its inception on large scale developments that result in the homogenization of the built environments and over the years, it has proved to be increasingly low cost and an industry vastly controlled by tour operators. These characteristics would attach appellatives of resort tourism as being a form of travel that is "culturally poor," "largely unsustainable" environmentally speaking, and deeply "commoditized" (Vainikka, 2013). For these reasons, mass resort economies in Mexico, but also in the Caribbean and beyond, have been denounced in tourism and development literature, as a continuation of plantation economies (see Patullo, 2011).[1] None of these discussions were, however, ones that Mexican businessmen and government officials were having at the moment. Tourism was still incipient as an economic sector, and its importance for national and global economies, still reduced.

The first large inversions in hotels and tourist infrastructure in Mexico date from the late 1940s as part of the presidency of Miguel Alemán (1946–1952) who is considered by most the first in realizing the economic potential of tourism for the national economy (Jiménez, 1993). Investments in tourism infrastructure followed the same revolutionary and modernizing goals stated in the 1938 National Tourism Committee and were enunciated in Alemán's Carta Turística, a state declaration of intentions in tourism matters. Investments were directed primarily toward two locations: the Mexico City area and the west coast port of Acapulco. They were part of the beginning of Import Substitution Industrialization (ISI) policies of state led development and a direct predecessor of a political corporatist system characterized by the one-party rule and "rubber stamp" legislature. That is, a strong president from a single dominant party, the Partido Revolucionario Institucional (PRI) with control over elections and no real opposition. This corporatist system was consolidated during the earlier decade with land reform and the nationalization of oil companies started by Lázaro Cárdenas (1934–1940), better known as the leader of the Mexican revolution, and it is generally described as a fundamental actor in the

1 Patullo studied the transformation of St Lucia's landscapes in the Caribbean into mass tourism by following the construction of the Jalousie Hilton resort and Spa today part of the Sandals Group. She pointed out: "What is important about the story of Jalousie is that it is so typical. There are countless other Jalousies in the Caribbean and the story of the Jalousies, taken together, provoke an essential debate about tourism: Who benefits? And how? It asks questions about sovereignty (when beaches and valleys become foreign fields). It presents dilemmas about economic well-being and how profits are distributed. It generates arguments about how tourism can (or cannot) protect the environment, and speculates about the social and cultural relations between visitors and hosts" (Patullo, 2011, p. 4).

process of Mexicanization (Terry et al., 2014). Tourism governance would gradually grow to be an integral part of this process of Mexicanization.

During Alemán's presidency, the first tourism destinations were created, Acapulco, Manzanillo, Mazatlán, Puerto Vallarta, Cabo San Lucas, Cozumel, Isla Mujeres, Veracruz, Mérida, Guadalajara, and México City, better known as Mexico's traditional tourism destinations. In 1949, Alemán's wrote the first federal Tourism Law. During the next six years, under the presidency of Adolfo Ruiz Cortines (1952–1958), Mexico starts promoting tourism inside the country too and to invest in both roads and rooms to cater to both national and international tourists (Jiménez, 1993).

Tourism's importance for governance would increase in state actions from this point on. The first Department of Tourism was established under the presidency of Adolfo López Mateos (1958–1964) who also created the first Trust Fund for the Development of the Tourism Sector (Fondo de Garantia y Fomento del Turismo FOGATUR) which would become a central actor in tourism governance in the decades to follow. The main goal of the Trust Fund in this moment was mainly to coordinate the promotional activities of the sector, just as its predecessor had done. During López Mateos and the next legislature, with Gustavo Díaz Ordaz (1965–1971), tourism governance remained restricted to the state promotion of tourism investment and infrastructure development, and inversions remained centralized in the areas of Mexico City, Acapulco, and now also, the US Border Region.

According to Wilson (2008), it was precisely in the mid-1960s that Mexico began to emphasize explicitly tourism development as part of the nation's economic development strategy. As she puts it, tourism governance in Mexico had three main goals at this moment: earning foreign exchange, creating employment, and diverting internal migration towards tourism development poles. The institutionalization of the National Trust Fund for Tourism Infrastructure in 1967, later to become, in 1974, the National Fund for Tourism Development (Fondo Nacional de Fomento al Turismo – FONATUR), would point at the central role that tourism would start to have as part of national development policies in the following decades (Clancy, 2001). Slowly but steadily tourism increased its presence in executive decisions relating to the development of the Mexican economy.

9.4 Tourism Poles: Modernization on the Basis of Profitability and the Rise of the Resort Model (1970s-late 1980s)

In 1973, Mexico City and Acapulco still received about 60% of foreign tourists visiting Mexico and it was these places that still concentrated most, if not all of the state promotional efforts. It is also in this decade, that a spread of the tourism industry to other parts of the country boomed. As Michael Clancy (2001, p. 121) noted, it is the decade of the 1970s that Mexico became "one of the most popular destinations in the world and tourism a major component of the Mexican economy."

The diversification of tourism destinations in Mexico coincided with the global democratization of travel. It was done through large-scale planned investments in particular destinations considered as "poles." After a three-year study by the Banco de Mexico to identify possibilities for expanding Mexican exports, the Mexican government created the so called Integrally Planned Tourist Centers (CIPS in Spanish). These were large tourism development projects created from the ground up in some of the poorest areas of the country with large coastal areas: Cancún on the Caribbean coast of the Yucatán Peninsula; Ixtapa-Zihuantanejo, close to Acapulco in the state of Guerrero; Los Cabos and Loreto on the western peninsula of Baja California; and Bahías de Huatulco in the southern state of Oaxaca. In all these "poles," as Clancy (2001, p. 132) noted, "the state played a strongly interventionist role in 'pushing' tourism through planning, providing infrastructure, and acting as entrepreneur and banker." Cancún and Itxapa specifically were made possible through large international loans by the Inter-American Development Bank and the World Bank who financed large-scale airports, modern sewer and water facilities, electricity, and other amenities. In these places, (Clancy 2001, p. 121) notes, "because the new resorts were built from the ground up, the tourist bureaucracy became the governing power within the area." And as such it has remained.

It is important to spell out some of the reasons that the state gave to choose the Caribbean coast of the Yucatán Peninsula to develop the first tourism pole amongst other areas in Mexico because these have become replicated onwards. These reasons were spelled out in the INFRATUR feasibility study of tourist potential of the region. Looking at the specifics of this case shows why and how tourism was conceived and why it would gain strength as a civilizatory and modernist project.

The INFRATUR feasibility study for the CIPS was done by bankers and with a banker's point of view (Pelas, 2011). It was a US$2 million study done through a computerized system that aimed to find locations where to construct from the ground up to enhance regional development in the country. Perfect weather all year round, cultural assets nearby such as archeological remains and a population in need around that could easily be moved as labor force were amongst the most salient demands. The Yucatán Peninsula at that moment, and as the plan noted, "possessed an army of under-employed or irregularly employed workers, since the demise of henequen and chicle, and these workers lived close to some of the most beautiful marine environments in the Caribbean. Rapid tourist development would bring them both together" (Redclift, 2005, p. 87).

Cancún developed mostly during Echevarria's presidency (1970–1976) who explicitly conceived the CIPS, and in particular Cancún, as the means to economic development per se, with development as a cultural category understood in pure modernization terminology. That is, various stages in economic development to be achieved through big investment in infrastructure and technologies, to be followed by sociopolitical change and then socio-cultural progress (Nederveen Pieterse, 1996). In practice, the establishment of the so-called 1974 Cancún Project was done in five phases: separating the tourist zone from the new city; a sanitizing project; a project of electrification; the opening up of telecommunications network and building and construction. In its development, the scale and risk of investments assured complementary private investments which was also enabled by a regional history of making money from fiber production and from speculative land deals (majorly in Quintana Roo) as well as by the steady consolidation of the tourism industry at an international scale.

In 1974, the creation of the World Tourism Organization (WTO) consecrated tourism internationally as "a white industry," as an "industry without chimneys" (Mowforth & Munt, 2015). Financing Cancún made sense not only in Mexico, but also, abroad. The idea was clear, state planned tourism poles revitalized regional economies through a trickle-down effect and this directly contributed to the national economy. Mexico would become a leading world example, in particular through Cancún's success. According to Clancy (2001)

> a check of Mexican data for 1976 reveals that the tourism industry accounted for a larger share of GDP than did the mining industry, the chemical industry, the automobile industry, or the electrical equipment industry. [Tourism] was about the same size as the petroleum industry.

Since then, tourism's importance for the state never stopped to grow.

Tourism's centrality as a source of foreign currency was identified and expressed in earnest directly attached to discourses of national and regional development during López Portillo's (1976–1982) presidency. The focus at the moment was to improve the infrastructure in tourist regions, mainly the transportation and communication sectors, and conversations about opening these sectors to private national and international investments were consolidated. During López Portillo's presidency, the Mexican government, vastly looking at Cancún's take off, would made an explicit effort to regulate tourism development in the country under a modernist ideology that prioritized scientific and technical expertise. The elaboration of the first Federal Act for Tourism Development is an example (Truett & Truett, 1982) of the materialization in public policy of the Mexican state's realization of the generalizable centrality of this sector in future economic growth.

Yet, the Federal Act for Tourism Development mostly focused on tourism's employee training practices and hence focused only on the operational level of tourism ventures, ignoring the management level and exemplifying the tentative and inward-looking development strategies of tourism's policies at large. This inward looking into tourism governance showed a lack of vision of tourism as a social, cultural, and spatial practice with ecological and political ramifications. This inward looking has continued until today.

In Cancún, for example, resort tourism advanced through the enclosure of the beach and resources for a few and hence, through the design of segregated spaces that were not meant, by design, to be accessible for all. Enclosing the Caribbean beach for tourism happened both through walls, fences and gates, but also through more immaterial practices of policing bodies and their movements in space. Enclosures created a segregated landscape that reproduces class and race discrimination making it hard for local populations to view, access and enjoy the beach, which is public land in Mexican law (Córdoba Azcárate, 2020). Besides, unfettered tourism urbanization without any ecological regulations soon proved to be disastrous for the environment. Receding mangrove forests, exposed corals due to sand dredging for construction, water pollution due to unregulated sewage, and hotels run-offs to the ocean are amongst the destination most common threats.

The global awakening of environmental awareness in the next decades will bring attention to some of these problems as well as a shift in development and tourism discourses at international and national scales. Unfortunately, changes in the way tourism governance would be practiced on the ground would still have to wait.

9.5 Tourism as a National Aid: Productive Diversification of Tourism Activities and the Rise of Alternative Tourism (1980s–1990s)

During the decade of the 1980s, Mexico stands out as the second most indebted nation in Latin America. The 1982 debt crisis and Miguel de la Madrid (1982–1988) mark the explicit turn of Mexico to neoliberal policies, a shift that would have a radical effect in the way tourism governance operates.

In this decade, Mexico is in full consonance with other Latin American countries and deeply pressed by foreign debt, with high inflation rates, and a myriad of social demands not met with existing programs that continue to be largely infused with modernization ideals from the last decade. In this conjuncture, De La Madrid's turn to neoliberal policies were the materialization of a state politics of austerity. "Crises" he said in his memoirs "are, paradoxically, the true opportunities for change" (Carmona Dávila, 2020, p. 2). And by change he considered not only the economic dimensions but the social and cultural ones, too. In 1983 the Mexico's National Development Plan set four main objectives: to preserve and fortify democratic institutions; to get out of the crisis; to recover economic growth capacity; and to initiate the qualitative changes in economic, political and social structures with a major leit motif "the moral renovation of society." Tourism was an industry well suited to guide this modern aesthetics of progress and marketization. Further investment "a la Cancún" in tourist infrastructure development was the means to achieve this goal.

In De la Madrid's Primer Informe de Gobierno (1 sept. 1983) tourism's relevance "for economic and social development" was expressed in earnest. In his words, tourism's importance "lies in its capacity to generate productive jobs, to contribute to a balanced regional development, to stimulate the rest of economic sectors and to strengthen the cultural identity of our 'pueblo'." To express support for the activity De la Madrid drafted the first National Tourism Program (1984–1988) whose intention was "to order tourism development, make effective the right to rest, regulate jobs, regional development and the general balance of payments."

De la Madrid's turn to tourism under state's austerity rules put an end to the old Banco Nacional de Turismo whose functions were taken by the Fondo Nacional de Fomento al Turismo (FONATUR) as well as to the Consejo Nacional de Turismo which was transformed into the contemporary Secretaría de Turismo (SECTUR) (see figure 9.1). At the forefront of this newly created Secretariat, in charge of tourism activities still today, De la Madrid would put Antonio Enriquez Savignac, one of the main inventors of Cancún. In the following decade, Savignac would become the director of the World Tourism Organization marking a fundamental continuity in ideologies across political scales which would be pivotal in Mexico's stabilization as a world tourism leader.

According to tourism data from 1986 and 1989, the CIPS model implemented the decade earlier was a success and Cancún soon became the most popular destination in Mexico and Latin America. At an institutional level, the economic success of CIPS during this decade meant the triumph of statist and top-down development strategies, or what is the same, the consolidation of the leading role of the Mexican state in conceptualizing, planning, financing, and administering tourism integral centers (Torres & Momsen, 2005). Largely aided by NAFTA's relaxation of trade barriers after 1994, investments in tourism development generated jobs, predominantly in construction and the service industry in and around Cancún and the other CIPS. These investments brought roads, airports, and universities which together helped to slow down the country's ongoing massive labor migration to the United States and over the years they have also created an emergent urban middle class.

Informed by Cancún's statistical success, the last years of the 1980s and the decade of the 1990s are characterized by the productive diversification of tourism activities under a strengthened neoliberal gospel. Yet, tourism stories of success in official development discourses started to contrast sharply and widely with the lived experiences of service workers who could not reach ends meet, and the rapid ecological deterioration in and around beach tourism destinations which were no longer possible to evade. Tourism as modernization through the growth poles strategy in short, was amassing evidence that in reality, it meant the exacerbation of social differences via the subordination of agriculture to tourism and the production of an unbalanced and asymmetric dependent development landscape: with coastal urban tourist resorts like Cancún representing the core exploiting and feeding up on rural areas considered peripheries (Castellanos, 2010; Torres & Momsen, 2005).

Table 9.1: Tourism Governance in Mexico: timeline of tourism's inclusion into national development plans and tourism models with an example from the Yucatán Peninsula "Table by author".

	1940s–1950s	1960s–1970s	1980s–1990s	2000s	2012–present
International Development Discourses	Tourism enters state governance	Development as Modernization	Sustainable Development	Development with Human Face	Community Empowerment/ social justice
Tourism in National Development Plans	Minimal to non-existent	Liminal Role	National Aid	National Priority	Ambivalent and contested role
Tourism Models in National Geography	Punctual Beach resort tourism	Tourism poles Beach resort Tourism	Productive Diversification Alternative cultural and natural tourism	Tourism as local engine Experimential tourism Peublos Mágicos	Indigenous Tourism vis-à-vis Mega tourism projects
Tourism in the Yucatán Peninsula	Regionally bound, no state support	Cancún, sea, sun and sand tourism	Ecotourism in protected areas and archeological tourism	Hacienda Tourism Culinary Tourism Spiritual Tourism New Age Tourism	Indigenous Solidarity initiatives Tren Maya

Besides, entering NAFTA, was followed by a strong devaluation of the peso by more than 100% (Biles, 2004) and meant a sharp decline in real income plus the Mexican government absorbing millions in debt and taking over 12 banks that were responsible for about 20% of all debts (Biles, 2004). The devaluation of the peso meant further relaxation of restrictions on foreign intervention, and the actual beginning of a foreign intervention era along with the rise of tourism as a main "state aid," or anchor, in national development plans.

Two large crises at the end of the 1980s would mark a shift from modernization on the basis of profitability as a tourism governance model, towards sustainability in both development and tourism goals, at least, in discourse and paper. On the one hand, the 1985 Mexico City earthquake were more than 5000 people died and which put Mexico City as a tourism destination on hold, and hurricane Gilbert in 1988, which totally devastated Cancún. Both disasters would bring long lasting impacts on tourism planning and governance at a national and regional levels and the Mexican state's effort to diversify its tourism offer started on the ground. In practice, this meant, a relative movement of inversions from the tourism poles and Mexico City towards other areas of the Mexican geography. Fundamentally, it also meant a shift in development discourses and in tourism's role in development discourse and practice.

The consolidation of the international environmental movement in the early 1990s and the worldwide proliferation of natural protected areas hand to hand with the rise of sustainability discourses, would help to frame these shifts. It is in this precise moment, that The World Commission on Environment and Development (1987 p. 15) used sustainability as a concept for the first time. The Commission defined sustainable development as "development that meets the needs of the present without compromising the ability of future generations to meet their own needs." Sustainable tourism, operationalized under the auspices of Agenda 21- a non-binding action agenda emanating from the UN Conference on Environment and Development in Rio de Janeiro in 1992 with recommendations for national, regional, and local levels was supposed to mean the consolidation of sustainable strategies and culturally responsible forms of travel. It was intended toward catering to consumers that did not identify as tourists, but who aimed to learn and contribute while touring, that is, travelers.

Contrary to mainstream mass/beach resort tourists that looked for an extension of home while travelling, travelers in the early decade of the 1990s were said to be on the move and motivated for a "quest for authenticity" (MacCannell, 1973). What this meant in terms of tourism governance was the acknowledgement that a shift from tourism as recreation in beaches, to tourism as appreciation and education of culture and nature was in the making. This quest for authenticity did not mean, however, the re-organization of the way in which decisions on tourism would be made. It meant a displacement from staging destinations as "paradisiacal beaches" to staging them as "wilder" and/or more "traditional." In Mexico, this meant an institutional shift from the attention to coastal areas toward attention to inland areas, as well as the systematic channeling of state funds, for the first time, to both archeological remains and Natural Protected Areas.

However, the shift towards sustainable tourism and the embrace of sustainable development happened in Mexico at the same time as heightened structural adjustment programs, which meant the rapid economization of conservation state-run enterprises through alternative tourism (Mowforth & Munt, 2015). Only in 1988 Mexico was one of the top five countries in terms of biodiversity. It accounted for over 17.9 million hectares of its surface covered with over 150 designated natural protected areas.

Carlos de Salinas (1988–1994), the main figure leading the Mexican "neoliberal mirage," led the radical economic re-structuring and guidelines of classical economic liberalism in Mexico that would had long lasting impacts for tourism governance. In practice, this meant the proliferation of technocrats in the government and the removal of most governmental controls on foreign trade and investments. Hence, whilst for example, pink flamingos, nesting and breeding at the Biosphere Reserve Ria Celestún in the Gulf Coast of the Yucatán Peninsula and only five hours away from Cancún, would become, side by side with sandy beaches and Maya archeological remains such as Chichen Itza or Tulum, a brand for national sustainable development as well as the epitome for a renewed tourist Mexico in the world, in terms of tourism governance, planning, and land management, however, the discursive shift towards sustainability was mostly that, a change in the register of words, or as they put it in Mexico, "papel mojado." In practical terms, nothing changed much from the previous phase. Novel tourism destinations, such as Celestún, kept being refashioned by state led tourism alternatives through market-driven engineering efforts that advanced through land enclosure and the privatization of natural and cultural resources for the enjoyment of a few (Córdoba Azcárate, 2020).

Several key national policies point at tourism as the main avenue to develop and modernize within this "green" neoliberal new gospel. The 1988–1994 National Development Plan for example, channeled for the first time, large investments in tourism infrastructure and huge tourism projects in National Parks and National protected areas, including the Biosphere Reserve Ria Celestún started. Likewise, the 1991–1994 National Program of Tourism Modernization, for example, conceived as a federal policy under the National Development Plan, would explicitly state among its main objectives the promotion and marketing of natural resources providing investments in infrastructure development and its modernization across the Mexican geography.

The creation in 1995 of the Alliance for Tourism under the 1995–2001 National Plan for the Development of the Tourist Sector, explicitly stated "to take care to cut regulations to ease development of resort areas, make it easier for foreign investment in the tourist areas, improve infrastructures and (for the first time in national policy) promote ecotourism" (press release: "Zedillo's plans to boost tourism," 1995).

First operationalized as a specific way of travelling in 1990 by the International Ecotourism Society (TIES, 1990) ecotourism is defined as "responsible travel to natural areas that conserves the environment and improves the well-being of local people." The Biosphere Reserve Ria Celestún in the Yucatán Peninsula acted as a laboratory of state practices in their implementation of ecotourism not just as a form of travel but as an economic option at a local, regional and national levels. Almost as the new CIPS, natural protected areas such as this one, would be directly targeted as primordial areas capable of generating divisas, create employment, promote public and private inversions and help towards a productive diversification of the economy. The prominent place that the beach resort had in earlier decades was to be joined by the natural enclave through the promotion of ecotourism.

In practical terms, the main strategy used to promote Natural protected Areas was part of Mexico's just launched Fondos Mixtos de Promocion Turistica, a mixture between federal, state level, and private entrepreneurs of the hotel sector to promote tourism activities in Mexico City and the tourism poles created the decade earlier, particularly, Cancún. In reality, there was continuity in the way tourism governance happened, and a displacement to incorporate more areas than the CIPS to the tourist mix. These areas though, such as the Biosphere Reserve Ria Celestún would still considered to be "assets" for the national economy.

It was not until President Zedillo de Leon (1994–2000), however, that in compliance with 1992 Rio Summit, Mexico started to actually regulate ecotourism as a sustainable development alternative in natural protected areas. During his presidency, FONATUR incorporated sustainability into official discourses (Chávez-Cortes & Alcantara, 2010). The 1995–2000 National Tourism Program referred for example, to the strengthening of the competitiveness and sustainability of the Mexican tourism product that will help to create employment, foreign exchange, and the promotion of development in established tourist destinations, in particular Cancún and recently established ecotourism centers such as Celestún.

During Zedillo de Leon, tourism is officially consolidated as a "national aid" (1995 press release). Despite discursive rhetoric aiming at sustainable options, tourism was still conceived as an aid to foster foreign currency and market imperatives. Heading celebration of World Tourism Day 1998, Zedillo would put it himself as follows:

> at a time when international financial markets are plagued by intense instability, internal and external tourism can provide fundamental support for attaining the maximum economic growth possible under such circumstances (. . .) in the first half of this year, more than one billion dollars were generated by tourism activities. It is estimated that this figure will rise to eight billion dollars by the end of 1998, to surpass the amount generated by oil sales (. . .) Mexico is eight in the world in terms of its number of international visitors, and sixteenth as regards the attraction of foreign currency. Our tourism strategy therefore seeks to bring us increasingly closer to world standards regarding the generation of foreign currency for each tourist entering Mexico.

This renewed discursive centrality of tourism in national development plans, came accompanied by major international efforts not only to curate environments according to scientific and technocratic understandings of nature, but also, by efforts to modernize culture, and particularly, Indigenous cultural practices. In the Yucatán Peninsula for example, ecotourism and cultural tourism were highlighted for the first time in regional official development discourses as the main avenues to develop/modernize "the Maya," and help "this peripheral region" towards progress. A direct result of these visions is for example, The Organization Mundo Maya born in 1992 as a transnational organization formed by Belize, El Salvador, Guatemala, Mexico and Honduras with the intention to create new tourism products related to Maya culture and to strengthen existing ones side by side with sustainable development policies.

In Mexico, Mundo Maya covers an area of 241,784 km2 and involves the states of Chiapas, Campeche, Yucatán, Quintana Roo, and Tabasco, areas that have been defined by the state as primordial in terms of federal policy where to develop tourism as a national priority. The main intention of Mundo Maya is still to strengthen a newly created region through "the mixed participation of public and private sectors as well as local communities" (Sectur, 2015: 91). For many inland Maya communities, Mundo Maya meant in practice that modernization plans through tourism, were to be materialized as processes of coercive assimilation like in the “old times,” For Kunchmil for example, a rural inland Maya village studied by Castellanos, sate led assimilation practices were experienced as a regression to “a dark past” (p. 24), “a return to slavery and the future loss of individual and collective rights to property and labor.” On an everyday basis, these efforts consisted in “re-clothing the Maya in western outfits and dragging them to consumer markets” (Loewe, 2010), most notably through insertion in the wage economy through tourism and labor migration to Cancún.

In complete consonance with national development plans, the 1995–2001 Yucatán’s State Plan, under the presidency of Victor Cervera Pacheco also embraced sustainable tourism discourses and ecotourism as “the real way to travel Yucatán,” making room for the first time to displace institutional attention and investments from Cancún’s tourism growth pole model to nationally and internationally declared natural protected areas. At a regional level then, it was in the mid-1990s when we are going to see systematic promotion of archeological Maya vestiges and the consolidation of the city of Mérida as a cultural capital of the world. Other examples of tourist attractions curated under these policies are the pre-classic Maya sites of Oxkintok and the remodeling of inland haciendas as Maya museums and boutique hotels (Córdoba Azcárate, 2020; Breglia, 2006; Loewe, 2010).

9.6 Tourism as National Priority: Tourism as Engine for Inclusive Development Strategies (2000s)

The 2000s signs the consolidation of tourism as the favorite national development engine in Mexico. Known in international development circles as the decade of the United Nation’s “Millennium Development Goals,” this decade marks the institutional shift from tourism as a national aid, to tourism as a national priority in Mexico.

The UN Millennium Development Goals stand for a generalized preoccupation with heritage protection and the consolidation of discourses of human and culturally responsible development policies; the institutionalization of poverty alleviation, community development strategies, and pro-poor and indigenous development strategies. It is in this decade when the UNESCO (2005, p. 4) calls for a serious consideration of the cultural aspects in both development and tourism practices, one that emanates from sustainability debates in earlier decades but which are going to point explicitly at the human and cultural aspects of them both: “Cultural diversity is a rich asset for individuals and societies” the UNESCO put it, “the protection, promotion and maintenance of cultural diversity are an essential requirement for sustainable development for the benefit of present and future generations.” During this decade, international actors and states are going to devote an increased attention to tourism and people’s lives and world’s codes of ethics and regulations as the way forward. A Global Code of Ethics for Tourism is born and Taleb Rifai, UNWTO Secretary-General's words in a public speech in 2013, are representative of the spirit:

> UNWTO is guided by the belief that tourism can make a meaningful contribution to people's lives and our planet. This conviction is at the very heart of the Global Code of Ethics for Tourism, a roadmap for tourism development. I call on all to read, circulate and adopt the Code for the benefit of tourists, tour operators, host communities and their environments worldwide.

The Mexican state's privilege of tourism as the main engine for more inclusive and culturally respectful development strategies must be contextualized under this larger umbrella. Tourism, at least in discourse, should aim not only to modernize and not only include local populations and natural resources in a sustainable way, but also to give back and empower. This discursive shift from tourism as vehicle for regional modernization to national priority, coincides with the government of Fox Quesada (2000–2006) and a moment where Mexico became the largest economy in the world where foreign interests were controlled by an overwhelming majority, with foreign financial institutions representing 75% of all deposits and loans (Biles, 2004).

The 2001–2006 National Development Plan conceived four main criteria for national development: "inclusion, sustainability, competitiveness and regional development". It explicitly stated that "tourism is a priority for the State, therefore it is determined to ensure its competitiveness." Determined to make tourism in Mexico a national priority and to turn Mexico into a leading tourist destination, Fox put it in an interview: "We will look for the development and strengthening of the tourist offerings to consolidate national destinations and to diversify the national tourism product, using Mexico's enormous wealth of natural and cultural resources to our advantage" (Fox, press release 2001).

The National Program for Tourism in Fox's six years of presidency became the governance tool to do so. In the Program, the president stated that tourism activities should be considered as "one of the decisive factors in the increment of opportunities, a better distribution of income and in the use of a concept of sustainability of natural and cultural resources." The visions of the program were to make Mexico a leading country in tourism by 2025, since it will have diversified its markets, products and destinations, and its firms will be competitive at the domestic and international level. Tourism then, is recognized as "playing a key role in economic development and it will have grown with full respect for the natural, cultural and social environment, contributing all the while to enhancing national identity."

In the tourism industry, this decade is characterized by a meaningful shift towards more "existential forms of authenticity" (Wang, 1999). Procuring for these forms of travel means catering not only to specific objects or spaces but to potential existential states of being that are to be activated by tourist activities in particular meaningful locations or moments. A turn to the past and to spaces or cultural difference is going to rule the tourism industry in the generation of these experiences. As a result of state efforts and this international shift in forms of travelling, Mexico witnessed the proliferation of heritage cities, colonial cities, culinary and gastronomic tours, religious and art-oriented tourism, premium tourism, boutique hotels, golf, spas, and nautical tourism. Indigenous luxury haciendas in inland Yucatán serve once more, as representative of the decade's tourism governance model, a form of tourism that happens in historically and regionally meaningful architectures and that enable the re-production of particular lifestyles.

Luxury Hacienda Tourism was launched in inland Yucatán in the 2000s among discourses of empowerment for Indigenous communities, in particular, Maya women. It was the philanthropic effort of a well-known Mexican banker side by side with the state to turn to travelling as a lifestyle that appreciates tradition and historical content. To do so, they selectively remodeled a series of henequen haciendas into boutique hotels in inland Yucatán. However, and despite discourses of community empowerment, participation and historical preservation, tourism policy and planning and tourism ventures on the ground as this one, are still fully fashioned according to the extractive logic articulated in Fox's words above. "Tourism is development" as Sectur (2001) put it, where development is still conceived as modernization and civilization of cultural difference for market purposes and above all, a tool to attract foreign capital. Initiatives as this one run the risk to become modern forms for the reproduction of old forms of servitude and racial imagination in so far as the luxury experience means placing Indigenous bodies back into productive spaces where they were exploited.

Perhaps the major example of tourism development in this period beyond the Yucatán region is that of Pueblos Mágicos. This federal initiative was launched in 2001 to promote more towns and villages as tourism destinations and according to their "cultural uniqueness". This uniqueness is measured in terms of experiences the visit to these places can offer to the visitor (see Velázquez García, 2013). To date Pueblos Mágicos is amongst the country's fastest growing tourist niches (see Ramírez Pérez, 2021).

By the mid-2000s, the financialization of the economy, that is, the increasing importance of financial markets to the workings of the economy (Davis & Suntae, 2015), meant in tourist regions as the Yucatán Peninsula but also across the Mexican geography, the increase of obstacles to small businesses and individuals when soliciting credit in the formal sector forcing low and moderate income families to turn largely to the informal sector (Biles, 2004). This decade for example, is going to witness the consolidation of mixed institutions to attend to the needs of the majority: formal institutions like Cajas Populares, Monte Piedad (Mexico's national pawnshop), and Banco Azteca, but also informal ones based on occasional lending, regular money lending; tied credit and group lending-practices whose entanglement with tourism dynamics has become a matter of everyday practice.

Calderon's presidency (2006–2012) elevated tourism as a matter of national priority and public policy priority declaring at the end of his mandate, 2011 as the Year of Tourism in Mexico. Most well-known by the signature of the 2011 National Tourism Accord, Calderon's presidency is going to fully promote tourism along the old adage of modernization theory. With the commitment of turning Mexico into one of the top five tourism destinations in the world, Calderon's government saw the National Tourism Accord as an effort to reinforce tourism as a trigger of national development. Tourism he said, "unites us beyond our differences," and it that spirit, he committed state policy to:

> Increase connectivity and facilitate tourism; Construct, maintain and improve tourist infrastructure and promote urban organization; Boost tourist promotion in Mexico and abroad; Encourage public and private investment and encourage financing for the tourist sector; Boost the competitiveness of destinations and tourist businesses; Diversify and enhance the supply of tourist services with better quality destinations, products and services; Encourage the integration of national production chains; Offer the best service and promote a culture of tourism; Promote regulatory changes to benefit the sector and to promote the sustainable development of the sector.

Recognized by the World Tourism Organization with the Charter for Solidarity with Tourism Award, Calderón was the first head of state to join the World Tourism Organization (UNWTO) and World Travel and Tourism Council (WTTC) joint campaign highlighting the importance of travel and tourism to global growth and development (see UNWTO, 2011), which emphasized Mexico's centrality in matters of tourism governance at an international level. In his speech receiving the award, Calderon ratified something that had already been said: "For Mexico, tourism is a national priority and a political priority. It is an essential activity for promoting growth and raising living standards [. . .] The sector is also key for promoting the regional development we seek and for doing so sustainably."

By 2010, Mexico was the 10th country receiving more foreign tourists in the world (22.3 million annually) with annual revenues of US$1.8 billion. It is the only Latin American country within the 15 principal tourism economies of the world (OMT 2010) and where tourism is the third source of national income, following oil and remesas from the US (according to Banco Mexico). However, according to Datatur 2012, 50% of the country tourist offer is still concentrated in eight destinations. DF, Riviera Maya, Cancún, Acapulco, Guadalajara, Los Cabos, Monterrey, and Puerto Vallarta. 65% of foreign tourists in Mexico were hosted in sun and sand destinations. 77% of those chose three destinations: two of the 1970s CIPS, Cancún (29%) and Los Cabos (10%), and a completely new tourist region born next to Cancún, the Riviera Maya (38%).

The OECD regional report for the Yucatán Peninsula, stated that although tourism policy "has focused on greater planning and promotion, development of infrastructure, and expanding the supply and diversity of tourist destinations while recognizing the importance of sustainability, the tourism industry in the Yucatán Peninsula, focused primarily on Cancún and colonial cities (Mérida) (. . .) has generally failed to generate many benefits for rural areas" (OECD, 2007). In the state of Yucatán, more than 60% of tourism infrastructure spending, employment, production, consumption, and investment are concentrated in Mérida and Progreso. In the state of Quintana Roo, over 90% are still concentrated in Cancún. "Paradoxically," the OECD reports, rural areas of the region offer a diverse array of tourist attractions but "the economic impacts of the hundreds of thousands of annual visitors to Yucatán's archeological sites are basically limited to admissions (tickets) and souvenir sold by non-local informal vendors, with rural locations capturing very few economic benefits and populations essentially serving as sources of cheap labor for Cancún" (2007, p. 132).

9.7 Tourism's Ambivalent Role: In-between Defunding and/or Pivotal Element for Social Justice (2012-present)

With Peña Nieto (2013–2018) tourism remained at the forefront of national development plans, at least in discourse. "We will open Mexico to the world," he said, "and the world will visit Mexico. Tourism is a fundamental component of our economy and a key driver of national development" (press release). However, and despite discursive support, the reality is that since 2012, the Mexican state has started a steady divestment in tourism projects, and most importantly, in institutional capacity building.

The 2013–2018 National Development Plan corroborated the commitment of the state to enable public-private financing and investment in "projects with tourism potential," by means of: encouraging access to credit; increment in inversions to entrepreneurs and tourism sector; increment in inversions towards consolidation of tourism offer in 1970s poles; and through local capacitation. The National Development Plan included a Programa Sectorial de Turismo which aimed to "specialize in six segments: sun and beach tourism, cultural tourism, ecotourism and adventure tourism, health tourism, sports tourism and special interest or luxury tourism." This program also enunciated the need to create a greater synergy between the action of the private sector and the government, "that will give us the opportunity to enhance the tourist sector for the country's economic development" and it promoted tourism as a means to "cooperate and build with state and local governments, facilitate private investment and encourage entrepreneurship and the development of micro, small and medium businesses, and diversify our product range." Reference to tourism governance is plagued with words such as "sustainable efficiency," sustainable growth, sustainable economic relationships; participative planning, inclusive tourism, sustainable and fair development, and the idea of "social tourism as an element of justice."

However, tourism has been largely absent in Peña Nieto's main legacy, the Pacto por Mexico and it has been overtly shadowed by discussions on energy, education, and political reforms. During his presidency, tourism conversations have been overtly guided by discussions on the benefits of the activity and still measured according to the number of jobs generated and comparisons on the PIB per capita of major urban tourism centers. Tourism has bounced back to comfortably be only a sector of the economy and not also, a set of generative social, cultural, political, ecological, and spatial practices.

One of the characteristics of this period has been a more proactive involvement of regional policy in tourism governance, something that has happened along the same discursive referents of tourism being only a sector or the economy. In Yucatán for example, the 2012–2018 Development Plan of Yucatán State specifically highlights tourism as "viable alternative towards the diversification of Yucatán's economy" pointing directly at the need to consolidate the region as a "modern tourism destination" with more quality through more diversification. In particular, the plan aims to consolidate Yucatán as a National Premium Destination through the direct investment in boutique tourism, such as luxury hacienda hotels, and one of the fastest growing niches of the tourism market in this decade worldwide. The state development plan highlights an increment in private inversion and also the need to professionalize the service sector labor force. In doing so, the state is reproducing the inwards looking take on tourism governance displayed in national development plans in earlier decades as well as potentially deepening local conflicts by pointing at the need to educate and train bodies as docile workers for the industry. This is clearly visible in ventures such as hacienda hotels where Indigenous workers are trained to carve out European gardens in tropical landscapes and tend to European culinary and healing techniques on shifts of 24 hours.

Despite Peña Nieto's less emphatic impulse to tourism in national policy, in 2016, the Mexican Tourist Board announced 35 million international tourists visiting Mexico, which represents a 9% growth with respect to 2015 and almost doubling the global tourism industry average growth situated at 3.9%. Among the reasons listed for the continuous growth in tourism in Mexico are the country's appeal to luxury travelers, its cuisine, nature, and art as well as the ease of multi-destination trips- amongst which the Yucatán Peninsula climaxes.

The region has become once again a privileged laboratory for tourism governance in Mexico's new presidential moment, with Manuel López Obrador, best known as AMLO (2018). In his inaugural speech, AMLO celebrated "tourism, culture and development" as part of a mega infrastructural development project, Train Maya, intended to reactivate the economy of the Mexican South East. Through Tren Maya AMLO has rescued tourism from the background position it had during Peña Nieto's presidency. Yet, its re-incorporation to the forefront of national development strategies has been far from uncontroversial and the old ghosts of modernization are to date deeply haunting this project.

Choosing Chetumal, in the Yucatán Peninsula, to present the National Tourism Strategy 2019–2024, AMLO has already made a performative statement highlighting the central role that the activity, at least in discourse, will have in his presidency.

Among the plans stated in the National Tourism Programs are to build Tren Maya; to develop less visited destinations by granting more licenses and permits for tourism business as well as by launching a domestic program to entice domestic tourism (which will have a unique strategy called Sonrisas por México (Smiles for Mexico) and will offer low-income Mexicans the chance to travel free); to improve working conditions for tourism works in some of Mexico's most popular tourism beach destinations-Acapulco, Los Cabos, Puerto Vallarta, and Playa del Carmen; to improve safety in tourism destinations by having the Secretaiat of Tourism to work hand in hand with the National Defense Secretary and the Security Secretary and the Navy, as well as to invest in advertisement campaigns of Mexico abroad, and not only in the USA but mostly in the United Arab Emirates, Japan, Korea, and parts of Europe, as France, and Italy.

Yet, in 2019 and 2020, The Ministry of Tourism has been the agency with the greatest drop in resources and the Senate of the Republic has approved the disappearance of the Tourism Promotion Council of Mexico (CPTM), or Mexico Tourism Board, responsible for coordinating, designing and operating the strategies in the sector for the country. Moreover, opposition to the Train Maya keeps gaining momentum amongst intellectual elites and Indigenous groups both in Yucatán, Chiapas and Guatemala who see the project as yet another ambitious neoliberal exercise in the dispossession of land, resources, and cultural traditions in the name of tourism and Western centric ideas of progress. What tourism governance will look in this period is still to be seen. Yet, something is clear: the overarching preponderance of the state in tourism planning and policy as well as a heavy and acritical reliance on tourism as a driver of economic growth and progress via foreign investments remains almost intact.

9.8 Conclusion

Tourism governance in Mexico has adopted different forms over time but it has shown several important continuities. One of the continuities that defines tourism governance in Mexico, is for example, the interventionist role of the state in development and planning. Such an interventionist role is revealed amongst other things, in the state acting as entrepreneur and banker in the development of tourism infrastructures across the landscape (Bennet & Sharpe, 1980; Clancy, 1999, 2001) and always side by side with the business community and private initiatives. Another important continuity in how tourism governance works in Mexico, is an existing underlying tension between official development discourses of sustainability, participation, empowerment, and social inclusion, which are celebrated in state development plans in accordance with international regulations, but which clash at a local level with extended and contested practices of neoliberal governance. This governance is materialized in the widespread privatization of resources, the enclosure of land for tourism, the selective inversion in targeted areas and the disciplining of a large wage labor force.

The result has been an institutionalized legitimation of tourism as an economic tool side by side the re-engineering of the country as a collection of fragile landscapes of and for consumption. I call these landscapes fragile because the Mexican state's embrace of international tourism funds for regional development purposes has enabled infrastructure growth, but it has also implied the imposition of the UNWTO's largely managerial understanding of nature and culture under heritage protection policies as nothing more and nothing less than economic assets. This managerial understanding of nature and culture has been facilitated in earnest by tourism coupling with nationalistic development narratives and modernization ideals about social progress and indigenous control that have generalized representations of different regions in the country over the years as fertile lands in crisis, populated by untapped resources and needed of intervention. This association in the realm of development discourses has legitimized extraction logics in the implementation of tourism ventures on the ground. This is evident for example in the way in which tourism governance has involved different alliances between government, private, transnational, and philanthropic actors that have nonetheless resulted in similar social and spatial segregation practices as the one envisaged by design in the earlier CIPs.

Inquiries into tourism governance acquire a renovated interest in the context of current financial and economic crisis, climate change and its associated risks like the disruption of international economic exchanges, the spread of political instability, the reduction of standards of living and the further advancement of environmental degradation. In such a context, models for more communal governance are needed. Such models would place decisions on land and labor in the hands of the workers and communities that inhabit and live in the areas targeted by the state as ideal areas for tourism development.

References

Aramberri, J. & Butler, R. (2005). *Tourism Development: Issues for a Vulnerable Industry.* Channel View Publications.

Bennett, D. & Sharpe, K. (1980). The State as Banker and Entrepreneur: The Last-Resort Character of the Mexican State's Economic Intervention, 1917–76. *Comparative Politics*, 12(2): 165–89. doi:10.2307/421700.

Berger, D. (2006). *The Development of Mexico's Tourism Industry: Pyramids by Day, Martinis by Night.* New York and Basingstoke: Palgrave Macmillan Press.

Biles, J. (2004). Globalization of Banking and Local Access to Financial Resources: A Case Study from Southeastern Mexico. *The Industrial Geographer*, 2(2): 159–173.

Bunten, A.C. (2008). Sharing Culture or Selling Out? Developing the Commodified Persona in the Heritage Industry. *American Ethnologist* 35(3): 380–95. www.jstor.org/stable/27667498.

Butler, R. (2006). Tourism and Indigenous peoples. Taylor & Francis.

Breglia, L. (2006). *Monumental Ambivalence: The Politics of Heritage. 1st ed. Joe R. and Teresa Lozano Long Series in Latin American And Latino Art and Culture.* Austin: University of Texas Press.
Castellanos, B. (2010). A *Return to Servitude: Maya Migration and the Tourist Trade in Cancún. First Peoples, New Directions in Indigenous Studies.* Minneapolis: University of Minnesota Press.
Carmona Dávila, D (2020). Memoria Política de Mexico. De la Madrid Hurtado Miguel. Accessible at: https://www.memoriapoliticademexico.org/Biografias/MMH34.html. Date accessed, 30th October 2020
Córdoba Azcárate, M. (2020). *Stuck with Tourism. Space, Power and Labor in Contemporary Yucatán.* Oakland: University of California Press.
Chambers, E. (2009). *Native Tours: The Anthropology of Travel and Tourism.* Long Grove, Ill: Waveland Press, Inc.
Chavez-Cortes & Alcantara Maya. (2010). Identifying and Structuring Values to guide the choice of sustainability indicators for tourism development, *Sustainability*, 2(9): 3074–3099.
Clancy, M. (1999). Tourism and development – Evidence from Mexico. *Annals of Tourism Research*, 26(1):1–20.
Clancy, M. (2001). *Exporting Paradise: Tourism and Development in Mexico. Tourism Social Science Series.* New York: Pergamon.
Dallen T. & Nyaupane, G. (2009). *Cultural heritage and tourism in the developing world: A regional perspective*, Routledge.
Davis, G. F., Suntae, K. (2015). Financialization of the Economy. *Annual Review of Sociology*, 41: 203–221.
González Vicuña, V. (2013). *Securing Paradise.* Duke University Press.
Jiménez, A. (1993). *Turismo. Estructura y desarrollo*, Mexico: McGraw Hill.
Loewe, R. (2010). *Maya or Mestizo? Nationalism, Modernity, and Its Discontents. Teaching Culture: UTP Ethnographies for the Classroom.* Toronto: University of Toronto Press.
Magaña-Carrillo. (2009). La política turística en México desde el modelo de calidad total: un reto de competitividad Economía, *sociedad y territorio*, 9(30): 515–544
MacCannell, D. (1973). Staged Authenticity: Arrangements of Social Space in Tourist Settings. *American Journal of Sociology*, 79(3): 589–603.
Mowforth, M. & Munt, I. (2015). *Tourism and Sustainability: Development, Globalisation and New Tourism in the Third World.* Abingdon, Oxon; New York, NY: Routledge.
Nunez, T.A. (1963). Tourism, Tradition, and Acculturation: Weekendismo in a Mexican Village. *Ethnology*, 2(3): 347–52. doi:10.2307/3772866.
OECD (2007). Territorial Reviews: Yucatán, Mexico. Retrieved from: https://www.oecd.org/cfe/regional-policy/oecdterritorialreviewsYucatánmexico.htm. Date accessed: 30th October 2020
OECD (2018) Mexico, Tourism Trends and Policies 2018. OECD Publishing, Paris.
Pattullo, P. (2011). *Last Resorts: The Cost of Tourism in the Caribbean.* London: New York: Monthly Review Press.
Pelas, H.R. (2011). Tourism development in Cancún, Mexico: an analysis of state-directed tourism initiatives in a developing nation. Thesis MA Georgetown University.
Pieterse, J.N. (2009). *Development Theory.* Second edition. Los Angeles; London: SAGE Publications Ltd.
Ramírez Pérez, D. (2021) Las marcas del pueblo. Tesis para optar al grado de Doctor en Antropología Social. Colegio de Michoacán, A.C. Centro de Estudios Antropologógicos.
Redclift, M. (2005). A Convulsed and Magic Country: Tourism and resource Histories in the Mexican Caribbean. *Environment and History*, 11(1): 83–97.
SECTUR. 2011. Agendas de Competitividad de los Destinos Turísticos de México. Secretaria de Turismo de Mexico & Universidad de Quintana Roo.

Sharpley, R. & Telfer, D.J. (2015). Tourism and Development: Concepts and Issues Channel View Publications.
Smith, K. & Robinson, M. (2006). Cultural Tourism in a Changing World Politics, Participation and (Re)presentation. Channel View Publications.
Terry, E., Fallaw, B., Joseph, G. & Moseley, T. (2014). *Peripheral Visions: Politics, Society, and the Challenges of Modernity in Yucatán*. Alabama University Press.
Torres, R. & Momsen, J. (2005). Planned Tourism Development in Quintana Roo, *Mexico: Engine for Regional Development or Prescription for Inequitable Growth? Current Issues in Tourism*, 8 (4): 259–85. https://doi.org/10.1080/13683500508668218.
Truett, L.J & Truett, D.B (1982). Public Policy and the growth of the Mexican Industry 1970–1979. *Journal of Travel research*, 20(3): 11–19.
Vainikka, V. (2013). Rethinking Mass Tourism. *Tourist Studies*, 13(3): 268–86. https://doi.org/10.1177/1468797613498163.
Velázquez García, M. (2013). La formulación de las políticas públicas de turismo en Mexico. El caso del programa federal Pueblos Mágicos 2001–2021. *Diálogos Latinoamericanos* (21): 89–110.
Wahab, S. & Pigram J. (1997). *Tourism, Development and Growth: The Challenge of Sustainability*. Routledge.
Wang, N. (1999). Rethinking Authenticity in Tourism Experience. *Annals of Tourism Research*, 26(2): 349–70. https://doi.org/10.1016/S0160-7383(98)00103-0.
Weaver, D.B. (2013). Asymmetrical Dialectics of Sustainable Tourism: Toward Enlightened Mass Tourism. *Journal of Travel Research* 53 (2): 131–140.
Wilson, Tamar Diana, 2008. Economic and Social Impacts of Tourism in Mexico. *Latin American Perspectives* 35 (3): 37–52.

Amna Al-Ruhelli & Rashid Al-Hinai

10 Oman

10.1 Introduction

Oman stretches from the south-eastern corner of the Arabian Peninsula, overlooking Arabian Gulf in the northeast to the Sea of Oman and Arabian Sea to the southeast. Oman has shared borders with the Kingdom of Saudi Arabia (KSA), United Arab Emirates (UAE), and Yemen. In addition, Oman is featured by its various geographical characteristics that created a wide range of environmental and ecological diversity (Robertson, Searle, & Ries, 1990). Oman is known by its natural landscapes that included natural reserves, wadis (valleys), mountains, caves, deserts, sand dunes, beaches, islands, water springs, beach lagoons, and canyons (Kwarteng & Vijaya, 2009; Subramanian et al, 2010).

Administratively, Oman is divided into 11 governorates. Muscat is the capital and located on the north eastern coast. Al-Batinah North and Al-Batinah South are to the west of Muscat. Ad-Dhakhilyah, Ad-Dahirah and Al-Wusta are to the south. Ash-Sharqiyah North and Ash-Sharqiyah South are to the east. Musandam and Al-Buraimi are on the border with UAE, and finally, Dhofar represents the southern part of the country. The government is monarchy-based and centralized. The Sultan is the head of the state and the head of the Council of Ministers (the Prime Minister) as well as the head of all supreme councils (Allen, 2016). According to Rabi (2002), the policy making in Oman takes two deferent paths: top-down and bottom-up that include inputs, revisions, and feedbacks from several authorities prior adopting any policy. Based on the Fundamental Statute of the State that was issued by the Royal Decree No. 101/96, the laws and supreme decisions in Oman are ratified directly by the Sultan himself and come out in a form of royal decrees. Whereas other orders, executive regulations, and ministerial decisions are issued directly by the concern authorities.

In relation to the demographic condition, Oman has small and young population. According to 2017 indicators. The population of Oman estimated as 4.6 million, divided into 2.5 million Omani nationals and 2.1 million foreign citizens (NCSI, 2018). The capital of Oman (Muscat) is the most populated area with a population density of 365 person/km2 whereas the average population density in the whole country is less than 15 person/km2. According to Al-Hinai (2018), the population growth in Oman (among nationals) is 3.3% and 65% of the population is under the age of 29 years.

Furthermore, Oman is one of the developing countries that has inflating economy. Although the oil sector plays a prominent role in the economy of the country, the GDP indicators show that there is a continuous increase in the petroleum activities. For example, the total GDP in 2000 was estimated around US$19,500 million

https://doi.org/10.1515/9783110638141-010

(the contribution of petroleum activities that time was around 50%). In 2010, the GDP increased to nearly US$57 billion and petroleum activities represented 47%. In 2017, the GDP reached US$70 billion and the contribution of petroleum activities decreased to nearly 29% (NCSI, 2018) (Figure 10.1). The decrease in petroleum sector contribution toward the GDP reflects Oman 2020 long-term vision to reduce the reliance on petroleum sector.

In Oman 2040 new vision, tourism was placed as one of the main sectors for diversifying the economy of the country. The country has several historical, cultural, and natural assets that could be efficiently utilized and served in order to form a decent tourism destination and attract more foreign and local investment in this sector. Due to the uniqueness features of the country, it is very important here to elaborate on tourism governance and the related development aspects. Thus, the main objective of this chapter is to provide an overview and give some insights on tourism governance and tourism development in Oman.

10.2 Shaping Tourism Development in Oman

10.2.1 Tourism Governance Before the Year 2000

To understand how the tourism industry is shaped in Oman, it is very important to show the history related to tourism governance. Until the 1960s, Oman had suffered from social conflicts and political and economic degradation and instability. The year 1970 represents the fundamental stepping-stone for the modern development in Oman across all sectors, including tourism. During the first half of 1970s, the tourism sector was attached to the information sector; first under an independent directorate called General Directorate of Information and Tourism from 1972 to 1973, and after that it expanded to be a ministry called Ministry of Information and Tourism until to the end of 1974 (Ministry of Information, 2018).

Then tourism sector was detached from this ministry and relocated as a small department under Ministry of Commerce and Industry. In 1989, this department was expanded to be a general directorate under the same ministry (Royal Decree 112/89) that had regulated this sector until nearly 2004. Moreover, the first ministerial undersecretary office for tourism sector was established in 1993 to steer tourism development from a higher level (Royal Decree 77/93). The dramatic shift in the tourism sector occurred in the 2000s, first by establishing the Ministry of Tourism in 2004 (RD 61/2004) and then forming a governmental company called Oman Tourism Development Company in 2005 to be the main Oman governmental arm for tourism development. Consequently, a ministerial steering committee between concern authorities was structured in 2006 (Royal Decree 117/2006) to enhance and facilitate tourism development and provide a sufficient legislative and administrative

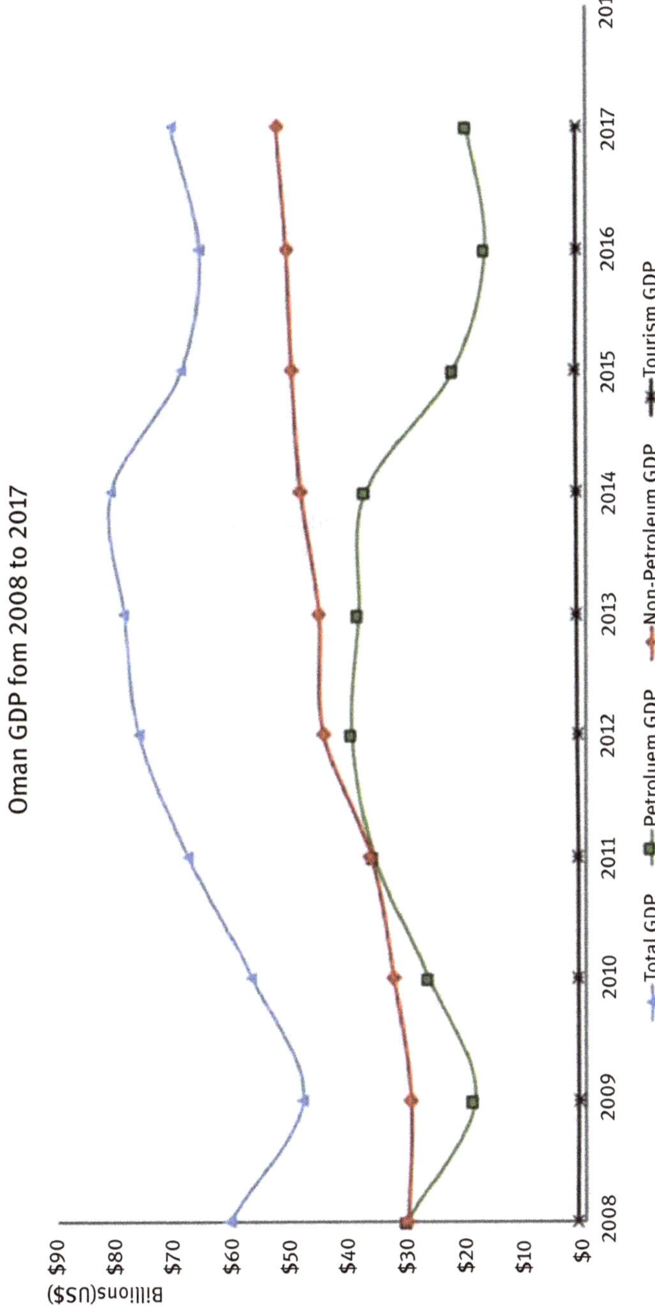

Figure 10.1: Oman GDP and the distribution of petroleum and non-petroleum activities including a reflection to tourism sector. (Source: collected information from Oman National Center for Statistics and Information NCSI, 2018).

support from a higher level. This committee acted for 4 years before it got abolished in 2009. Recently in 2020, major changes occurred in the government structure which include reducing the number of governmental bodies and providing more agile and efficient government. Therefore, for enhancing the utilization of historical and archaeological assets for the tourism uses, tourism sector was merged with heritage sector, forming a ministry called Ministry of Heritage and Tourism (Royal Decree 91/2020), where heritage sector was regulated previously by a separate ministry called Ministry of Heritage and Culture (Figure 10.2 shows the organizational structure of the ministry). In addition, Oman Tourism Development Company was linked to the newly formed Oman Investment Authority to strengthen the government investment base in general and boost the investment in tourism sector in particular (Royal Decree 61/2020).

Until now, the main two authorities responsible for tourism development are the Ministry of Heritage and Tourism, and Oman Tourism Development Company. Nevertheless, tourism industry is a complex industry and there are many authorities interact with tourism development in Oman such as Ministry of Housing and Urban Planning, Environment Authority, Ministry of Commerce, Industry and Investment Promotion, Ministry of Transport, Royal Oman Police, and regional municipalities.

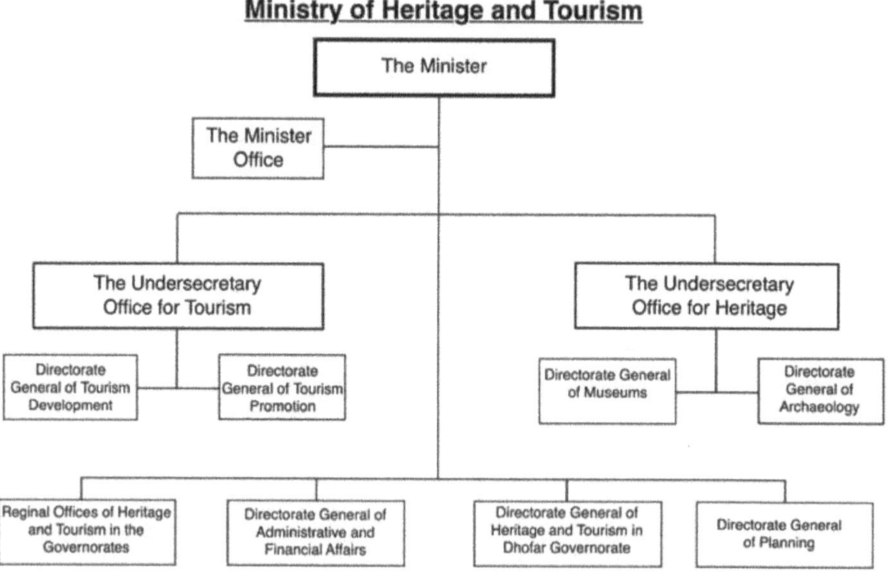

Figure 10.2: The organizational structure of the Ministry of Heritage and Tourism.

At the level of polices and strategies, Oman is perceived to be very cautious during 1990s in promoting tourism and in establishing tourism infrastructural projects

(Wippel, 2016), as the country was sensitive about disturbing its natural environment and the Omani culture (Inskeep, 1994). According to Al-Masroori (2006), mass tourism was planned to not be practiced at the country. Even though tourism in Oman became official in the 1980s and the government started to transfer some of the governmental lands for tourism use, like the public lands in Al-Sawadi area, the tourism projects were still very limited. For example, by the end of 1980s, the accommodation facilities in the country did not exceed 30 units with nearly 2000 hotel rooms in total (Winkler, 2007). In general, during 1970s and 1980s there was low attention from the government to the tourism sector because its focus was directed mainly to the oil sector and developing the infrastructure services of the country. However, in the late 1980s the government established a supporting system and started to allocate some fiscal fund for enhancing touristic, economic, and industrial investments (Royal Decree 40/87). During the same period, a system for organizing and regulating travel and tourism offices was established (Royal Decree No. 12/88). These offices were formulated as a type of tourism activities and were governed under the Commercial Companies Law and the Commercial Register Law of 1974. Both of these steps came in parallel with establishing the General Directorate of Tourism in 1989 as mentioned earlier (Ministry of Commerce and Industry, 1998).

In 1996, Oman announced its first national long-term vision or strategy "Oman 2020," which focused on areas such as privatization, education, and diversification of economic resources and set targets and goals to be achieved by the year 2020 among the various development sectors. Oman 2020 set a target for tourism sector to achieve a contribution of 5% of the national GDP by the year 2020 (Oman, 2007; Subramoniam, Al-Essa, Al-Marashadi, & Al-Kindi, 2010) which was that time estimated as less than 1% (Winkler, 2007) and recently (2017) reached to around 2.6%.

10.2.2 Tourism Governance After the Year 2000

The year 2000 represented the start of the dramatic change in the country's tourism policy. In 2002, the Tourism Law was issued and since has become the main legal framework that regulates and controls tourism sector. In addition, in 2004 Oman became officially a member of UN World Tourism Organization (UNWTO, 2018), which was followed by establishing the Ministry of Tourism. In 2005, a National Tourism Development Plan was formulated to help guide the integration of tourism as a sustainable socio-economic aspect. The plan has required all tourism projects to reflect Oman's historic, cultural, and environmental heritage (Ministry of Tourism, 2005). It started with a goal of creating singular tourism attractions for Oman's old towns and settlements to experience its history and cultural heritage and to attract more responsible tourists. Oman actively avoided promoting costal and mass tourism that focus on sun, sand, and sea (Al-Masroori, 2006).

However, with the need to increase tourism contribution to the national GDP, Oman has shifted toward mass tourism. For instance, since the second half of 2000s, the country started to allocate many areas to establish integrated tourism complexes (ITCs) along the coastal area such as Al-Mouj, Jebel Sifa, and Salalah Beach ITC Projects. Furthermore, in 2006, the country issued the system related to the foreign ownership in ITCs (Royal Decree No. 12/2006), which clarifies these types of projects and offers incentives to foreign owners, such as a free-hold ownership for the units within ITCs and a permanent residence in the country. Until this point foreign ownership of real estate units is not allowed in the country outside ITCs excluding for citizens of Gulf Cooperation Countries. In 2010, the ministry established a Development Control Plan Framework as a guiding document for investors to establish ITCs which clarified the procedure, criteria, regulations, and legislation related to this type of projects. In addition, since 2013, the ministry has updated and expanded the classification programs and standards, which originally developed prior the establishment of Ministry of Tourism, to include many types of accommodation facilities (hotels, hotel apartments, campsites, eco-lodges, etc.) and tourism services (travel and tourism activities, marine activities, leisure and entertainment activities, organization of festivals and shows, etc.) to maintain decent quality and productivity among tourism sector (Ministry of Tourism, 2010).

On the strategic side, the ministry launched in 2016 the Oman Tourism Strategy 2040 to guide tourism development for the next two decades (Ministry of Tourism, 2016). The strategy defined goals and targets to be achieved by the year 2040 such as:

A. Increase the direct contribution of tourism sector in the national GDP from around 2.6% to be between 6% and 10% by 2040.
B. Provide 80,000 deferent types of hotel rooms to meet the satisfaction of all tourist groups (there are currently around 20,000 rooms).
C. Secure more than 500,000 jobs in the tourism sector (direct and indirect). According to WTTC (2017), there are currently less than 100,000 jobs associated with this sector.
D. Enhance the participation of SMEs in tourism development to include at least additional 1200 enterprises.
E. Increase the number of national and international tourists to be more than 11 million per year by the year 2040. In 2017, the number of international tourist arrivals was around 2.4 million and the forecast of 2040 is almost 5.5 million per year.

The strategy emphasizes four main areas: (1) developing experiential system, which includes developing Oman competitive advantage, enhancing natural and cultural resources, and restoring historical areas and traditions; (2) enhancing tourism infrastructure and utilities to facilities tourist's access to the various tourism attractions; (3) enhancing planning and management related to tourism and provide the essential programs and regulations to successfully manage the implementation of the strategy; and (4) providing various types of accommodation facilities and resorts.

According to the strategy, the total volume of investments in tourism for the strategy period is prospected to be around 18.9 billion OMR (approximately US$50 billion). The strategy defined also that around 88% of this investment is going to be through the private sector and the remaining 12% will come through the government (Ministry of Tourism, 2016).

Consequently, since 2017, the Ministry of Tourism started to conduct regional development tourism masterplans for the priority clusters and regions as defined by the strategy. These masterplans focus on determining the tourism identity, attractions, and potential of each region, and thus providing lands, projects, and activities to enhance tourism development and experience in the areas that have high tourism attractiveness. The preparation of Dhofar Governorate masterplan is already completed and there are currently four running masterplans' studies for Muscat, Musandam, Ash-Sharqiyah South, and Ad-Dakhiliyah Governorates. By the year 2020, tourism activates start taking place at more accelerated rate as various tourism development and projects were established (Al-Hinai, 2018). The newly formed Ministry of Heritage and Tourism is currently working in updating the Tourism Law and reviewing the legal and administrative instruments related to the sector. In addition, and as an immediate action related to COVID-19 pandemic, a recovery plan was prepared in the ministry (as acknowledged, but not published to date) to overcome the short and medium term influences expected from this pandemic and also to provide more support and incentives to the private sector in order to face the pandemic and its consequences. Furthermore, this recovery plan pays more attention to local and regional tourism which is the main target area for the coming development phase. On the investment side, a new law for foreign capital investment was issued in 2019 to facilitate the investment in all sectors and provide more flexibility and privileges to attract foreign investors and foreign capital (Royal Decree 50/2019).

10.2.3 Tourist Visa Policy

In terms of Oman's openness to tourism, the first change in tourist visa policy occurred in the early 1990s and the country started to expand its visa regulations to include tourists (Winkler, 2007). As a result, the visa numbers issued for tourism purposes reached 47,000 visas in the year 1995. Nowadays, the executive regulations associated with the country Law of Residence of Foreigners clarifies that the Royal Oman Police (the authority responsible for immigration and visas) has conducted several positive enhancements in the visa system, which is characterized currently with more inclusiveness and effectiveness (Royal Oman Police, 2018). Although Oman now offers a wide range of visa categories, there are three main types associated to tourism: a single entry 10-day visa, a single entry one-month visa, and a multiple entries one-year tourist visa which are usually valid to be used within one month from the date of issue. The new regulations allow citizens of more than 70 countries to benefit from the enhanced e-visa direct process or to apply through the official representative of the country, such as Oman embassies' worldwide offices. However, citizens of other countries can also enjoy visiting Oman by applying for tourist visas through the local travel and tourism offices approved by Royal Oman Police. Nevertheless, the Oman government also implements two forms of joint visas: one with Dubai government and the other with Qatar government. These allow any tourist to visit two destinations (Oman and Dubai or Oman and Qatar) with one single visa. By these efforts, according to Ministry of Tourism statistics, the number of annual tourist visas issued has increased and reached to more than 123,000 in the year 2017. As an exemption from the visas system, citizens of Gulf Cooperation Countries are excluded and could freely enter to the country without restrictions (Al-Hinai, 2018).

10.3 Tourism Dynamic and Realities

In Oman the tourism sector is considered one of the accelerated economic sectors, as reflected by the recent indicators. For example, from 2000 to 2017 a dramatic shift has occurred; the number of accommodation facilities increased from 100 to 367; the hotel rooms increased from 5,300 to 20,500; and the related direct jobs increased from 5,500 to more than 14,000 (NCSI, 2018). Although tourism direct contribution to Oman GDP is still very low and not exceeding 2.6%, there is already some indirect positive influence from this complex industry on other sectors such as transport providers, food and privilege enterprises, and retail businesses which create benefit to local communities. Table 10.1 shows some of the recent indicators related to tourism development in Oman.

Table 10.1: Tourism indicators in Oman (data source: collected information from NCSI).

Year	2009	2011	2013	2015	2017
Inbound Tourists (million)	1.52	1.02	1.40	1.91	2.40
No. of Hotels	224	248	282	318	367
No. of Hotel Rooms (thousand)	10.5	12.2	14.4	16.7	20.5
Tourism Added Value (million OMR)	505	521	653	758	719
Tourism Direct Contribution to GDP (%)	2.7	1.9	2.1	2.8	2.6

10.3.1 Characteristics of ITCs and Limited Tourism Projects

During the past two decades the tourism projects in Oman developed under two different forms or categories. The first is the integrated tourism complexes (ITCs) and the second is the limited tourism projects. Integrated tourism complexes (ITCs) are mega projects and include touristic and real-estate elements. According to the relevant regulations, the minimum area of this category is 200,000m2. Since the year 2005, the country has launched many ITC projects to speed up tourism developments and attract foreign investors (Ministry of Tourism, 2010). Most of these have strategic locations on the coast of Oman. However, it is observed that the overall development progress of these projects is not meeting the aspirations, especially where some are partially completed and operated, such as Al-Mouj and Jebel Sifah projects, while some are suspended and others have not yet started.

Conversely, the limited tourism projects are small to medium scale projects and include mainly touristic elements such as accommodation and tourism activities. The maximum area of this category is usually not exceeding 200,000m2. This category includes a wide range of tourism project types, for example resorts, hotels, eco-lodges, campsites, and tourism entertainment facilities, and as a result, most of the tourism projects implemented fall under this category. Unlike ITCs, the overall completion progress of this category is noticed to be higher, although there are also some failures and delays in the implementation of these projects (Ministry of Tourism, 2010). Table 10.2 demonstrates the main features and differences between ITC projects and limited projects, as extracted from the tourism regulations and legal instruments:

Table 10.2: Tourism ITC projects and limited projects characteristics (data source: collected information from Ministry of Tourism).

Category	Tourism ITC projects	Tourism limited projects
Associated land area	> 200 thousand m^2	< 200 thousand m^2
Annual rental fees of government touristic lands	– 0.5 OMR in Muscat Governorate and Salalah city and 0.2 OMR for other regions for land area < 1 million m^2 – 0.05 OMR for land area > 1 million m^2	– 0.5 OMR in Muscat Governorate and Salalah city – 0.2 OMR for other regions
Land rental fees calculation bases	50% of the acquired land	50% of acquired land or the % of built-up area (which is maximum)
Exemption period from rental fees	5 years from signing the usufruct contract with the government	

Table 10.2 (continued)

Category	Tourism ITC projects	Tourism limited projects
Contract period	50 years (extension possible)	– 25 years for low investment projects. (extension possible) – 50 years for high investment projects (extension possible)
Permitted built-up area	30% of the total land	According to the individual land details
Preparation period and the period to start commencing projects	Up to 3 years (based on the type and scale of the project)	
Residential units allowed	Yes (up to 50% of the components, in some cases some exception are given to the investors with a maximum ratio of 4:1, residential:touristic components)	No
Foreign ownership and related incentives	– Free hold ownership – Permanent residence for owners and their families	Not allowed
Restrictions level	High (these projects are subjected to higher-level approvals). Foreign ownership is not allowed in some areas	Low
Tourism investments on private lands	Allowed but subjected to an official assessment and approval for both types	

10.3.2 Approval Process Related to Tourism Development

In relation to the preparation phase of projects, the process of obtaining all of the necessary official approvals, fulfilling the requirements of concern authorities on the masterplans and detail plans, and obtaining the building permits of projects usually consumes a considerable amount of time due to the multiplicity of procedures and regulations implemented. Figure 10.3 shows the process of obtain an approval for a tourism project in Oman.

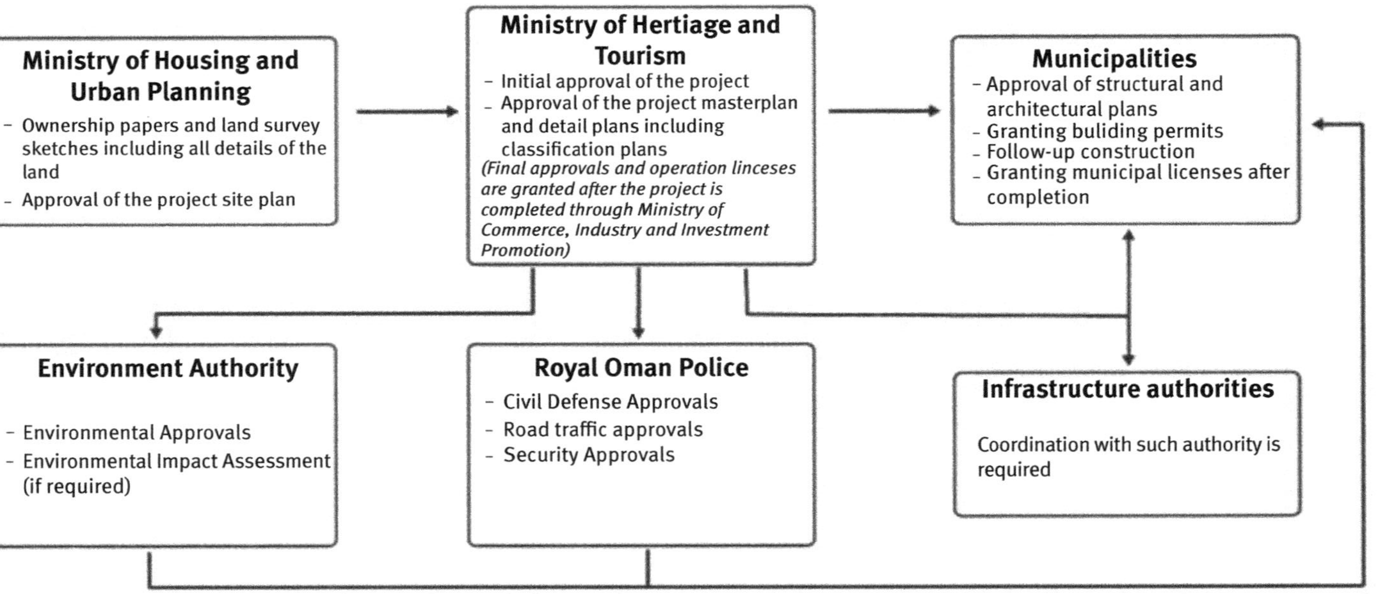

Figure 10.3: Approval process for tourism projects in Oman.
Note: Ministry of Commerce, Industry and Investment Promotion is currently developing the Invest Easy Digital Platform which acts as One-Stop-Shop for all approvals and reduce paper work and time required to generate approvals.

10.3.3 Participation of Private Sector in Tourism Development

The Oman government has acted from different angles and through different authorities to participate in tourism development and encourage the involvement of private sector administratively (via providing regulations and legal instruments) and physically (via participating in tourism spatial development and providing land and sites to be developed by the private sector). Thus, the government has participated in developing some resorts and accommodation facilities, improving the landscapes and features of some cities, adding some tourism infrastructure and activities, restoring and managing some of the historical and archeological sites, and improving and managing of natural reserves and natural and geological sites (Al-Hinai, 2018). Perhaps, due to the limited administrative and financial resources and the insufficient experience related to tourism, the overall governmental performance might be still behind the country's aspirations and targets.

In relation to the contribution of private sector in tourism development, recently there has been high interaction from this sector to enhance tourism. However, because of limited local experience in the diversity of tourism products and the short-term vision in developing tourism projects, it is obvious that most of the projects developed in the last period are mostly accommodation facilities with some amenities. According to the latest statistics and indicators, the number of hotel facilities, for example, has expanded from 287 (14,000 rooms) as in 2013 to 412 facilities (22,000 rooms) by the end of 2018 with additional 151 projects are still under construction (Ministry of Tourism, 2019). This dramatic increase affects the general occupancy rate negatively where it reduced from 48% in 2013 to 38% in 2018 (NCSI, 2019). In relation to the agencies working in travel and tourism, there was, for example, 678 active agencies in 2013. This number has expanded to 1239 in 2018. Additionally, there was 54 registered tour guides in 2013 and this number has expanded to 394 in 2018 (Ministry of Tourism, 2019). These numbers show the positive participation of private sector in tourism development.

10.3.4 Challenges in Tourism Development

Despite of the overall noticeable development achieved by tourism sector, since 2005 there have been many considerable delays on developing tourism projects as well as failures in some cases. As an expert point of view, the delay in Oman tourism development could be associated with several reasons such as:

A. The long and complicated governmental procedure in obtaining the necessary official approvals that most of the investors are inconvenient with,
B. The insufficient regulations and mechanisms that support the joint-governmental work and the fast-changing planning and development regulations and rules in the country,

C. The influence of economic crises on public and private investments (for example 2008–2009 economic crises and 2015–2017 oil prices crises),
D. The influence of the surrounding geo-political implications since 2011,
E. The weakness of tourism market in some periods,
F. The financial deficit and lack of readiness of some investors, and
G. The limited involvement and participation of local communities.

Due to all of these, and since 2016, the country has started to take more active steps to build a strategic path for the future development of tourism. Oman Tourism Strategy 2040 is considered nowadays as the main facilitator of this sector. In addition, tourism has been highlighted as a priority in the action plans of the main national strategies and policies, such as Oman Ninth Five-year Development Plan 2016–2020, Oman 2040 Strategy, and Oman National Spatial Strategy, which were to be launched by 2020. Simultaneously, in 2016, the country launched a high-level national program called Tanfeeth (this Arabic term stands for the English word "implementation"). This program was established for diversifying the economy and enhancing Oman GDP with non-petroleum products and entitled all authorities in the country to collaborate. A separate governmental unit was also established for the purpose of managing and supporting the implementation of this program. In each sector the program defines initiatives. In tourism, there are many initiatives provided to facilitate and accelerate its development. One of these is the preparation of the tourism masterplans. There are other important initiatives, such as privatizing the management of nature sites, enabling adventure activities, developing a year-round calendar of events, introducing e-visa and facilitating of new markets, and activating tourism development fund (THR Innovative Tourism Consultants, 2016). In 2019, the former Ministry of Tourism launched and started working on 26 strategic initiatives which are handled recently by the Ministry of Heritage and Tourism. These initiatives include improvements and enhancements in many areas such as; legislative framework and organizational structure, planning framework, investment procedures, tourism promotion management and procedures, tourism quality and competitiveness framework, human capital inclusiveness, and intelligence information system. All of these efforts are seeking to achieve the prospected goals and maintain a decent future tourism destination.

10.4 Tourism Attractions and Projects

Oman, as one of oldest civilizations in the Arabian Peninsula, has a long history and has played a sensitive role in the ancient Silk Maritime Trade Roads as a trade junction between China and India and other continents (Owtram, 2004). Due to its historical weight, the country heritage is rich with cultural and archeological diversity. There are numerous forts and castles, archeological sites, old villages, and architectural values. For example, Bahla Fort (in Ad-dakhiliyah Governorate), Ancient City of Qalhat (in Ash-Sharqiyah South Governorate), and Land of Frankincense (in Dhofar Governorate) are UNESCO registered historical and archeological sites (UNESCO, 2018).

In addition to the above-mentioned attractions, there are many sites across Oman conveying the beauty and lifestyle of old towns and villages. For instance, the site of Majan, which means the "mountains of copper," is located in the north of Oman and dates back to the third millennium BCE (Weisgerber, 1983). The site contains a settlement connected to ancient great Majan civilization. Another site is the Ubar which is known as the "lost city," which is located in the southern part of Oman. Ubar site is also known for its frankincense trade that has been exported and produced at the south of Oman since the first century CE (Fisher & Fisher, 1999). In the interior of Oman, Harat al-Bilad in Manah provides a taste of the old lifestyle and oasis settlements in Oman. Tourist can visit this abandoned oasis settlement, experience the tribal patterns, and the settlement structure and architecture (Bandyopadhyay, 2004). Moreover, Omani society is relatively a conservative society in relation to its lifestyle and customs and it is famous for its high level of hospitality. It enjoys also ethical, religious, and ideological tolerance, which distinguishes it as society in the region.

Furthermore, Oman is also known for its nature-based, eco-friendly, and adventure tourism. It has many nature reserves and a wide range of natural attraction that promote adventure tourism activities, including diving, beach sports and activities, fishing, trekking, mountain biking and climbing, desert and mountain safaris, bird watching, camel and horse riding). For its unique environment the National Geographic in 2018, listed Oman as the world's ninth most exciting destination to be visited for the year 2019.

Since the establishment of the former Ministry of Tourism, many projects (ITCs and limited tourism projects) took place across the country. The ministry efforts had concentrated on transforming Oman into a nature-based and eco-friendly destination that respects the concept of sustainability as highlighted in all policies and strategies. ITC projects, for example, have been adapted and created to express those efforts. These types of projects claimed to be eco-friendly and sustainable as they are inspired by Oman's natural environment, and through reflecting the heritage and traditions of the country on their urban and architectural forms (Nebel, 2016).

Al-Mouj, as an ITC example, is a project in the heart of Muscat Governorate that provides a waterfront community with 6km stretches of pristine natural beach. The project featured by capturing Oman's diverse cultural and natural lifestyle. The project architecture incorporated the traditional Omani houses, while following the modern style of world-class developments. This project offers golf course, high quality hotels, and a variety of residential, retail, and business units. It provides a distinguished resort lifestyle with friendly urban environment. Jebel Sifah, as another example, is also considered an ITC project. The project masterplan contains around 2700 luxurious residences, four hotels, gulf course, retail areas, and restaurants. This project is partially completed and operated (around 30%). That which is operation provides experiences that combine sea, sky, mountains, and desert all in one place (Ministry of Tourism, 2016).

On the other side, there are many exclusive resorts distributed throughout the country as limited tourism projects. One of the well-known eco-friendly projects is Six Senses Ziggy Bay Eco-resort in Musandam Governorate. Although this 5-stars luxurious beach eco-resort is extremely expensive, it is considered as one of the unique models inspired by Oman old architecture and lifestyle and reflects a living picture of a project that respects culture, nature, and the people in the area at the same time. Anatatra Al-Jabal Al-Akhdar Resort, as final example, is a world class mountain-type resort that resides on a beautiful mountain-edge in Al-Jabal Al-Akhdar touristic mountain (Ministry of Tourism, 2016).

10.5 Conclusion

It is concluded from the discussion that tourism in Oman did not get the sufficient governmental support and prior to the 1990s. However, since that decade tourism development has taken a positive direction. Specifically, since the year 2002, the government focus on tourism has been by issuing the Tourism Law in 2002 and then forming the Ministry for Tourism and joining UN World Tourism Organization in 2004, then merging tourism with heritage sector in the Ministry of Heritage and Tourism in 2020. It is also clear that, since the establishment of the former Ministry of Tourism, a continuous enhancement in the field of tourism policies, strategies, and regulations has been witnessed in the country. By now we can say that the government of Oman has formed a clear vision and path toward tourism development and integration to promote a sustainable and eco-friendly tourism destination. For example, in 2016, both of Oman Tourism Strategy 2040 and Tanfeedh National Program clearly defined the tourism sector as one of the main targeted sectors in need of development in order to enhance and to contribute toward the national economy of the country.

Although, Oman 2020 Vision placed a target for tourism sector to contribute by 5% in Oman's by the year 2020, this target is no longer possible to be achieved. Currently the tourism sector contributes about 2.6%, which is considered as an acceptable share in comparison with the efforts made by the government and the consequences of the international economic crisis and the surrounding political implications. However, recently, Oman has achieved a dramatic change in the tourism field, as the recent indicators clarified, and the country is still seeking to develop a decent sustainable and eco-friendly future destination. What Oman requires in the next phase is to focus on enhancing the tourism policy, upgrading the regulations and working mechanisms, attracting reliable investments, involving local communities and private sector in tourism planning, and developing sufficient tourism infrastructure to enable tourists and visitors to access tourism natural and cultural attractions. It is important also to put in place a management program for the sector in order to respond to the requirements of the international tourism market.

References

Al-Hinai R. S. (2018). *Engineering Stewardship in Spatial Planning and Sustainable Communities in Oman: Analysis, Guidance and a Case Study Connected to Tourism*, Vienna University of Technology, Austria.

Allen Jr., C. H. (2016). *Oman: the modernization of the sultanate*. Routledge.

Al-Masroori, R. S. (2006). *Destination Competitiveness: Interrelationships between destination planning and development strategies and stakeholders' support in enhancing Oman's tourism industry*. Griffith University, Australia.

Bandyopadhyay, S. (2004). Harat al-Bilad (Manah): tribal pattern, settlement structure and architecture. *Journal of Oman Studies*, 13, 183–263.

Fisher, J. & Fisher, B. (1999). The use of KidSat images in the further pursuit of the frankincense roads to Ubar. *IEEE transactions on Geoscience and Remote Sensing*, 37(4), 1841–1847.

Inskeep, E. (1994). *National and regional tourism planning: methodologies and case studies*. Routledge.

Kwarteng, A. Y., Dorvlo, A. S., & Vijaya Kumar, G. T. (2009). Analysis of a 27-year rainfall data (1977–2003) in the Sultanate of Oman. *International Journal of Climatology*, 29(4), 605–617.

Ministry of Commerce and Industry (1998). Decision 154/98 related to the classification of hotel and tourism facilities, Ministry of Commerce and Industry, Muscat.

Ministry of Information (2018). Brief History. Retrieved from https://www.omaninfo.om/english/" https://www.omaninfo.om/english/. Date accessed: October 16, 2018

Ministry of Tourism (2005). *National Tourism Development Plan*, Ministry of Tourism, Muscat.

Ministry of Tourism (2016). *Oman Tourism Strategy 2040*, Ministry of Tourism, Muscat.

Ministry of Tourism (2019). *Annual Statistics 2018*, Ministry of Tourism, Muscat.

National Center for Statistics and Information (2018). Statistical Year Book 2017 Data, NCSI, Muscat.

National Center for Statistics and Information (2019). Statistical Year Book 2018 Data, NCSI, Muscat.

National Geographic (2018). National Geographic Travel Reveals the Best Trips of 2019, Retrieved from http://press.nationalgeographic.com/2018/11/27/national-geographic-travel-reveals-the-best-trips-of-2019/. Date accessed: September 18, 2018

Nebel, S. (2016). Tourism and Urbanization in Oman: Sustainable and Socially Inclusive? Under Construction: Logics of Urbanism in the Gulf Region, 55–69.

Owtram, F. (2004). A modern history of Oman: Formation of the state since 1920 (Vol. 30). IB Tauris.

Rabi, U. (2002). Majlis al-Shura and Majlis al-Dawla: Weaving old practices and new realities in the process of state formation in Oman. *Middle Eastern Studies*, 38(4), 41–50.

Robertson, A. H., Searle, M. P., & Ries, A. C. (Eds.). (1990). *The geology and tectonics of the Oman region (Vol. 49, No. 1, pp. 3–25)*. London: Geological Society.

Royal Decree No. 101/96. Promulgating the Fundamental Statute of the State (Issued 6 November 1996). Official Gazette 587.
Royal Decree No. 112/89. Establishment of Tourism General Directorate under Ministry of Commerce and Industry (Issued 30 December 1989). Official Gazette 422.
Royal Decree No. 117/2006. Structuring a Ministerial Committee for Tourism (Issued 18 November 2006). Official Gazette 828.
Royal Decree No. 12/88. Issuing the System of Travel and Tourism Agencies (Issued 6 February 1988). Official Gazette 377.
Royal Decree No. 40/87. In Relation to the Fiscal Fund for the Private Sector in the Fields of Industry and Tourism (Issued 10 June 1987). Official Gazette 361.
Royal Decree No. 50/2019. Issuing the Law of Foreign Capital Investment (Issued 7 July 2019). Official Gazette 1300.
Royal Decree No. 61/2004. Establishment of Ministry of Tourism (Issued 9 June 2004). Official Gazette 769.
Royal Decree No. 61/2020. Establishment of Oman Investment Authority (Issued 7 February 2020). Official Gazette 1344.
Royal Decree No. 77/93. Appointing an Under-secretary for Tourism Sector under Ministry of tourism and (Issued 7 December 1993). Official Gazette 517.
Royal Decree No. 91/2020. Amending the Name of Ministry of Heritage and Culture to Ministry of Heritage and Tourism (Issued 19 August 2020). Official Gazette 1353.
Royal Oman Police (2018). *Decision No. 129/2018 related to the amendments of some clauses of executive regulations of the Low of Residence of Foreigners*, Muscat: Oman.
Subramoniam, S., Al-Essai, S. A. N., Al-Marashadi, A. A. M., & Al-Kindi, A. M. A. (2010). SWOT analysis on Oman tourism: A case study. *Journal of Economic Development, Management, IT, Finance, and Marketing*, 2(2), 1–22.
UN World Tourism Organization (2018). Member States, Retrieved from http://media.unwto.org/members/states. Date accessed: November 20, 2018
UNESCO (2018). Oman, Retrieved from https://en.unesco.org/countries/oman. Date accessed: October 20, 2018
Weisgerber, G. (1983). Copper production during the third millennium BC in Oman and the question of Makan. *Journal of Oman Studies*, 6(2), 269–276.
Winckler, O. (2007). The birth of Oman's Tourism Industry, *Tourism*, 55(2), 221–234.
Wippel, S. (2016). Port and Tourism Development in Oman: Between Economic Diversification and Global Branding. Under Construction: Logics of Urbanism in the Gulf Region, 101–118.
Oman, D. (2007). Development of community-based tourism in Oman: challenges and opportunities. Tourism in the Middle East, 188–214.
Ministry of Tourism (2010). Development control plan framework for tourism development projects. Retrieved on August 20, 2019 from https://omantourism.gov.om/wps/wcm/connect/mot/2ca1ff80446dc55cb406bfbf0c53e49e/DCPF.pdf?MOD=AJPERES&CONVERT_TO=url&CACHEID=2ca1ff80446dc55cb406bfbf0c53e49e
THR Innovative Tourism Consultants. (2016). Oman Tourism Strategy: Executive Summary. Document No. 309 Retrieved on August 24, 2019 from https://omantourism.gov.om/wps/wcm/connect/mot/4bd8ab5a-f376-44b0-94f0-812a31bd0b99/ENGLIGH+EXECUTIVE+SUMMARY+.pdf?MOD=AJPERES&CONVERT_TO=url&CACHEID=4bd8ab5a-f376-44b0-94f0-812a31bd0b99

Magdalena Banaszkiewicz & Sabina Owsianowska

11 Poland

11.1 Political and Historical Context

The development of tourism in Poland in the last two decades should be considered in the context of changes in the transformation period, which created new conditions for mobility and restored the region to a global map of tourist destinations. Although it is not the case that behind the proverbial "Iron Curtain" there was no travel at all, tourist traffic was drastically reduced in comparison with the choice of forms and destinations available in the West in the compared period (Giustino et al., 2013; Keck-Schajbel, 2012, 2015; Sowiński, 2005). Trends and directions of tourism development observed in Central and Eastern Europe after the collapse of the Soviet Union are largely similar (Banaszkiewicz et al., 2017; Jafari, 2008; Hall et al., 2006; Vukonic, 2006). On the one hand, we can talk about the structural dimension (transformation of institutions and entities responsible for tourism development in individual countries), and on the other, about the specificity of the tourism market offer, inter alia development of cultural tourism, primarily city break, turn towards local identities and ethnic revival in these tourist products based on ethnicity rather than homogenizing globalism dissonances related to the commercialization of heritage, including socialist heritage (Banaszkiewicz, 2016; Owsianowska & Banaszkiewicz, 2015).

After the collapse of the Soviet Union in Central Europe there was a desire to emphasize one's own identity, which was reflected in the so-called persuasive cartography (Smith, 2002; Ziegler, 2002). Derek Hall (2017, p. 32), deconstructing the notion "Central and Eastern Europe" in geopolitical terms, writes that by the end of the 20th century the notions of "Cold War" and "Eastern Europe" had practically disappeared from use, but the term "Central Europe" quickly and clearly marked its presence both in everyday language and in the dictionary of tourism marketing, which became the basis for presenting the region not only as a gateway or a bridge between East and West, but also as a central point of Europe. At the same time, in view of the rise of travel itself, there has been a dynamic development of research and studies on tourism, with the presence of the regional scientific community in the international arena becoming more and more visible. It is true that Stanislav Ivanov, Bulgarian tourism researcher and editor-in-chief of the *European Journal of Tourism Research*, taking into account the low representation of scientists from Central and Eastern Europe in the scientific councils of the 50 most important magazines in the field of tourism and the disproportionately low number of their publications, noted in the discussion at the TriNet group (3.10.2016): "The Iron Curtain in the former communist countries of Central and Eastern Europe in the field of tourism/hotel industry studies has not yet fallen!" However, the increasing number of initiatives to change this situation is difficult

https://doi.org/10.1515/9783110638141-011

to ignore. Transformation processes and state of research in the region are the subject of special issues of *Journal of Tourism and Cultural Change* – "Tourism in (Post)Socialist Eastern Europe" (Banaszkiewicz et al., 2017), *Folia Turistica* (2015), as well as the book *Anthropology of Tourism in Central and Eastern Europe: Bridging Worlds* (Owsianowska & Banaszkiewicz, 2018).

It is worth mentioning that one of the most characteristic elements of the legacy of the socialist system is a strong link between tourism, recreation, and physical culture/sports, in practical, but also scientific and didactic terms. For this reason, the first faculties of tourism and recreation (e.g. Poznań, Kraków) were established in academies (universities) of physical education in the 1970s (e.g. Banaszkiewicz et al., 2017; Owsianowska & Banaszkiewicz, 2018). Despite systemic changes, today they still play an important role in tourism research, expert work for the socio-economic environment and personnel education, including the development of higher education programs. They also provide a reference or template for universities and faculties/institutes with a different profile (e.g. economic), offering studies in tourism and recreation, management in tourism, and so on. The interdisciplinary nature of tourism studies determines the content of curricula and selection of instructors. Lecturers come from various disciplines (geography, law, economics, anthropology, sociology, and philosophy) or represent the sciences of tourism. Tourism studies have so far not gained the status of a separate scientific discipline (there is also no agreement on the grounds for their autonomy) (cf. Alejziak & Winiarski, 2006). This affects the financing of research or academic career development opportunities. Despite the limitations, cooperation is undertaken within the framework of national and international projects: research, publishing, as well as linking business with the academy. The transfer of knowledge, between theoreticians and practitioners, is facilitated by scientific journals, e.g. *Folia Turistica* (which has been published continuously since 1990), *Turyzm/Tourism*, or *Turystyka Kulturowa/Cultural Tourism*.

The aim of the chapter is to show the specificity of the tourism market in Poland in terms of history, structure and challenges. Moreover, it describes the main types of tourism that represent the mainstream of the Polish tourism market. The discussion of tourism development is contextualized in relation to the state's tourism development strategy as well as trends resulting from global and local determinants.

11.2 State and Regional Government

The links between tourism and physical culture/sports, mentioned above, had an impact on institutional issues also after 1989. In the years 1991 to 1999, the Office for Sports and Tourism was the main body of the central tourist administration. After the liquidation of the Office, and since 1999 tourism was for a short time within the structures of the Ministry of Transport and Marine Economy, and since

June 21st, 2000, the Ministry of Economy. In the years 2000 to 2005 the Ministry of Economy and Labour was in charge of tourism, in 2005–2007 the Department of Tourism in the Ministry of Economy. In July 2007, the department was transferred to the Ministry of Sport, which was renamed the Ministry of Sport and Tourism. However, from 2020, Department of Tourism is subject to the Ministry of Economic Development, Labour and Technology. At the regional and local level, the development of tourism is dealt with by voivodeship and marshal offices, poviat offices, and communes (Alejziak, 2009; Kozak, 2009; Zawistowska, 2014).

The general system of tourism management in Poland, modeled to a large extent on Western European solutions, but adapted to the specificity of the local market, was developed at the beginning of the 21st century. The tasks of the state in the field of tourism policy concerned mainly the promotion of tourism. A network of national tourist organizations (NTOs) and national tourist administration (NTAs) has been created. On January 1, 2000, the Polish Tourist Organization, under the supervision of the Department of Tourism, commenced its activity (see Figure 11.1 and 11.2). Its main task is to promote Poland as a country attractive for tourists. In 16 voivodships there are regional (ROT), and in smaller territorial units, local tourist organizations (LOT), in accordance with the principle of decentralization, dealing with the area of a commune or poviat.

Figure 11.1: Poland: National Tourist Administration (NTA). (Source: authors' elaboration).

The promotion of Poland in the world is the responsibility of foreign branches (ZOPOT), which operate in 14 countries: Austria, Germany, Sweden, the Netherlands, Belgium, Great Britain, Italy, France, Spain, Ukraine, and Russia, and outside Europe in the USA, Japan, and China (from October 2015). It is worth noting, however, that the scope of activity of foreign branches may extend beyond the office host country (e.g. a branch in Austria promotes on the Hungarian, Austrian, and Swiss markets, in the USA on the North American and Canadian markets, and in Sweden – on the Scandinavian markets). The geographical distribution of ZOPOT correlates with the most important countries for Polish inbound tourism. This applies to nine out of 10 European markets and four out of five non-European markets (cf.).

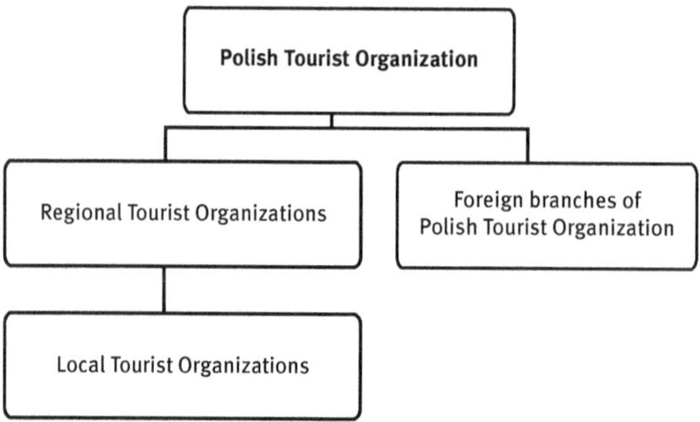

Figure 11.2: Poland: National Tourist Organization (NTO). (Source: authors' elaboration).

Currently, the Tourism Department within the Ministry of Economic Development, Labour and Technology is responsible for programming the development and shaping legal and economic mechanisms in the field of tourism. The task of the Department is to indicate priorities and determine the directions of promotional activities on the domestic and foreign markets; development, implementation and monitoring of programs related to tourism, including, for example, those focused on tourism management or tourist safety. With regard to the tourism economy, the Department deals with the mechanisms of regulating the tourism market, especially in the area of development of entrepreneurship, improvement of quality, protection of consumers of tourist services. It conducts matters that arise from the tasks of the minister in charge of tourism; it also participates in the process of creating Community law on tourism and deals with matters that arise from Poland's membership in the EU and other international organizations (e.g. UNWTO) and structures (Visegrad Group V-4). Of particular importance in the activities of the Department of Tourism are tasks related to education of personnel for tourism, corresponding to the requirements of the changing market, including those set out in the Act of December 22nd, 2015, on the Integrated Qualifications System. Another area of the Department's activity is research, including statistical studies showing national and international tourism traffic.

11.3 Accession to the EU: Consequences for the Development of Tourism

The accession of new countries to the European Union (and later the Schengen area) was another breakthrough (Ferfet, 2008; Hall et al., 2006). Cooperation with the "old Union" had been encouraged in the pre-accession period: cross-border structures were created, so-called "euroregions," within which cooperation was being promoted in respect of transnational strategies for destination development, integrated promotion, harmonization of standards, staff training, and subsidizing local initiatives. Applications for EU grants had to meet certain criteria corresponding with trends in European and worldwide tourism. This way, the tasks of sustainable development were carried out, best practices and solutions were implemented (such as Greenways Trails), including promotion of natural and cultural heritage.

Accession to the European Union, although important for political and economic reasons, has not led to a radical shift in the direction of tourism itself. In communication from the Europe (2010). Communication from the Commision to the European Parliament, the Council, the European Economic and Social Commitee and the Committee of the Regions - Europe, the world's No.1 tourist destination - a new political framework for tourism in Europe /*COM/2010/0352 final*/; Eur-Lex - 52010DC0352 - En - EUR-Lex (europa.eu).

> tourism is a major economic activity with a broadly positive impact on economic growth and employment in Europe. [. . .] As an activity which impinges on cultural and natural heritage and on traditions and contemporary cultures in the European Union, tourism provides a textbook example of the need to reconcile economic growth and sustainable development, including an ethical dimension. (Europe, 2010)

It is clear that, it is a document outlining a strategy for the tourism policies of individual Member States rather than setting out a common EU tourism policy.

As far as Poland's tourism policy is concerned, it is based on the one adopted on November 24th, 2017. The Act on tourist events and related tourist services (Journal of Laws, item 2361), replacing the Act of August 29th, 1997, on tourist services, which for two decades was the basic legal act regulating the activities of tour operators, tourist brokers, tourist agents, tour guides, tour managers, and hotel service providers in Poland. The Act of 2017 implemented into the Polish legal order the provisions of Directive 2015/2302/EU of the European Parliament and of the Council of November 25th, 2015, on tourist events and related tourist services, amending Regulation (EC) No 2006/2004 and Directive 2011/83/EU of the European Parliament and of the Council and repealing Council Directive 90/314/EEC. The Act was also aimed at improving the existing legal solutions, in particular as regards protection of travelers against the effects of insolvency of tour operators and entrepreneurs and facilitating the purchase of tourist services.

The instability of the tourism market and the simultaneous appearance of many small operators caused problems related to the solvency of tour operators, which in recent years have intensified. In the years 1999 to 2014, in Poland, there were 62 cases of activating funds from financial security in connection with the bankruptcy of a tour operator (Wanat-Połeć & Sordyl, 2015); hence from 2016 there is the Touristic Guarantee Fund, which is an additional pillar of insurance, providing funds to cover the costs of return to the country and reimbursement of payments for unrealized and interrupted tourist events for customers of travel agencies, which will lose liquidity and declared insolvency.

11.4 Tourism in Numbers

The year 2017 brought 8% growth in tourism in Europe with 538 million international tourist arrivals, accounting for 40% of the world's total. The dynamics of tourist traffic in Central and Eastern Europe is not as spectacular as in Asia and is far from the position of leaders such as France or Spain, but we can see a systematic increase in the interest of tourists in this area. The European Union Tourism Trends report, elaborated by The European Commission and the World Tourism Organization (UNWTO) distinguishes within the Central and Eastern European subregion so-called EU destinations and Extra-EU destinations, but does not refer to Central Europe per se. According to the recent report, in 2016 Central and Eastern Europe recorded a 4% increase in arrivals (127 million) and a 6% increase in receipts (€48 billion), though arrivals to five countries (Bulgaria, Poland, Hungary, Romania, and Croatia) grew even faster (at a rate of 8%, while receipts grew 10%, reaching €29 billion) (The European Union Tourism Trends, 2018). In 2016, international arrivals to Poland grew 4% to 17 million, while tourism earnings grew 10% to €10 billion. This is connected with two significant events in 2016: Wroclaw was one of the 2016 European Capitals of Culture, while Kraków hosted World Youth Day.

According to the World Economic Forum (The Travel and Tourism Competitiveness Report, 2018) for 2017, Poland was ranked 46th. Among the other countries of the region, Estonia (37), the Czech Republic (39), Slovenia (41), and Bulgaria (45) ranked slightly higher. However, it is important that Poland's position is strengthening systematically (in 2011 it was 49th, in 2009 only 58th). In the ranking of the magnitude of effects generated by the tourism industry (expressed in real prices in terms of dollars) in the years 2005–2015, Poland was ranked 12th among 28 EU Member States, although it was only 25th in terms of the contribution of tourism to the GDP of the country. This means that tourism still does not contribute to the economic development and quality of life of the population to the same extent as other sectors of the national economy. The total contribution of Travel & Tourism to GDP, including wider effects from investment, the supply chain, and induced income impacts was PLN 89.9 billion in 2017 (4.5% of GDP) and has been expected to grow by 3.6% to PLN 93.2 billion (4.5% of GDP) in 2018, though it was forecasted that by 2028 it would be 5.5% of GDP (Travel & Tourism Economic Impact, 2018).

Studies conducted by the Central Statistical Office in Poland show that the rate of development of tourism in Poland remains at a constant level. In 2017, 83.8 million foreigners (including 18.3 million tourists and 65.5 million same-day visitors) came to Poland (4.1%, 4.4%, and 4.0% more than in 2016 respectively). Nearly half of the arriving foreigners was in the age group of 35–54 years (48.2%), with people aged 25–34 years constituting a large group (18.8%) also. In 2017, expenditures of foreigners visiting Poland amounted to PLN 56.7 billion, which was 3.6% more than in 2016. In 2017 the largest number of tourists was recorded in Małopolskie and Mazowieckie voivodships (5.1 million including 1.5 million foreigners and 4.9 million).

Warszawa (6.2 million) and Kraków (5.3 million) occupy the position of leadership if the number of overnight stays is taken into account (Turystyka, 2017).

However, the growing popularity of Poland among foreigners has its negative effects. In the last few years the prices of tourist services have been growing rapidly, which makes Poland's competitiveness for domestic tourists lower than in the past. An attempt to reverse this trend is an attempt to encourage domestic exploration through a system of one-off discounts. The project "Poland, see more – weekend at half price," which was a joint action of the Polish Tourist Organization and the Ministry of Sport and Tourism, organized twice a year (in March and October), proved to be a success. This is a nationwide promotional campaign, to which entrepreneurs from the tourism industry submit their offers and tourists use them at reduced prices. The fifth edition, in autumn 2018, was attended by more than 165,000 tourists who took advantage of the offers of 757 partners. Research commissioned by the Polish Tourist Organization shows that Poles praise not only low prices, but many of them value the opportunity to spend time with their families and believe that "Poland, see more – weekend at half price" activates society. The advantages of the campaign are also noticed by entrepreneurs, for whom it is important that they can additionally earn in the so-called low season, the vast majority of whom announce participation in subsequent editions. They consider as the most important: promotion of their activity (87%) and reaching new clients (78%) (Polska Zobacz Więcej, 2018).

11.5 Types of Tourism

As shown by research conducted by the Polish Tourist Organization aimed at segmentation of the domestic tourism market, taking into account the various expectations of departure, trip planning strategies, and forms of activity undertaken during the trip, four segments presenting different tourist attitudes were distinguished: leisure seekers (45%), family holidaymakers (25%), explorers (19%), and all-inclusive tourists (11%). Some destinations of domestic tourists coincide with foreign recipients, but the main difference lies in the predominance of rest over sightseeing. The most popular forms of tourism practiced by foreigners in Poland include heritage tourism, city breaks, rural tourism, pilgrimage tourism, and medical/health tourism.

11.5.1 Heritage Tourism

Heritage tourism, regardless of whether it concerns material or immaterial heritage, is one of the most important types of global cognitive mobility (Light, 2015; Timothy, 2011; Timothy & Boyd, 2003). In Poland, heritage resources include both centuries-old architecture (e.g. historical cities with a preserved urban layout from the Middle

Ages), an increasingly modern museum offer (including institutions such as the Solidarity Centre in Gdańsk), but also a number of phenomena that are used in the form of cultural event tourism (folklore festivals, inclusion of the Kraków crib making industry on the UNESCO List of Intangible Cultural Heritage). Annual surveys conducted by the Polish Tourist Organization confirm that the majority of foreigners associate Poland with history and rich heritage (94%) and probably their travel plans most often include cities and UNESCO sites. The development of this type of tourism is facilitated by the creation of thematic trails, such as the Wooden Architecture Route or the Silesian Industrial Monuments Route belonging to the European Route of Industrial Heritage network. Culinary tourism is an increasingly important form of cultural tourism developing in Poland based on the unique heritage of the region. In 2018, the European Academy of Gastronomy awarded for the first time the title of European Capital of Culture, which went to Kraków (the city defeated Lisbon in the final). One of the cyclical events that have become a permanent fixture in the calendar of European gastronomy festivals is the Dumpling Festival, organized in mid-August in Kraków.

11.5.2 Pilgrimage Tourism

Pilgrimage tourism also deserves special attention when describing the cultural potential of Poland. According to Jackowski et al. (2014) there are between 500 and 800 Catholic holy sites in Poland. Since the early Middle Ages, Poland has been an important point on the pilgrimage map of Europe. It is one of the easternmost countries to which the Camino de Santiago route reached. Already in the Middle Ages, Jasna Góra, the Pauline monastery famous for its miraculous painting of the Virgin Mary, was treated as a center of worship on a supra-local scale. Pilgrimages to Jasna Góra are one of the most important religious phenomena in Poland, and Poles as an ethnic group constitute almost one fifth of Christian pilgrims in Europe. For centuries, Kraków has been the undisputed leader. Since the end of the 16th century, the saying "If there was no Rome, Kraków would be Rome" has been known. Kraków is considered to be a center of religious tourism not only because of the remains of locally worshipped saints, such as St. Stanislaus and St. Hedwig, but above all because of the popularization in the 1990s of the Divine Mercy cult associated with the activities of St. Faustyna Kowalska and the sanctuary in Łagiewniki and the elevation to the altars of the "Pope-Pole" by John Paul II. In 2016, almost 1.5 million pilgrims from all over the world participated in World Youth Day in Kraków. Since 2017, the International Congress of Religious Tourism and Pilgrimages has been organized in Kraków. It is attended by about 200 tour operators and tour agents from 30 countries and has already become the most important event in religious tourism branch in Central and Eastern Europe (International Congress of Religious Tourism and Pilgrimages, 2020).

11.5.3 City-break Tourism

In the last decade, the market of city break tourism in Poland has been particularly strengthened. This phenomenon is conditioned by the expansion of low-cost airlines and the spread of non-hotel forms of accommodation (mainly through Airbnb). Low prices of transport and accommodation together with competitive prices of services and the cost of food caused that in cities such as Kraków, Warsaw, Gdańsk, or Wrocław were overrun with tourists coming for just two to four days. Recently, the messaging of this form of travel has provoked questions about the negative effects of tourism, especially in the form of the progressive gentrification of city centers (Kruczek, 2018).

11.5.4 Rural Tourism

Poland's tourist attractiveness in the area of rural tourism should also not be overlooked. It results from the preservation of the traditional agricultural landscape and rural architecture, as well as the authenticity of folk customs and traditions, which enrich the offer with a wealth of unique experiences, including religious traditions, for example Corpus Christi processions, and agricultural traditions, for example harvest festival. The most attractive areas for the development of rural tourism with elements of ethnic tourism are Kashubia, Podlasie, and Podhale.

11.5.5 Medical/Health Tourism

Medical tourism is not obvious, although a more and more visible development trend in Poland. On the Medical Tourism Index (Medical Tourism Index Global Ranking, 2016–2017) Poland is the only representative from the region of Central and Eastern Europe. This is an important niche when, according to MTA, the value of the global medical tourism market amounts to US$100 billion per year. Moreover, it is a form of travel less susceptible to seasonal fluctuations than typical holiday tourism. The analysis of the phenomenon of medical tourism from the supply and demand side carried out by Białk-Wolf and Arent (2018) indicates that the number of medical tourists coming to Poland is about 155,000, of which the largest group (75,000) are people using dental services, 48,000 are health resort tourists, 22,000 choosing from a wide range of aesthetic medicine services, and 10,000 use hospital services. The development of this branch of tourism can be significantly stimulated thanks to appropriate government programs, although so far no systemic framework for the development of this niche has been developed in Poland.

11.6 Image of the Country and Tourism Promotion

One of the challenges for Poland and other countries of Central and Eastern Europe after 1989 was to change the image and break the stereotypes perpetuated for decades. In the 1990s, the promotion of national tourism was not centralized and based on long-term development plans. Before accession to the EU, one of the promotion priorities was to emphasize that European values are – and have always been – important for Poles. After 2004, the main challenge was to increase Poland's presence in foreign media and to convince about the attractiveness of the country as a destination. Qualitative image research conducted among residents and tour operators from potential emission markets, such as France, Germany, Great Britain, and Ukraine, confirmed that the image of Poland is "ambiguous, difficult to define." Despite the positive associations related to the beauty of nature, cultural values, dynamically developing economy, and figures of famous Poles, for example Fryderyk Chopin, Maria Skłodowska-Curie, John Paul II, Lech Wałęsa, the country was not perceived as an attractive destination. The exception, according to German tourists, were city break trips and leisure tourism by the sea and lake. Intensive efforts were necessary to create a new, recognizable image of the country (Johann, 2014; Kruczek & Walas, 2010; Ageron Polska, 2011; Owsianowska, 2011, 2014, 2017b).

In the promotional activities prior to 2010, the PTO (2005) promotional campaign entitled "Welcome to Poland" (the so-called commercial with "Polish plumber") stands out from the rest of the world. The controversial and quite radically shattering fixed ideas about Poland, however, was assessed negatively in the fundamental issue of building the country's brand. Prepared at a time when the shape of the European Constitution was being discussed, it aroused interest by skillfully engaging in the public debate on issues such as the uncontrolled influx of "cheap labor" from the new Member States. It provocatively refers to stereotypes and the figure of a handsome "plumber," who, contrary to expectations, stays in the country and invites you to visit, presenting the main attractions of Poland. A measurable effect was an increase in the number of tourists from France and other Western European countries. The spontaneous and accidental nature of the campaign and the lack of a long-term strategy meant that the benefits of the campaign were not used properly.

A project of long-term activities appeared a few years later, on the occasion of the Polish Presidency of the EU in 2011 and the organization (together with Ukraine) of EURO 2012. These events provided an excellent opportunity to initiate a comprehensive image campaign. Based on the research preceding it, four groups of countries were identified, potential emission markets: the first of them was dominated by the image of Poland as "little modern, slightly backward, religious, agricultural and focused mainly on tradition and history;" for the second, it was a country with a "double face," traditional and religious, and at the same time modern, free market, attractive for investors (cf. Ollins, 2008); for the third, Poland was a democratic country with a dynamically developing economy and good work organization; while the respondents from the fourth group did not have much information about Poland.

In 2009, an agreement was concluded for co-financing of the project under the Operational Program Innovative Economy "Promoting Poland's tourist assets" – Let's Promote Poland Together 2009–2015. In 2010, a cruise of the sailing ship Fryderyk Chopin took place, and the campaign "Promoting Poland's tourist assets" was carried out. "Polska. Move your imagination" started at ITB in Berlin in 2011. Its aim was to show a modern, dynamic country with a diversified tourist offer. Films and computer animations, multimedia presentations and events created the desired image of Poland and reflected competing narratives about the country and its inhabitants. On the one hand, they showed attachment to tradition, on the other hand, they emphasized openness to cultural changes, for example expressed in lifestyles, the growing role of women, or acceptance for other people and groups, including ethnic and social minorities. Creating the image of Poland focused on the most important values associated with the main Polish cities (Warsaw, Gdańsk, Kraków, Poznań, Wrocław, and Łódź) and regions (for example, Masuria), history, eminent people, customs, and so on. The central value is hospitality, which is the essence of creation also in subsequent campaigns: "Feel Invited" and "Come and find your story" (this so-called fairy tale campaign used fiction – fairy tale, literary, film – to promote the country; cf. Owsianowska, 2017b).

In 2012, the PTO concluded an agreement with the Ministry of Economy for co-financing of the "I like Poland!" project, which was to reach recipients in China, India, and Japan, in order to improve Poland's competitive position on Asian markets and create new development opportunities for the Polish tourism industry.

According to a report by the European Travel Commission (ETC), Poland is one of the European countries that spend the least on tourism promotion: in 2016, the budget of the Polish Tourist Organization amounted to €9.7 million (and was 36% lower than in 2015). Promotional actions carried out by the Polish Tourist Organization were effective thanks to cooperation with external entities (representatives of the tourism industry, regions, cities, diplomatic missions, etc.) and consolidation of financial resources (www.pot.gov.pl, 2020).

11.7 Europeanization, Multiculturality, and Dissonant Heritage

The fundamental changes, already in the period preceding Poland's accession to the EU, concerned not only the adjustment of national legislation and tourism development strategy. The adoption of EU institutional and legal arrangements, with regard to tourism, also concerned the promotion of European values. Obtaining EU funds, for example capital and tools for modernization and innovation of the region, made it easier to catch up with several decades of neglect in the sphere of economy and socio-cultural life. In the material sphere, this means investments in infrastructure, restoration of monuments, know-how for managing destinations and attractions. In the sphere of promotion, Europeanness appears, for example, in the slogans "This is the most European of the Polish regions" or "This is where the European Union begins" (Owsianowska, 2017a). For example, the areas of eastern Poland – apart from the Masurian lakes – which are absent from the above-mentioned actions, became the subject of a project devoted to the promotion of sustainable development of tourism. "The Beautiful East" campaign, addressed to domestic and foreign tourists, notably in Germany and Ukraine, presented the tourist values of the macro-region, covering five voivodships: Warmińsko-Mazurskie, Podlaskie, Lubelskie, Świętokrzyskie, and Podkarpackie. It offers mainly products of active and specialized tourism, sightseeing tourism, culinary tourism, and rural areas. Promotion of Eastern Poland as a place of active tourism and multicultural region was to help to use the tourism potential of the peripheral areas of the country. Tourism is treated as a remedy for a lower level of economic development, and due to the cultural and tourist policy of the EU, the areas of eastern Poland gain color and attractiveness thanks to the reconstruction of the multicultural world before World War II.

The issue of ethnic, national, and regional identity is closely related to the issue of belonging to European culture and to a community of supranational scope, which is also a source of ideological conflicts. Eurosceptic opinions refer, among other things, to differences in worldview, which is largely due to the attachment to the Catholic Church declared by the majority of Poles and the fear of secularization associated with the Western European system of values, or to the sovereignty of the state. Another controversial area is the approach to history and remembrance of difficult events that have not been "processed" by local communities – de facto heirs to the difficult heritage of multiculturalism, ethnic conflicts, and the Holocaust. As a result, the lack of discussion and action that would help to settle the controversial past and to understand the needs and concerns of all stakeholders is causing resistance to some initiatives. An example is the display in tourist products of the values of different cultures in areas that are currently mono-ethnic. On the one hand, it is treated as an attraction of the offer and an opportunity to get to know other as former neighbors, this includes Jews, Lemkos, Ukrainians, and Germans, to restore and cultivate the memory of them. The promotional message exposes the multiculturalism of the described regions, the richness of traditions, such as culinary traditions through slogans "mosaic of cultures" and "borderland of styles." On the other hand, the reconstruction for the tourism industry of a world that existed in the past, sometimes arouses (irrational) fear, which is skillfully exploited and fueled by populists and nationalists as a source of hostility towards strangers or ressentiment. However, despite the above-mentioned dilemmas and the complicated political situation after the parliamentary elections in 2015, the support for Poland's membership in the EU was high with over 80% of the population in favor (Center for Public Opinion Research, 2016, 2019).

The creation of the country's image refers to ideological differences, which influence the choice of promoted cultural heritage resources. After 2015, the way of approaching some controversial points in the history of the country has changed, including the most recent one, including questioning the role of Solidarity activists, e.g. Lech Wałęsa in 1980s, and re-evaluation of the achievements in the years 1989–2015. One may ask whether initiatives such as the Subjective Bus Line and the Lech Wałęsa Workshop, would nowadays gain the recognition from the PTO as the certified tourism products (cf. Owsianowska & Banaszkiewicz, 2015). In the officially forced version of the past, the history just after the end of the WWII is also subject to revision. For instance, celebrations connected with the so-called cursed soldiers refer to a simplified vision of their role in the postwar Poland and soldiers responsible for genocide against civilians and ethnic minorities are included in the group of commemorated persons. This process of rewriting of the difficult past is reflected in the imposition of a new interpretation on cultural institutions subordinate to the Ministry with regards to, inter alia, the exhibitions at the Museum of the Second World War in Gdańsk or an attempt to intervene in the European Solidarity Centre program in exchange for a subsidy from the state budget.

In the promotion of regions, ideological discussion may also concern traditions and local legends which sources seem unacceptable to Catholicism, according to some representatives of governmental administration. The reason may be a too "pagan" a symbol (i.e. a witch) in the logotype, storylines and promotional campaigns of one of the regions (Świętokrzyskie). In fact, these examples are an expression of the voices of a strong minority in Polish society. Most Poles reject extreme, absurd, and fanatical views and are able to easily reconcile patriotic values and attachment to tradition with a sense of belonging to the European community, with the pursuit of modernity, tolerance, and openness to others. However, in the approach to heritage and in competing promotional strategies, as if through a lens, social divisions reflect the legacy of a complicated history, including 123 years of partitions (1772–1918), the difficult process of building an independent state in the interwar period, interrupted by the outbreak of World War II, and the experience of two totalitarianisms in 1939–1945 and 1945–1989. The dynamics of recent years clearly reveals internal fractures, crises, and the renegotiation of identity and community narratives, also through tourism discourse.

11.8 Education for Responsible Tourism

Along with the development of tourism, the demand for qualified staff has increased. Higher education at the bachelor, master, and doctoral levels (from 2019 within the so-called doctoral schools) is conducted in the structures of various universities: from private universities related to tourism, through the already mentioned academies of physical education, to economic universities, pedagogical universities, and various faculties of general universities. There are similar dilemmas related to teaching as in other countries, namely the need to balance business and more humanistic approaches, as well as vocational and philosophical aspects (cf. Tribe, 1997). Personnel education for tourism should combine theory and practice, but the way of implementing this postulate ultimately depends on the profile of the university or faculty where such faculties as tourism and recreation, international tourism, management in tourism, or intercultural relations in tourism are implemented.

After Poland's accession to the EU, the education system was reformed and adapted to European standards through the introduction of the so-called Bologna system. In the field of tourism, as in other fields, it is oriented towards learning outcomes, validation, and certification. Qualifications can be acquired in (1) a formal way in schools, universities, and other educational institutions; (2) non-formal ways through various training, internships, and courses; and (3) informal, related to learning in professional situations, everyday life practices, through media coverage. In 2014, The Educational Research Institute in Warsaw and teams of experts have undertaken actions to create a Sectoral Framework of Qualifications for Tourism (SQFT), including occupations in the hotel, catering, and tourism sector (HO-GA-TUR). In the first quarter of 2015, the first stage of creation of SQFT was completed, then the descriptions of qualifications as well as the principles of validation and certification were prepared. Their implementation was to improve the quality of personnel by increasing the transparency of qualifications and identifying key areas of competence and adapting human resources education and training programs to the needs of the industry.

Tourism education in Poland has a long tradition dating back to the 19th century, formalized for the first time in the 1930s and consistently developed in the post-war realities. The breakthrough that took place after 1989, both in the sphere of mobility in the region and in the tourist economy, opened new directions of education, taking into account local specificity and global trends and tendencies. Challenges related to changing conditions in the labor market and factors affecting the organization of travel, stay, and leisure patterns require not only knowledge, skills, and competencies strictly related to tourism services from people employed in the industry. Climate change, dynamic political and economic situation, refugee crisis, among other nota bene phenomena inscribed in ideological disputes not only in Poland; they are a part of reality on both sides of the guest-host relationship. This means that staff should be expected to have a good orientation in a broader social, political, economic, or environmental context, critical thinking, the ability to assess the reliability of information sources, as well as the implementation of theory in practice (Owsianowska, 2018). Actions for the sustainable development and promotion of responsible tourism, in line with the provisions of the Global Code of Ethics for Tourism (UNWTO), are an important element of human resources education, but their importance should be given greater prominence. In addition to the programs implemented as part of formal education, it is worth mentioning such initiatives as post-turysta.pl (2013), whose creators conduct workshops for various groups of recipients.

11.9 Conclusion: Tourism Development in 2020 and Beyond

The main objective of the "Marketing strategy of Poland in the tourism sector for the years 2012–2020" was to indicate the most important directions and forms of actions that should be taken within the framework of tourism policy in order to improve the international recognition of Poland as an attractive and hospitable country for tourists and having competitive tourist products of high quality. The study describes trends in the global tourism industry and tourism policy of the European Union. It also includes an analysis of Poland's tourism competition, a description of the basic indicators of inbound tourism and the potential of emission markets. The selection of products and target groups, tools, and instruments were also indicated, and above all the vision, mission, strategic objective, and operational objectives were described. Poland exists as a brand, but it is still not very distinctive, which, according to Anholt (2007), places it in the group of "new and boring" countries, mainly due to its association with history, which does not appear to be attractive for tourists in itself. For this reason, Poland treats urban and cultural tourism (understood as "5A" – attractions, amenities, accommodation, access, atmosphere) along with meeting industry products, and the strategy for promotion of Poland recommends to use the brands of cities, which in many markets are stronger than the brand of the country as a whole.

The low level of coordination of the tourism industry development policy in Poland is best evidenced by the fact that public expenditure on tourism amounted to PLN 820 million, which was only PLN 61 million from the central budget (Polish Economic Institute 2020, p. 39). The COVID-19 epidemic has forced radical measures to save the tourism industry. According to Statistics Poland (2020a), in the first half of 2020, the number of crossings of the Polish border was 86.9 million. Compared to the same period of the previous year, lower border traffic was recorded, both for foreigners (by 41%) and Poles (by 38%). In the summer of 2020, the situation was saved by domestic tourism, also developing under the influence of the protectionist policy of the Polish government (including a tourist voucher for families to spend on accommodation in the country), even though, in July and August 2020 the number of tourists using accommodation facilities decreased by 29% compared to the same period last year (Statistics Poland 2020b) The value of the losses to be incurred by the Polish tourism industry in connection with the pandemic is difficult to estimate. Aid programs prepared by the Polish government – the so-called anti-crisis shields (including additional benefits, exemption from paying social security contributions, non-returnable subsidies for micro and small businesses) are perceived by representatives of the tourism industry as far from being sufficient. The new post-epidemic realities will require a redefinition of the role of the tourism sector in the economy, taking into account the far-reaching social and cultural changes resulting from "forced non-mobility." Undoubtedly, those destinations that a year ago considered actions aimed at limiting the negative consequences of mass tourist traffic (mainly Krakow or Zakopane) are now facing a completely opposite phenomenon – the lack of tourists, which has a dramatic effect on the local economy. Earlier interest in Poland was to a large extent due to spontaneous search of tourists themselves, attracted by the price attractiveness of the stay, rather than to a well-considered tourism policy of the state, within the framework of which coherent promotional and marketing activities would stimulate interest in diversified groups of tourist products. Perhaps, the current epidemic crisis may be one of the factors stimulating a more sustainable development of tourism, combining in a coherent and consistent way the economic, social, environmental, institutional, and political orders.

Acknowledgement: The chapter was prepared within the framework of a research project No. 156/BS/KTiR/2018, funded by the Ministry of Science and Higher Education in Poland, Research potential.

References

Ageron Polska, (2011). Badania wizerunkowe Polski i polskiej gospodarki w krajach głównych partnerów gospodarczych [Image research of Poland and Polish economy in the countries of main economic partners]. Report commissioned by the Ministry of Economy. Ageron Polska, Warszawa.

Alejziak, W. (2009). Determinanty i zróżnicowanie społeczne aktywności turystycznej [Determinants and social diversity of tourist activity]. Studia i Monografie AWF Kraków nr 56.

Alejziak, W. (2016). Multi-Level Governance jako instrument planowania rozwoju i zarządzania obszarami recepcji turystycznej [Multi-Level Governance as a tool of tourism planning and he management of tourism destinations]. Turystyka i Rekreacja – Studia i Prace. 16, 9–31.

Alejziak, W. & Winiarski, R. (2006). Tourism in scientific research in Poland and worldwide. WSiZ Rzeszow-AWF Krakow.

Anholt, S. (2007). Competitive Identity: New Brand Management for Nations, Cities and Regions. Palgrave Macmillan.

Banaszkiewicz, M. (2016) The "embodiments" of Stalin in the tourism landscape of Moscow. *International Journal of Tourism Anthropology*, 5(3–4), 221–234.

Banaszkiewicz, M. & Owsianowska, S. (2018). Anthropological research on tourism in Central and Eastern Europe. In S. Owsianowska & M. Banaszkiewicz (Eds.) *Anthropology of Tourism in Central and Eastern Europe*. Bridging Worlds. Rowman and Littlefield.

Banaszkiewicz, M., Graburn, N. & Owsianowska, S. (2017) Tourism in (Post)socialist Eastern Europe. *Journal of Tourism and Cultural Change*, 15(2), 109–121.

Białk-Wolf, A.& Arent, M., Turyści medyczni w Polsce w kontekście międzynarodowym, http://cejsh.icm.edu.pl/cejsh/element/bwmeta1.element.desklight-587f337d-d130-48a3-9ecc-e78680fe29d2/c/06-Turysci_medyczni.pdf.

Centre for Public Opinion CBOS (2016). Polska w Unii Europejskiej. Komunikat z badań [Poland in the European Union. Research report], vol. 31, https://cbos.pl/SPISKOM.POL/2016/K_031_16.PDF.

Centre for Public Opinion CBOS (2019). 15 lat członkostwa Polski w Unii Europejskiej. Komunikat z badań [15 years of Polish membership in the European Union. Research report], vol. 59, https://cbos.pl/SPISKOM.POL/2019/K_059_19.PDF.

Europe (2010). Communication from the Commission to the European Parliament, the Council, The European Economic and Social Committee and the Committee of the Regions - Europe, the world's No 1 tourist destination - a new political framework for tourism in Europe /*COM/ 2010/0352 final*/; EUR-Lex - 52010DC0352 - EN - UER-Lex (europa.eu)

Giustino, C. M, Plum, C. & Vari A. (2013). *Socialist Escapes: Breaking Away from Ideology and Everyday Routine in Eastern Europe*, 1945–1989. Berghahn.

Hall, D. (2017). Tourism in the Geopolitical Construction of Central and Eastern Europe (CEE). In D. Hall (Ed.) *Tourism and Geopolitics: Issues and Concepts from Central and Eastern Europe*. CABI.

Hall, D., Smith, M. & Marciszewska, B. (2006). Tourism in the New Europe: the Challenges and Opportunities of EU Enlargement. CABI.

International Congress of Religious Tourism and Pilgrimages, https://icortap.com/

Jackowski, A., Bilska-Wodecka, E., & Sołjan, I. (2014). Pilgrimages and religious tourism in Poland in the 21st century – current situation and perspectives for development. Zeszyty Naukowe Uniwersytetu Szczecińskiego. Ekonomiczne Problemy Turystyki. 4, 253–270.

Jafari, J. (2008). Eastern Europe's Tourism: Old Wines and New Bottles. In Ferfet, K. (Ed.) *Tourism in the New Eastern Europe*. Warsaw College of Tourism and Hospitality Management.

Johann, M. (2014). The Image of Poland as a Tourist Destination. *European Journal of Tourism, Hospitality and Recreation*. 4, 143–161.

Keck-Szajbel, M. & Stola, D. (Eds.) (2015). Crossing the Borders of Friendship: Mobility across Communist Borders. *Eastern European Politics and Societies and Cultures*. 29(1), 92–95.

Keck-Schajbel, M. (2012). Shop around the Bloc: Trader Tourism and its Discontents on the East German-Polish Border In Bren, P., Neuburger, M. (Eds.) *Communism Unwrapped: Consumption in Cold War Eastern Europe*. Oxford University Press.

Kozak, M. (2009). Turystyka i polityka turystyczna a rozwój: między starym a nowym paradygmatem [Tourism and tourist policy and the development: between the old and new paradigm]. Wydawnictwo Scholar, Warszawa.

Kruczek, Z. (2015). Sektorowa Rama Kwalifikacji w Turystyce i jej znaczenie w kształceniu i certyfikacji kadr turystycznych [The Sector Qualifications Framework for Tourism and its Meaning in Education and Certification of Tourism Staff]. In Rapacz, A. (Ed.) *Gospodarka turystyczna w regionie [Tourist Industry in a Region*]. Wroclaw Univ. of Eonomicx Publ., 396–404.

Kruczek Z. (2018) Turyści vs. mieszkańcy. Wpływ nadmiernej frekwencji turystów na proces gentryfikacji miast historycznych na przykładzie Krakowa [Tourists vs. Inhabitants. Impact of overtourism on gentrification of historical cities on the example of Krakow]. Turystyka Kulturowa, 29–41, http://turystykakulturowa.org/ojs/index.php/tk/article/view/956/810.

Kruczek, Z. & Walas, B. (2010). Promocja i informacja w turystyce [Tourist promotion and information], Proksenia.

Kuźmicki, M. & Wasilewska, J. (2015). The Image of Poland as a Tourist Destination in the Country's Main Economic Partners. Annales Universitatis Mariae Curie-Skłodowska, XLIX (3), 91–98, doi:10.17951/h.2015.59.3.91.

Light, D. (2010). Heritage and Tourism In Waterton, E. & Watson. S. (Eds.) *The Palgrave Handbook of Contemporary Heritage Research*. Palgrave Macmillan.

Medical Tourism Index Global Ranking 2016–2017, https://www.medicaltourismindex.com/destination/poland/

Marketing strategy of tourism in Poland for 2012–2020 (Marketingowa strategia Polski w sektorze turystyki), https://www.pot.gov.pl/pl/o-pot/plany-i-sprawozdania-pot

Ollins, W. (2008). *The Brand Handbook*. Thames & Hudson.

Owsianowska, S. (2011). Tourism Promotion, Discourse and Identity. *Folia Turistica*, 25(1), 231–248.

Owsianowska, S. (2014). Stereotypes in tourist narrative. *Turystyka Kulturowa/Cultural Tourism*, 4, 1–21.

Owsianowska, S. (2017a). Tourist narratives about the dissonant heritage of the borderlands: The case of South-eastern Poland. *Journal of Tourism and Cultural Change*, 15(2), 167–184.

Owsianowska, S. (2017b). "Come and find your (love) story". Remaking the image of Poland as a tourist destination. Via Tourism Review, http://journals.openedition.org/viatourism/1772; DOI: 10.4000/viatourism.1772

Owsianowska, S. (2018). Lifelong learning and multiculturality: A Polish case. In V. Cuffy, D. Airey, G. Papageorgiou, *Lifelong learning and tourism*. Routledge.

Owsianowska, S., & Banaszkiewicz, M. (2015) Dissonant Heritage in Tourism Promotion: Certified Tourism Products in Poland. *Folia Turistica*, 37, 145–167.

Owsianowska, S., & Banaszkiewicz, M. (2018). Anthropology of Tourism in Central and Eastern Europe. Bridging Worlds. Rowman &; Littlefield.

Polska Zobacz Więcej [Poland. See more], https://www.pot.gov.pl/attachments/article/7428/PODSUMOWANIE%20BROSZURA7_final.pdf

Polish Economic Institute [Polski Instytut Ekonomiczny] (2020), Branża turystyczna w Polsce. Stan sprzed pandemii, https://pie.net.pl/wp-content/uploads/2020/05/PIE-Raport_Turystyka.pdf.

Smith, A. (2002). Imagining Geographies of the "New Europe": Geo-Economic Power and the New European Architecture of Integration. *Political Geography*, 21(5), 647–670.

Sowiński, P. (2005) Wakacje w Polsce Ludowej: polityka władz i ruch turystyczny (1945–1989). "Trio": Instytut Studiów Politycznych Polskiej Akademii Nauk.

Statistics Poland (2020a), Border traffic and expenses made by foreigners in Poland and by Poles abroad in the 2nd quarter of 2020, https://stat.gov.pl/en/topics/prices-trade/trade/border-traffic-and-expenses-made-by-foreigners-in-poland-and-by-poles-abroad-in-the-2nd-quarter-of-2020,14,3.html.

Statistics Poland (2020b), Occupancy of tourist accommodation establishments in Poland in July and August 2020, https://stat.gov.pl/en/topics/culture-tourism-sport/tourism/occupancy-of-tourist-accommodation-establish-ments-in-poland-in-july-and-august-2020,5,22.html.

The European Union Trends Report, https://ec.europa.eu/growth/tools-databases/vto/content/2018-eu-tourism-trends-report

The Travel and Tourism Competitiveness Report, http://reports.weforum.org/travel-and-tourism-competitiveness-report–2017/

Timothy, D.J. (2011). *Cultural Heritage and Tourism: An Introduction*. Channel View Publication.

Timothy, D.J., & Boyd, S.W. (2003). *Heritage Tourism*. Prentice Hall.

Tourism Development Programme until 2020, https://www.msit.gov.pl/download/3/12550/TourismDevelopmentProgrammeuntil20201f3c.pdf

Travel & Tourism Economic Impact 2018. Poland, WTTC, https://hi-tek.io/assets/tourism-statistics/Poland2018.pdf

Tribe, J. (1997). The Indiscipline of Tourism. *Annals of Tourism Research*, 24(3), 638–657.

Turystyka w 2017 roku [Tourism in Year 2017], https://stat.gov.pl/download/gfx/portalinformacyjny/pl/.../1/.../turystyka_w_2017.pdf

Vukonic, B. (2006). Turning Point in European Tourism Development. In Alejziak, W. & Winiarski, R. (Eds.) *Tourism in Scientific Research in Poland and Worldwide*. University of Physical Education in Krakow Press.

Wanat-Połeć, E., & Sordyl, G. (2015). Wzmocnienie ochrony konsumentów niewypłacalnych touroperatorów w Polsce a koncepcja utworzenia Turystycznego Funduszu Gwarancyjnego, Annales UMCS Lublin-Polonia, XLIX, 4, 633–648; DOI: 10.17951/h.2015.49.4.633.

Zawistowska H., Dębski, M. & Górska-Warsewicz, H. (2014). Polityka turystyczna. Powstanie – rozwój – główne obszary [Tourist policy. Creation – development – main areas]. PWE.

Ziegler, D. (2002). Post-Communist Eastern Europe and the Cartography of Independence. *Political Geography*, 21/5, 671–686.

Joana Almeida & Pedro J. Pinto

12 Portugal

12.1 Introduction

At the 2017, 2018, and 2019 World Travel Awards, Portugal won three consecutive awards for best tourist destination. The National Tourism Authority "Turismo de Portugal" also won the award for World's Leading Tourist Board in these same three years, alongside the 14th UNWTO Award (January 2018) in Public Policy and Governance. This raises the question, in what ways do these awards reflect good tourism governance? To understand the basis for this success, this chapter presents a synthesis of the governance of tourism in Portugal over time and the great changes that occurred at two key moments: 2007 and 2014.

Portugal, with a population of about 10.3 million (INE, 2019), is part of the world's leading region in terms of tourist arrivals: specifically, Mediterranean Europe. Located in the Iberian Peninsula at the extreme southwest of the continent, Portugal is composed of the mainland area – a territory whose borders are among the oldest and most stable in the world, having been in place for over seven centuries, in a country that has existed for almost nine centuries – and by two Atlantic Archipelagos: Madeira and the Azores. In its relatively small territory (92,212 km2 / 35,603 sq. mi), Portugal has a long (1,240 km) coastline along the Atlantic Ocean. It also has a wide diversity of landscapes, natural parks, historical, architectural, cultural and religious monuments, intangible cultural expressions, culinary traditions, wine regions, historical villages, spas, golf courses, ports and marinas, waterways, lakes, reservoirs with piers and moorings providing different, unique experiences within easy reach.

According to the World Economic Forum, in 2016 Portugal was the 14th most competitive tourist destination of 136 countries (WEF, 2017). Tourism is an extremely significant activity for the Portuguese socio-economy, accounting for about 16.5% of the country's total exports, in 2016, and 17.8%, in 2017 (Bank of Portugal, 2018), with travel and tourism representing the largest share in Portugal's total exports of goods and services.

Tourism is well recognized for playing a key role in economic activity, job creation, and as a source of export revenue and domestic value added. In 2016, tourism directly contributed to 9% of GDP and 8.8% of employment, far above of the average for OECD countries, where tourism directly contributes an average of 4.2% of GDP and 6.9% of employment (OECD, 2018).

Note: The writing of the present chapter was partially funded by project MetroGov3C - funded through Grant FCT - PTDC/GES-URB/30453/2017 of the Portuguese Foundation for Science and Technology.

https://doi.org/10.1515/9783110638141-012

The strategic relevance of tourism in Portugal is thus evident not only in making good use of the country's heritage and territory, but also as a way of promoting regional planning, receiving investment, generating wealth, creating employment, and boosting Portugal's image abroad. In the last 50 years tourism has contributed significantly to Portugal's economic growth and development, inducing a social restructuring and a cultural transformation within the country, and making it increasingly cosmopolitan. But despite this long-term significance, recent years mark a big change. In 2010, the number of international tourist arrivals was 6.8 million, growing to 18.2 million in 2016, and around 21.2 million in 2017 (UNWTO, 2018). In fact, 2017 was a historic year for the tourism sector in Portugal.

In terms of tourism destinations there has been investment in upgrading, innovation, sustainability, and internationalization, with the destination's facilities often winning awards. In 2002, the nineth edition of the World Travel Awards distinguished an accommodation establishment in Portugal for the first time. But since 2016, Portugal has been receiving a striking amount of tourism awards: Europe's Leading Destination, Europe's Leading Beach Destination; the "Turismo de Portugal" was deemed World's Leading Tourist Board, Europe's Leading Tourist Board and World's Leading Tourism Authority Website [online]; Lisbon was considered World's Leading City Break Destination, Europe's Leading Cruise Destination, and Lisbon's Port as Europe's Leading Cruise Port; the Algarve was singled out as Europe's Leading Beach Destination; Madeira as Europe's Leading Island Destination; Passadiços do Paiva (Arouca UNESCO Global Geopark [online]) was deemed Europe's Leading Tourism Development Project in 2016 and in 2017. Turismo de Portugal was awarded the first international distinction of Accessible Tourism Destination (ATD2019), launched by the UNWTO (15th World Award, December 2018), to recognize those destinations that are making laudable efforts so that they can be enjoyed by all tourists, regardless of their physical, sensory, or cognitive abilities.

These awards recognize the success of the Portuguese Tourism Governance, but there is also an international context and some national and even municipal policies and strategies that should be taken into account, and some warnings as to the long-term sustainability of this success.

12.2 Tourism Governance in Portugal: Historical Background

12.2.1 1900–1950 – First Steps

The first official organization for Portuguese Tourism – the Tourism Office, part of the Ministry of Development (Ministério do Fomento) – was created in 1911. This office commissioned studies on the sector (Cunha, 1997). Portuguese tourism was then based primarily on thermal resorts for domestic tourism and on Madeira and Lisbon for international tourism.

World War had I hard economic, political, and social consequences, that severely affected tourism. Nevertheless, some reforms were carried out and projects launched, such as the "Estoril Tourism Plan" and the creation of the Pousadas (inns) building program, which became a trademark of Portuguese tourism. In 1920, the Bureau of Tourism became part of the General Administration of Roads and Tourism (Ministry of Commerce and Communications). A few local commissions were set up to ensure the preservation of national heritage, cooperation in road improvements, tourism, hygiene, and local promotion (Cunha, 1997).

By 1950 the number of international arrivals to Portugal reached 76,307. During this 50-year period, despite growing awareness of the importance of tourism and the legislative effort (about 240 legal documents related to tourism were published), Portugal's tourism sector failed to reach the desired international prominence (Cunha, 1997).

12.2.2 1950–1973 – Small Steps

In the early 1950s, with the European countries in recovery from World War II, a new era was beginning for tourism. In Portugal, the tourist potential gained increasing recognition, associated with favorable weather conditions, picturesque landscapes, exceptional conditions for spa and beach tourism, and the hospitality of the people. However, the country had been ruled by a dictatorship since 1926 and a few equally acknowledged that Portugal was not prepared with regard to the hotel industry and that the backwardness of the country could hinder tourism development. Tourism was beginning to be perceived as a true industry, and tourism was emerging as a major economic activity. For the first time, the tourism policy was designed according to a multisector approach (Cunha, 1997).

Politically, Portugal's dictatorship was increasingly isolated from the world, and tourism was a way of promoting its political image and an increasingly important source of external revenue. The government drafted, in 1952, tourism statutes that would provide the framework for future Portuguese tourism development. Tourism areas were also created in the municipalities with beaches or thermal areas, among other attractions, and by 1952, there were 83 tourism zones administered by municipal commissions. Two years later, the status of "public utility" was created, allowing hotel companies access to a wide range of tax exemptions. Then, in 1956, the Tourism Law was published.

It was only in the 1960s that tourism was first referenced in a state planning instrument, in the Plano Intercalar de Fomento (Interim Development Plan) (1965–1967). Unlike the two earlier multi-year plans (1953–1958 and 1959–1964), the third Plano de Fomento (1968-1973) already considered tourism to be a strategic sector of economic development, with the understanding that it was the foreign market that lead the expansion of hotel capacity and tourist activities (Moreira, 2018). A Secretariat of State for the Council Presidency with specific competence in matters related to information and tourism was established in 1962. For the first time, tourism was being dealt with at the central government level. Over this period, the hotel capacity more than doubled from 24,000 beds in 1953 to 57,000 ten years later. Hotel occupancy was still dominated by domestic demand, which provided about 65% of total hotel nights (Cunha, 1997).

12.2.3 1973–1990 – Unplanned Tourism Growth

After a short period of decline associated with the 1973 energy crisis, Portugal demonstrated one of the highest tourism growth rates in Europe (along with Turkey and Hungary). This generated euphoria and also some business irrationality, attracting – especially to the Algarve – pure real estate development, deeply disconnected from tourist reality. "The real estate replaced tourism accommodation" (Cunha, 1997). Between 1979 and 1987, the Algarve's hotel capacity increased from 5,000 to 25,000 beds and informal accommodation reached 150,000 beds. With this came a degradation of the region's landscape: the natural resources that attracted tourists were themselves put in jeopardy by unplanned development. The Algarve held over 40% of the national tourism supply, and together with the Lisbon Coast absorbed 70% of the international tourism nights. The "sun & sea" product was the prime attraction, so at that time Portuguese tourism faced the challenge of diversification.

Following the 1974 revolution, significant changes were made to tourism governance. For the first time tourism was given its own independent Secretary of State, and a set of concrete measures to promote tourism development were implemented, including professional training, the creation of tourism regions, and the establishment of a public company that owned and managed the Pousadas. At the municipal level, though, tourism governance remained highly disorganized.

In 1985, Portugal surpassed 10 million foreign visitors for the first time, and with its entry into the EEC in 1986, strengthened its position vis-à-vis Europe as a prestigious and experienced tourist destination. The idea of tourism as a fundamental factor for development was reaffirmed. Tourism was providing employment and improving the quality of life of a considerable part of the Portuguese population, and revenues generated by this activity were a major contribution to the GDP.

Despite this consolidation, Portugal found it difficult for decades to define a medium and long-term policy for tourism (Costa & Vieira, 2014). The first Plano Nacional de Turismo (National Tourism Plan) was enacted and published in 1986 and remained in force until 1992, and it has been recognized as an important shift in tourism policies in Portugal (Milheiro & Santos, 2005). Under this plan, 19 regional tourist boards were created, following a policy approach to decentralize state power, land-use planning was valued, as well as investments, professional training, tourist entertainment, balneology and spas, and promotion aimed at diversifying markets and increasing revenues. Other destinations, besides the classic Lisbon-Algarve-Madeira triangle, were starting to be "discovered" both by domestic and foreign tourism.

12.2.4 1990–2000 – More Tourism Growth

The 1990s were a successful decade for tourism around the world, with a global growth in the number of tourists of about 55%. Portugal grew 50%, second only to Turkey (100%) among southern European countries. This increase was heavily dominated by the product "sun & sea" (Neto, 2013).

This decade was extremely important for structuring the Portuguese tourism offer. The investment in road infrastructure boosted tourist investment, especially in low-density territories where tourism was seen as an opportunity for the demographic and economic revitalization. There was a diversification of the accommodation offered, with rural tourism gaining popularity. The network of Pousadas also expanded, as did the number of golf courses. Cultural tourism was promoted across the country, and a number of tourist plans, such as the Recovery of Historic Villages Plans or the Cultural Tourism Increment Program, were created. Distinctive tourist developments were built as well as event venues, which led to a rise in international cultural, social, and sports events. In 1994, Lisbon was the European Capital of Culture and the city hosted the World Exhibition (EXPO'98) four years later. This decade saw the liberalization of the European Community airspace and a consequent increase in air travel. In the 1990s investments were also made in the management and promotion of Portugal, at home and abroad.

12.2.5 2000–2010 Stagnation, But Ambitious Goals

In the 2000s, while there was a global slowdown in tourism growth, Portugal was the only country in southern Europe and North Africa where tourism did not grow and did not benefit from the growth of Europe (Table 12.1). What could be the causes for such a halt, especially when compared to its direct competitors (Croatia, Greece, Turkey, and Spain)?

While September 11th, 2001, and the Economic Crisis of 2008/2009 could partly explain a global slowdown in tourism growth, Portugal was hit even harder than average. In fact, it is considered a lost decade for tourism in Portugal, because the successive governments did not realize that the country had reached the high point of maturity of a tourism growth cycle in the late 1990s, and because new realities, new products, and new markets emerged, with an increased competition and lowered prices. Portugal did not keep up with these changes.

Table 12.1: International tourist arrivals – 1990, 2000 and 2010. Source: Based on Neto (2013) and WTO (2012).

	International tourist arrivals				
	x1000			Variation (%)	
	1990	2000	2010	1990/2000	2000/2010
World	435	674	940	54.9%	39.5%
Europe	262	385	475	46.9%	23.4%
South Europe	90	133	169	47.8%	27.1%
Croatia	7	5.3	9.4	−24.3%	77.4%
Greece	8.9	13	15	46.1%	15.4%
Turkey	4.8	9.6	27	100.0%	181.3%
Spain	34	46.4	52.7	36.5%	13.6%
Portugal	8	12	11	50.0%	−8.3%

Paradoxically, the political objectives for the sector were quite ambitious, aiming in 2003 to position Portugal in the top ten tourist destinations (over 20 million foreign tourists); in 2004, a dedicated Tourism Ministry was created; in 2006 the National Strategic Plan for Tourism (PENT) was drawn up on the premise of a Top Ten "sustained growth above the European average" target, expecting 20–21 million tourists by 2015. The reality is that in 2015 Portugal would receive only 14.5 million foreign tourists. Neto (2013) offers a possible explanation: 1) The objectives set by the governments were unrealistic and reflected a lack of knowledge about the reality, the dynamics of tourism, and the structural changes that were emerging in tourism all over the world; and 2) the strategies defined conceal interests that had little to do with tourism.

The PENT was approved in 2007 and became the strategic framework for developing tourism in Portugal over a 10-year period. The goals of this plan were:

A. To reorganize the organic and institutional structure of tourism both nationally and regionally by investing in the simplification of tourism planning in Portugal, setting up 11 regional tourism bodies (five Regional Tourism Authorities and six Tourist Development Centers), along with two Regional Tourism Directorates in Azores and Madeira. Previously, there were 27 regional and local bodies in mainland Portugal and two in the Autonomous Regions responsible for the management of the destinations and product structuring; the external promotion of the regions was left to the Regional Tourist Promotion Agencies;
B. To diversify the offering, which is why 10 strategic tourist products were structured and developed for the first time, namely: cultural and landscape tours; sun & sea; nature tourism; city break; gastronomy and wines; integrated resorts and residential tourism; business tourism; golf; health and wellness; nautical tourism;

C. To promote and organize events that would project national destinations abroad;
D. To strengthen the public and private investment in tourism.

In addition to the five new tourism regions (Norte, Centro, Lisbon, Alentejo, and Algarve) and the five tourism development poles (Douro, Serra da Estrela, West, Alqueva, and Litoral Alentejano), the government determined in 2005 the creation of mechanisms for fast-tracking the procedures of projects recognized as being of potential national interest (PIN), aiming to reduce context costs associated with largest private investments. Projects qualified as PIN due to special economic, social, technological, energy, and environmental sustainability.

In 2012, over 50% of the 90 approved PIN projects were tourist developments. They shared the characteristic of consisting of detached houses scattered over areas of high natural value, often in land protected by national law. In this context, three fundamental issues led to the development of conflicts (Almeida & Silva, 2019):

A. The perception that the process unfairly favors large private developers;
B. Mistrust in projects' reported goals as reported by the tourist /real estate developers, namely regarding the value of the investment (in millions of euros) and potential job creation, that are used to justify exceptions to the planning laws; and
C. The fact that these private projects are in areas of high environmental interest, which should constitute a public good.

If, on the one hand, it is acknowledged that the reduced context costs for new private projects with high investment could be positive, on the other hand, the exceptions produce a strong feeling of distrust. In fact, given the generalized and growing distrust for the mechanism among stakeholders, from public entities responsible for the environment and land use planning, NGOs and the general public – who felt uncomfortable with "privileges" awarded to some large tourism developers by the State – and the associated media buzz, the PIN program was quickly perceived as a lose-lose solution for all involved parties, including developers.

Over time the growth of tourism has contributed to the emergence of land use conflicts and urban conflicts. If until 2012 the conflicts were mainly related to the building sprawl in areas of high environmental and landscape values, nowadays the main sources of conflict emerge in the historical centers of the main cities due to the exponential growth of tourism in Portugal that has occurred since 2014 (Figure 12.1).

12.2.6 The Recent Years

As shown in Figure 12.1, 2012 represents a turning point. The exponential growth is surprising, but we offer three main explanations for these very successful years for tourism in the country: good tourism governance (a topic further explored in the

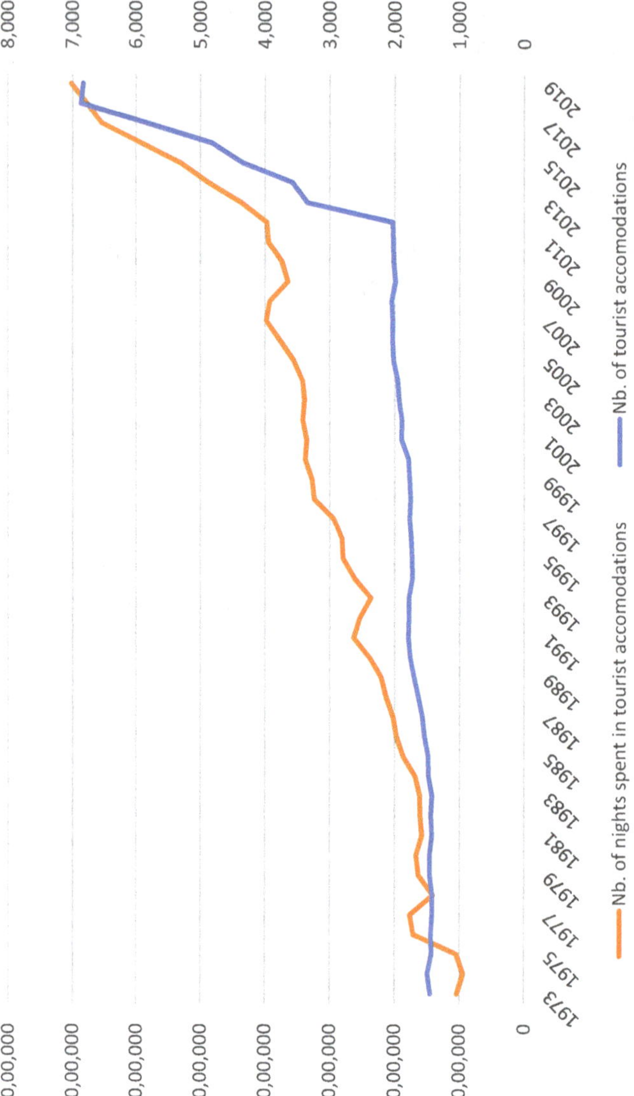

Figure 12.1: Number of overnight stays and number of tourist accommodation in Portugal 1964–2019. (Source: Based on data from INE – Portuguese National Statistics Institute/PORDATA).

next section), a favorable international context, and a set of consistent national and municipal policies.

In the international context we can highlight growing international mobility, explained by the proliferation of low-cost carriers and online accommodation booking platforms (Booking.com, TripAdvisor, Airbnb, etc.); the instability of other tourist destinations in direct competition with Portugal, such as Morocco, Tunisia, and Egypt; and the visibility provided by the amount of international tourism awards that Portugal, its cities, its destinations, and its institutional structures have won.

At the national level, several policies and laws were implemented, related with the attraction of more foreign investment: the non-resident (2009) and the Golden Visa (2012) programs and legislation related to the expansion of new forms of tourist accommodation: the urban Rental Law (2012) and the Short-term Rental Law (2014).

At the municipal level, investment in public space and the Urban Rehabilitation Incentive Programs, combined with national policies, were instrumental in increasing the attractiveness of the historic centers. Increasing numbers of tourists instilled these districts with a vibrancy that had been amiss for decades due to population loss and boosted struggling local economic activities. More recently, though, increasing tourism pressure over relatively compact urban neighborhoods has led to the emergence of conflicts with the resident population (Almeida et al., 2019).

The strong, fast growth of tourism in Portugal since 2014 has posed some problems for the destination and the components of the tourism system, in particular (Almeida, 2019; Moreira, 2018):

A. The shortage of qualified human resources;
B. The persistent low pay of many of the people employed in tourism;
C. A still marked concentration of tourist in traditional hubs;
D. Especially in the historic centers of Lisbon and Porto, there has been an increase in tourism pressure on the facilities and public infrastructure in cities, which is why some local authorities have tried to levy tourist taxes on overnight stays, with the revenue being invested in tourist attractions and/or the cleaning of public areas;
E. Local accommodation platforms have generated a huge increase in the number of units, but they are mostly concentrated in city centers and all too often in apartment blocks, where conflicts with residents can arise, as is especially evident in Lisbon and Porto. This is also one of the underlying causes for the rising cost of living in city centers, especially related to exponential growths in rent prices;
F. Conflicts have started to arise between residents and local authorities due to the operating hours of nightlife establishments, noise and disturbance of peace and quiet in public spaces where residential and leisure facilities are next door to one another;
G. Some streets in the historic centers of the large and some medium-sized cities are dominated by catering and souvenir outlets;

H. Tourists use the public transport services that connect the main tourist attractions constantly and intensively, especially in Lisbon and Porto;
I. There is constant traffic movement in the narrow streets where much of the important heritage is found, with occasional oversized transport in central areas. Traffic flow and parking difficulties have led Lisbon City Hall to restrict the occasional transport of passengers in some streets of the historic center.

In spite of this, over the last 50 years the tourist offer in Portugal has grown very significantly and, as the tourism product modernizes and diversifies – with the promotion of heritage, nature, gastronomy, or wine-based tourist products – the destination has become more distinctive. Innovation in accommodation, and professional training, have contributed to differentiate the offer. The country succeeded in projecting itself internationally by staging major events, and there were recent changes to the way the destination is promoted and marketed internationally, which coupled with improved and more affordable air accessibility, and the country's growing reputation as a safe and affordable destination, help explain the recent success.

The marked seasonality of tourist demand has been fading, so Portugal is starting to assert itself as a year-round destination. Tourism has helped employment, investment, exports, and with the development of the maritime economy, which were crucial to overcoming the deep economic crisis of the early 2010s. It has also spurred the regeneration of public spaces, the rehabilitation of buildings in the historic centers of Portuguese cities, and encouraged people to settle in low-density regions, contributing to territorial cohesion and balance. Currently, the efforts of the key players and stakeholders, both public and private, are converging to strengthen the competitiveness of the destination.

The growth of tourism in Portugal is so marked today that the costs and benefits of this activity need to be assessed, and those responsible for destination management and attractions must strive for sustainability and quality as the cornerstone of competitiveness.

12.3 The Current Tourism Governance in Portugal

12.3.1 General Overview

The Secretary of State for Tourism (SST) is placed within the Ministry of Economy. Turismo de Portugal I.P. (TP) is the National Tourism Authority. It is responsible for implementing tourism policy at national level and reports to the SST. TP oversees promotion, improvement, and sustainability of tourism activities, as well as training and investment. TP's mission is to: enhance and foster tourism infrastructure, promote human resources training, support investment in the tourism sector, coordinate domestic and international promotion, and regulate and inspect gambling activities.

Five Regional Tourism Bodies (Entidades Regionais de Turismo, ERTs) operate within Portugal (Figure 12.2). These are public law corporate bodies (including central government, municipalities, business and leisure associations, and major operators) with a specific territorial scope that act as destination management organizations with financial and administrative autonomy. They are responsible for domestic marketing and product development in close cooperation with TP, with whom they have a contractual relationship. In addition to these bodies, there are two Regional Directorates for Tourism in the autonomous regions of Madeira and the Azores.

There are also seven Regional Tourism Promotion Agencies (ARPT) which are non-profit, private associations that bring together the ERTs and private companies. They engage in international marketing in coordination with TP. TP and its partners ensure that marketing plans and campaigns supported by public and private funding align with the national strategy.

The total budget of TP was €244 million in 2016, of which half is derived from dedicated taxes on gambling, with the remainder from EU Structural Funds and other public funding sources. The promotional budget is around €45 million and includes expenditure on promotion at national and international level, as well as the co-financing of regional promotion abroad (via ARPT), to which private companies and regional tourism bodies also contribute.

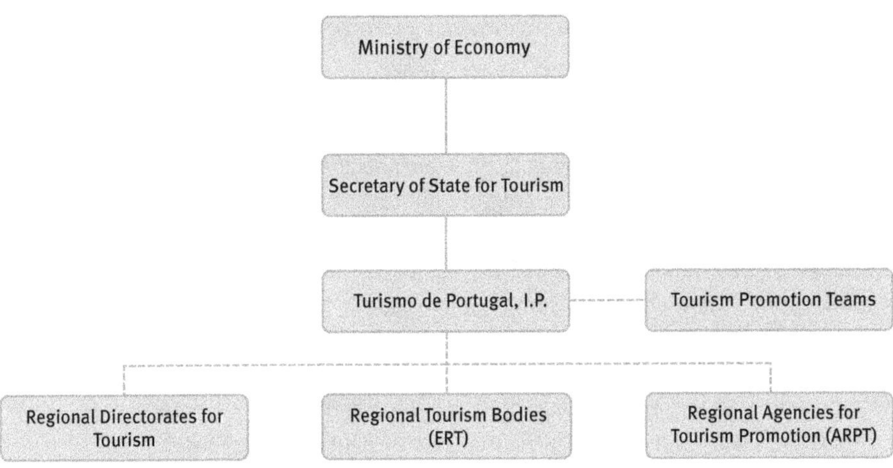

Figure 12.2: Portugal: Organizational Chart of Tourism Bodies. (Source: Adapted from OECD, 2018).

Portugal is experiencing challenges to remain competitive in the face of change, resulting in the need for continuous innovation. As a testimony of TP's continued efforts in innovation, it was awarded the 14th UNWTO Award (November 2019) for Innovation in Public Policy and Governance. This renewed effort in refreshing international promotion started in 2013 with the change in Portugal's international promotional campaigns: they stopped being institutional and started being digital. That year's campaign used exclusively digital media: Google (Google AdWords and Google Display), YouTube, and Facebook, as well as specific websites for certain markets. The website www.visitportugal.com, originally launched in 2004, was of great significance for this campaign. It enhanced the visibility of the destination and made it possible to reduce the costs.

In May 2017 a wholly digital campaign was launched by Turismo de Portugal in strategic European countries, as well as, Russia, United States of America, Canada, Brazil, India, and China. This international campaign had the tagline, "Can't Skip Portugal." Running in 2017 and 2018, the campaign consisted of four short films about Portugal that show some lesser-known aspects of the destination. Three of them are short films targeting three different segments. One features Jack, an entrepreneur who decides to take a relaxing break in contact with nature ("Can't Skip Freedom"), another shows us Chloe, a young French student who is seeking inspiration in a city ("Can't Skip Inspiration"), and the third is about Klaus, a German in his sixties who wants a change of life, a fresh start, living in the Portuguese countryside ("Can't Skip New Beginnings"). The fourth film brings these three characters together ("Can't Skip Us. Can't Skip Portugal"). The messages they convey are designed to promote the experiences and appeal to the emotions. "Can't skip . . . love, joy, smiles, life, opportunities, hope, new beginnings, challenges, and happiness." The campaign is based at a website (http://www.cantskipportugal.com/en), which lets visitors choose the destination according to how they feel and presents itself as an emotion travel guide.

In the present moment, sustainability is perhaps the most important challenge for Portugal. More specifically, how to remain competitive while promoting tourism as a tool for regional development, preserving the natural and cultural resources and authenticity of destinations, creating quality jobs, reducing seasonality and regional imbalances, and maintaining a good balance between residents and tourists. To address this challenge, the Ministry launched the Tourism Strategy 2027 (TS27), which establishes a strategic framework for tourism development in Portugal for the next decade. Development of the strategy involved large numbers of tourism stakeholders from across the country in an open and participatory process using web-based platforms and regional discussion workshops.

TS27 sets out a vision for the Portuguese tourism industry that:

> Affirms tourism as a hub for economic, social, and environmental development throughout the territory, positioning Portugal as one of the most competitive and sustainable tourism destinations in the world. (OECD, 2018)

This vision represents a broader perspective in terms of key tourism policy priorities, positioning Portugal as a sustainable destination; a cohesive territory, where the benefits of tourism are spread widely; an innovative and competitive destination; a country that values work and talent; an attractive destination to visit, invest, live, and study; an inclusive, open, and connected country; and an international benchmark in terms of the production of goods and services for the tourism industry.

Explicitly putting people at the core of tourism policy, Tourism Strategy 2027 established five strategic pillars (Turismo de Portugal, 2017):

A. Valorize the territory: using historical-cultural heritage and preserving their authenticity; promoting urban regeneration; improving product development to better match consumer needs; protecting natural and rural resources; promoting the importance of tourism in the context of the "maritime" economy.
B. Boost the economy: promoting the competitiveness of tourism businesses, through simplification and reduction of red tape and bureaucracy; attracting investment; developing the circular economy; fostering entrepreneurship and innovation.
C. Promote knowledge: improving the tourism professions; promoting the development of human resources; promoting continuous qualification of entrepreneurs as managers; dissemination of knowledge and tourism data and research; and promoting a smart destination strategy.
D. Generate networks and connectivity: improving air accessibility; developing mobility within the destination; promoting "tourism for all" from an inclusive point of view; involving society in the process of tourism planning and development; promoting networks and co-operation between tourism stakeholders.
E. Highlight Portugal: improving Portugal's positioning as an attractive destination to visit, invest, live and study; fostering the domestic market; promoting Portugal as a destination for congresses and events and for students/education; promoting the internationalization of tourism businesses.

For the first time, TS27 also sets out specific goals and targets for each of the three pillars of sustainable development. Economic goals cover overnight stays and tourism receipts; social goals cover seasonality, skills and qualifications, and residents' satisfaction; and environmental goals cover energy, water, and waste.

Enhancing destination value is a key area of focus in Portugal. For example, the Valorizar Program aims to support investment and enhance the quality and value of Portugal as a tourist destination. The program was created specifically to address seasonality and achieve a more balanced distribution of demand throughout the country.

12.3.2 The Regional Governance of Tourism: The Case of Lisbon

In 2007, the number of regional tourism areas was reduced from 19 to five regional tourism bodies, with the same limits as the NUTS II territorial units, including one for the Region of Lisbon (ERT-RL). These five entities are responsible for regional tourism development, in accordance with the national guidelines for tourism mentioned above. The number of tourism regions had for long been considered as excessive. Nowadays, each of the regions is managed as a whole by the same entity, where in the past several smaller entities managed specific tourism hubs within the region, often neglecting smaller hubs or failing to adequately promote less-travelled areas.

With the consolidation into a smaller number of entities, it was equally possible to concentrate human and financial resources not only at the regional scale, but also in their contribution to the collective effort of national and international tourism promotion. In the case of Lisbon, the ERT-RL consolidated promotion strategies for well-established "brands," such as the capital itself, Sintra, or the Estoril / "Portuguese Riviera," but was also able to identify and potentiate possible synergies with less-promoted tourism destination and complementarity between the different regional hubs.

The same legal document that established the five regions also set out the regulatory framework for the country's five main tourism development hubs, identified in the National Strategic Plan for Tourism (NSPT): Douro, Serra da Estrela, Oeste, Alqueva, and Litoral Alentejano. Besides these, Lisbon and Porto each have their own tourism promotion boards. These seven Regional Tourism Promotion Agencies (RTPAs) are non-profit, private associations that bring together the ERTs and private companies. They are therefore a sort of public-private partnership consortium, set out to involve and coordinate the action of public and private stakeholders. Lisbon's ARPT is the Associação de Turismo de Lisboa – Visitors & Convention Bureau (ATL).

ATL is an especially strong consortium among the RTPAs and is well-established as a tourism promotion board, having been created in 1997 and inherited tourism promotion competences in 2004 (ATL, 2019). It is participated by private stakeholders, such as hotel chains and travel agencies, but also by public entities, such as municipalities, which helps coordinate private investment and business decisions with local planning actions.

The ERT-RL, while patently a public body consortium, participated by the central government agency and by all municipalities in the region, also includes a board of consulting private entities, including economic development associations, tourism and travel industry associations, or even labor unions, among other private stakeholders (ERT-RL, 2019). These entities were active in the definition of the Regional Strategic Tourism Plan 2014–2019 (RSTP), which was accompanied by a specific governance model so as to ensure its implementation. Among the provisions therein, is the need to respect each entity's goals and competences. In broad strokes, the ERT-RL is tasked with the elaboration and frequent revision of the Strategic Plan, the coordination of public and private entities, and the channeling of national and European funding for the sector, whereas ATL is tasked with tourism promotion of the region (including tourism offices and information centers) and the coordination of actions by the private parties in tourism development, potentiating synergies among them – one typical example would be the coordinated efforts in affirming Lisbon as a major convention venue, or the promotion of new destinations within the region.

The RSTP also points out the importance of taking full advantage of the already available resources and experience, especially those of ATL, in developing tourism in the region, and in articulating the regional tourism strategy with that of each municipality. While the ERT-RL, through the RSTP, is tasked with the definition of Tourism Development Programs (focused, for instance, in specific sports activities, or nature-based tourism), it is then the ATL that is tasked with the management of most programs. The programs' implementation is overseen by a consulting board, that is tasked with verifying whether the program's objectives are being met and in promoting the alignment of these with the tourism strategies of all municipalities involved.

Each program is assigned a Manager (also referred to as an Enabler) and, whenever necessary, a supporting workgroup including the main stakeholders (public and private), so as to facilitate communication among interested parties. In a typical tourism development program, the ERT-RL is the entity responsible for the definition of goals and budget and the ATL supervises the implementation of the program through an enabler, promoting the cooperation among stakeholders, including the municipalities.

The RSTP is therefore a strategic plan focused on action/results, setting out the governance structure that is considered to be the most effective in achieving the goals of the different programs. By involving private and public stakeholders in the elaboration of the RSTP, and in its subsequent implementation, the Plan also further encourages future collaboration and the reinforces a critical mass of active and involved partners and establishes a mechanism of supervision focused on guaranteeing adequate coordination of interests and efforts focused on results.

The new territory of the ERT, encompassing the whole region of Lisbon, preserves some well-established sub-regional "brands" (Figure 12.3) but, more importantly, sends out an important signal that there are obvious synergies between tourism hubs (Figure 12.4) that were, in the past, promoted independently as if directly competing. Finally, the assumption that several municipalities have underdeveloped tourism strategies has led to the realization that it is paramount to analyze tourism potential in the region beyond the administrative boundaries separating municipalities, and finally assuming the importance of multi-municipal products (such as the Tagus River Arch around the Tagus Estuary).

12.4 Conclusion

The year 2017 was historic for the tourism sector in Portugal, but despite the wealth in tourism resources, Portugal has not always been a successful tourist destination. Was it good governance that determined the shift that led to its current success? This chapter has presented a synthesis of the tourism governance over the years in an attempt to answer this question.

Tourism was first addressed by the government in 1911, with the creation of the first official tourism body. If, at the time, the main tourist product were thermal

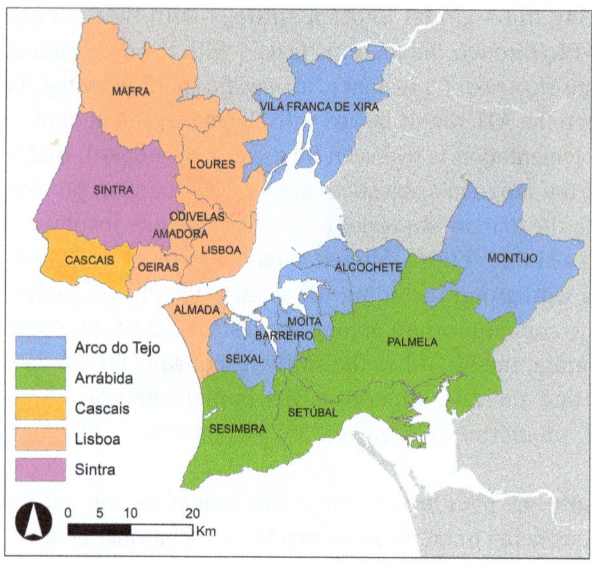

Figure 12.3: Lisbon Region Municipalities. (Source: Adapted from the Strategic Tourism Plan to Lisbon Region 2014–2019).

establishments, a century later sun and sea has become the biggest tourist attraction (concentrated on the Lisbon, Algarve, and Madeira triangle). Tourism governance has always been very centralized, led by central government bodies and often directly linked to large private investors. The regional tourism governance structure would only effectively be implemented after 2007, with the formalization of the Regional Tourism Entities (ERTs). The municipal level remained highly disorganized until mid-2010 and, while the importance of intersectoral cooperation in the public sector has been highlight since at least the 1980s, it only started occurring more frequently in the last decade.

While the 2000–2010 decade was a period of stagnation in tourism growth, it was during that period that the tourism governance at central government level received a major overall, and we may finally be seeing the results of the changes then made. This current decade has been characterized by the structuring of new tourist products, the increase, diversification, and dispersal of accommodation, and simplification of the creation of tourist enterprises and local accommodation facilities. Tourism entrepreneurship has become a business opportunity for small and medium-sized investors in the years following the economic and financial crisis. There are more incentives for companies to project themselves internationally and for the qualification of human resources. The focus seems to be on quality, sustainability, innovation, and, lastly, competitiveness, as a way of reducing state intervention. Forms of cooperation and collaboration, institutional and territorial partnerships and networks have been intensified, with various key players and stakeholders having

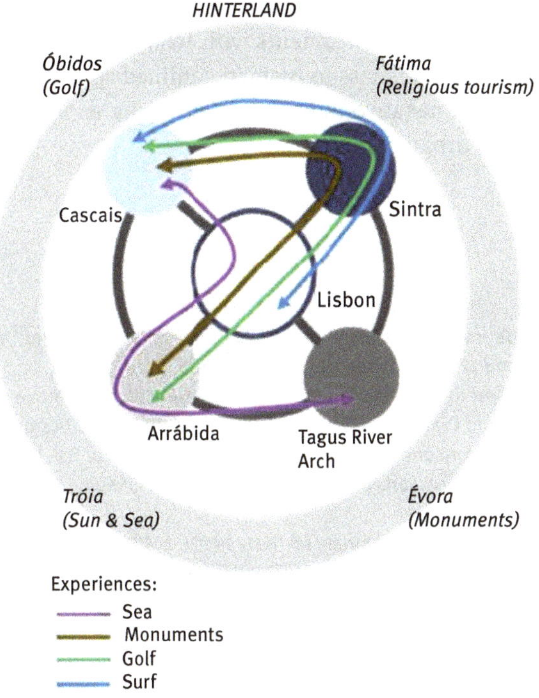

Figure 12.4: Centralities and regional experiences in the Lisbon Region. (Source: Adapted from the Strategic Tourism Plan to Lisbon Region 2014–2019).

been mobilized to set policies and help define strategies for tourism in the medium and long run. In short, there has finally been a political and social recognition of the importance of tourism. We would add that it is also a good example that deep institutional reform may take a while to bear fruit.

With the exponential growth of the number of foreign visitors, several municipalities finally woke up to tourism, which has led to investment in public spaces and tourism support infrastructure, and is one of the underlying causes for a major effort in rehabilitation of historic city centers. Greater cooperation between municipalities and the central government has emerged for the development of attractive tourism products, and a dynamic of public-private cooperation has developed, under the guidance of a new framework for the implementation of strategic tourist planning.

Immediately before the Covid-19 Pandemic, there were signs that the rate of the sector's expansion might be outrunning the capacity of city centers, and the municipalities that run them, to adapt to this process, adequately plan for it, and effectively regulate the peaks of occupation. The global Covid-19 pandemic has had a tremendous impact over global travel and tourism. Early signs of the recovery are very encouraging for the Portuguese tourism industry, as travelers have returned in large numbers whenever travel restrictions allowed. This increased tourist pressure is leading to a growing discontentment of some residents with what they perceive as a potential loss of quality of life, related to mass tourism in confined spaces. And so, tourism growth and its positive and negative impacts are emerging as a major topic within the context of spatial and urban planning.

References

Almeida, J. (2018). *Turismo & dinâmicas de transformação em bairros históricos, XIV Assembleia e XIII Fórum Internacional RUITEM, 11 and 12 June*, Lisbon University, Lisbon, Portugal.

Almeida, J. & Nunes Da Silva, F. (2019). Different understandings of "public interest" as a source of conflict. Portuguese Spatial Planning and Practice, in Eraydin, A. & Frey, K. (Eds.) *Politics and Conflict in Governance and Planning. Theory and Practice*. Routledge, New York. Pp. 93–111.

ATL (2019) Retrieved from: https://www.visitlisboa.com/pt-pt/sobre-o-turismo-de-lisboa. Dateaccessed: 26th October 2020

Banco de Portugal, 2018. "Annual Report - Activities and Financial Statements 2018". Banco de Portugal, Lisbon, Portugal. https://www.bportugal.pt/sites/default/files/anexos/pdf-boletim/relatorio_atividade_contas_2018_en.pdf

Costa, C. & Vieira A. L., (2014). Tourism planning and organization in Portugal, in Costa, C., Panyik, E., & Buhalis, D., (Eds.), *European tourism planning and organization systems: the EU Member States*, Channel View Publications, Bristol, p. 352-366.

Cunha, L. (1997). *Economia e política do turismo*. McGrawHill, Portugal.

ERT-RL (2019) Retrieved from: http://www.ertlisboa.pt/pt/institucional/quem-somos/. Dateaccessed: 26th October 2020

ERT-RL – Entidade Regional de Turismo da Região de Lisboa (2014) Plano Estratégico para o Turismo na Região de Lisboa 2015–2019. Retrieved from: http://www.ertlisboa.pt/fotos/editor2/lis_9970_04248_007_15.pdf. Date accessed: 26th October 2020

INE – INSTITUTO NACIONAL DE ESTATÍSTICA (2019). Demographic Statistics: 2018. Lisboa. Retrieved from: https://www.ine.pt/xurl/pub/358632586. ISSN 0377-2284. ISBN 978-989-25-0499-5. Date accessed: 26th October 2020

Milheiro, E., & Santos, J. P. D. (2005). O turismo em Portugal: Que passado? Que futuro. Aprender, 30, 119–25.

Ministério Da Economia e Da Inovação (MEI), (2007). Plano Estratégico Nacional do Turismo. Para o Desenvolvimento do Turismo em Portugal 2006–2015, Ministério da Economia e da Inovação e Turismo de Portugal, Lisboa, [online].

Moreira, C. (2018). Portugal as a tourism destination. Paths and trends. *Journal of Mediterranean Geography*, 130 [online]. Retrieved from: https://journals.openedition.org/mediterranee/10402. 10.4000/mediterranee.9909. Date accessed: 26th October 2020

NETO, V. (2013). *Portugal Turismo – Relatório Urgente*, bnomics, Lisboa.

OECD (2018), OECD Tourism Trends and Policies 2018, OECD Publishing, Paris. http://dx.doi.org/10.1787/tour-2018-en

Turismo De Portugal (TP), (2015). Turismo 2020 Plano de Ação para o Desenvolvimento do Turismo em Portugal, Turismo de Portugal, Lisboa, [online].

Turismo De Portugal (TP), (2017). Estratégia Turismo 2027, Turismo de Portugal, Lisboa, Available at https://estrategia.turismodeportugal.pt/sites/default/files/Estrategia_Turismo_Portugal_ET27.pdf

Umbelino, J., Boavida-Portugal, L., Ferreira, M. J., & Sousa J. F. (1993). O Inventário dos Recursos Turísticos de Portugal Continental, Inforgeo, 6, 93-99.

UNWTO – United Nations World Tourism Organization (2017). *Tourism highlights 2017*, Madrid: World Tourism Organization.

UNWTO (2012), Annual Report 2011, UNWTO, Madrid, DOI: https://doi.org/10.18111/9789284415366

WEF – World Economic Forum Wef. (2017). The travel & tourism competitiveness report 2017, Geneva: World Economic Forum. [online].

WTTC – World Travel & Tourism Council. (2017). Travel & tourism economic impact 2017 Portugal, London: World Travel & Tourism Council, Retrieved from: https://www.sgeconomia.gov.pt/ficheiros-externos-sg/wttc_portugal2018-pdf.aspx. Date accessed: 26th October 2020

Mar Loren-Méndez & Daniel Pinzón-Ayala

13 Spain

13.1 Introduction

This chapter presents a historical account of the management of tourism in Spain from the beginning of the 20th century, precocious in its international dimension, to the present day. Tourism in Spain has been understood as a development strategy as well as a tool serving to project a modern image of the country, with advancements in infrastructures playing a key role. While in the first half of the 20th century tourism was economically marginal with cultural heritage constituting the main tourist narrative, since the latter period of Franco's regime (1959–1975) the concentration of sun, sea, and sand mass tourism on the Spanish Mediterranean coast has become a key economic activity. The research presented here focuses on the Costa del Sol, in the province of Malaga. This western section of the Andalusian littoral is representative of the profound urban transformation that has taken place on the Spanish Mediterranean, with the relentless pursuit of policies to accommodate the emergent demands of mass tourism. This study reviews both the centralized state structure and the legal apparatus implemented to facilitate tourist development on the coast, demonstrating how tourism became a state priority. It then focuses on the present, showing how tourism has been managed in the context of democracy since 1975, confirming coastal mass tourism as an established product, highlighting decentralization as the main shift in government structure, and reflecting on the paradoxical inertia of the dictatorial period. Finally, the study identifies emerging tourist products, such as urban cultural tourism, and accounts for the success of Spain as a tourist destination, even surpassing the United States in 2017 for the first time. Precisely in connection with this indisputable success in the sector, the chapter finalizes by reflecting on the main challenges that the governance of tourism in Spain faces in the coming years.

13.2 The Beginning of Tourism: Internationalization, Modernity and Culture in the First Half of the 20th Century

In 1905, the Comisión Nacional del Turismo (National Commission for Tourism [NCT]) was the first organization connected to the tourist industry to be created in Spain (MInisterio de Fomento, 1905). It was created under the auspices of the *Ministerio de Fomento* (Ministry for Development) as, since its origins, tourism in Spain has been understood

https://doi.org/10.1515/9783110638141-013

as a development tool. The NCT mediated a strategy for the construction of a modern representation of Spain that could transcend the prevailing clichéd romantic image of the nation: a modern image of Spain was formalized through quality hotels, understood as diplomatic tools, capable of constructing the image of the country (Poutet, 1995). Such strategic policymaking aimed to welcome the foreign tourist and was generally applied in towns modeled as seaside health resorts. In 1911, the Comisaría Regia de Turismo y la Cultura Artística (Governing Commission for Tourism and Artistic Culture) was created, involved in promoting Spain abroad, reinforcing the concept of culture as the main tourist product of this period (Presidencia de Ministros, 1911).

In 1928, with Spain now under Primo de Rivera's dictatorship, the *Patronato Nacional de Turismo* (National Tourist Board) aimed to implement policies for the modernization of infrastructures. The approval in 1926 of the *Circuito Nacional de Firmes Especiales* (National Circuit of Special Road Surfaces), a plan to modernize the country's road infrastructure that facilitated access to cultural sites, highlighted the link between tourism and infrastructures while unveiling the cultural approach to Spanish tourist products and reinforcing its international vocation. A network of state-run hotels called Paradores was also promoted, gesturing once more to the internationalization of Spain. Despite these advances, no overall economic vision for tourism existed; rather, tourism was seen as a tool for the international projection of a modern image of Spain based on its culture. The organigram of the National Tourist Board would continue under the Second Republic (1931–1936), albeit in a more simplified version: between April and December 1931, the National Tourist Board was temporarily replaced by the *Dirección Nacional de Turismo* (National Direction of Tourism). The National Tourist Board was reinstated by the Decrees of the 4th of December 1931 and 12th of January 1932. Provincial delegations were eliminated, the three vice-presidencies of Art, Propaganda and Travel were unified under one vice-presidency (Moreno, 2010). Tourism was considered as one more dimension of the country's internationalization policies. In the 1920s and 1930s, tourism still fell below 0.1% of GNP.

Such was also the case at the beginning of Franco's dictatorship: despite the creation of the *Servicio Nacional de Turismo* (National Tourist Service) in 1938, and the subsequent *Dirección General de Turismo* (Department of Tourism) in 1939 (Gobierno del Estado, 1938; Jefatura del Estado, 1939), the state's aspirations for tourism were marginal, and "work proceeded without defining sectorial functions and objectives" (Loren-Méndez & Pinzón-Ayala, 2020). There was little international interest in Spain in comparison with other European countries such as France. One of the first interventions of Franco's Nationalists to attract foreign visitors was the organization of the *Ruta de Guerra del Norte* (War Route of the North), before the end of the Civil War (*Ministerio de Interior*, 1938).

In 1948, the United Nations Declaration of Human Rights proclaimed the right to rest and leisure, including periodic holidays with pay, ideas first initiated after the First World War (United Nations, 1948). The year 1949 saw the interest for Spain as a tourist destination accelerate for the first time, with the number of visitors higher than before the start of the Second World War. Throughout the 1950s, leisure trips helped re-establish normality in a damaged postwar European society. Meanwhile, between 1949 and 1959,[1] the strictly protectionist character of the Spanish government had softened, and the first economic liberalization policies were applied, substantiated at the end of the 1950s, and which would continue to characterize the last 15 years of Franco's dictatorship, an unprecedented speed of development for Spain. The country was able to take advantage of the postwar geopolitical situation by giving form to the European demand, thus forging the economic future of Spain, which has remained linked to tourism ever since.

At this time, foreign tourism still hinged on a curiosity for the exotic inherited from the romantic travelers of the preceding century, with the emerging model of the spa to complement the form of cultural tourism that prevailed in the international experience of Spain. Coastal tourism was eminently national, while the minority of foreign tourists continued to choose travel as a way of discovering Spain. Tourism anyhow was first practiced by the elite – the aristocracy – and, after the First World War, the emerging bourgeois and professional classes – and was concentrated in the cities. However, still no overall or comprehensive economic vision for tourism existed in this decade, with its instrumentation serving as a tool for the international promotion of the country prevailing. The image of Spain that was projected continued to be based on modernization and culture. The territorial transformation was still concentrated on infrastructures and the specific insertion of stand-alone hotels, roadway inns, and spas, with a special emphasis on cities.

During this decade, annual revenues from tourism increased in Spain, although the phenomenon was not visible to the general public, and its trace in geography was still not relevant:

> *The most significant change occurred in 1951 with the creation of the Ministry of Information and Tourism, largely concerned with information control, however, with little reference to tourism. As well as the organization and promotion of establishments and sites for tourism, priorities were to substantially increase the number of foreign tourists, develop the hotel industry, increase public and private investment, create areas of tourist interest, streamline procedures, and provide better training to workers in the sector.* (Loren-Méndez & Pinzón-Ayala, 2020, 5)

For the first few years, the state continued to be responsible for promoting tourist accommodation and services, as well as creating commissions, publications, and delegations at various levels.

1 This first phase known as the 'Segundo Franquismo' (Second Francoism) is referred to by Townson, N. (Ed.) (2007) as the national-catholic and corporatist phase: 1946–1959: "Between 1951 y 1958 the Spanish economy grew at the impressive annual rate of 4.35%, albeit starting from a very low threshold: the PIB for the year preceding the war, 1935, was not equaled until 1951, while the income per capita was not surpassed until 1935." (2007, XVIII–XIX).

Meanwhile, in an attempt to make the social dimension of tourism more effective, Britain and other northern European countries developed policies that imposed strict currency limits on citizens leaving the country. The level of exterior spending thus equaled citizen spending, leading to a model of low-income tourism. This produced a demand for low-cost accommodation requiring large areas of empty land to host so-called mass tourism. The Mediterranean coast was able to give form to that demand (Pack, 2006). In 1953, Spain developed its first *Plan Nacional de Turismo* (National Plan for Tourism), persisting with the notion that the main value of tourism was to counteract the negative international image of the Franco regime, giving only secondary importance to its economic benefits, such as the activation of industrial and commercial offshoots, or the entry of foreign capital (Velasco, 2005; 2008). The National Plan for Tourism was developed by a commission under the *Ministerio de Información y Turismo* (Ministry of Information and Tourism) and approved in July 1953. The plan was based on the 'Study for a National Plan for Tourism', developed in 1952 by the *Secretaría para la Ordenación Económico-Social* (Secretariat for Social and Economic Development), the first state document on the role of tourism in Spain. In 1954, the number of tourists reached a million; the figure doubled in 1957, then tripled in 1960 (Dirección General de Turismo, 1956; Fernández, 1991).

13.3 From a "Marginal" to a "Main" Economic Activity: Mass Tourism in the Mediterranean by the End of Dictatorship (1959–1975)

The end of the 1950s marked a turning point in the discourse of the dictatorship under Franco: Spain's identity, based on Civil War victory and self-sufficiency, became transformed within a framework of prosperity, peace, and progress, with the fundamental parameters of a pro-European attitude and tourism as central strategies (Chuliá, 2007; Loren-Méndez & Pinzón-Ayala, 2020, 6). After 20 years of postwar hardship and isolation, Spain began opening up to the international context, resulting in an economic liberalism that became compatible with the continued absence of ideological freedoms characteristic of military dictatorship (Barciela, C., López, M.I., Melgarejo, J., & Miranda, J.A., 2001). Tourism, alongside foreign investment and emigration, made it possible for a country internationally considered part of the so-called Third World to become the second fastest growing economy in the world after Japan at this time (Loren-Méndez & Pinzón-Ayala, 2020).

An international dimension to tourism, a modern image, and the development of depressed areas constituted the parameters of Spanish tourism that Franco's regime would maintain throughout this last phase between 1959 and 1975, the period that would ultimately define Spain as a tourist country. In this period, the areas clearly targeted for development were on the coast, leading to the identification of Spanish tourism as the quintessential Mediterranean beach holiday and leaving culture aside as a complementary tool in the construction of a national image that rendered modernity compatible with the most picturesque traditions.

The political and economic management of tourism was highly centralized. The purpose of such ministerial centralization was to ensure the fulfillment of norms, the quality of the development in tourism, and to guard against betraying the values of selected sites (Jefatura del Estado, 1963b). Centralization was criticized by both the local authorities and the more conservative members of parliament, who both saw the growth of tourism as a threat to the continuation of the regime. The stake in centralization and, therefore, the homogenization of construction in tourist territories, was complemented by a decentralizing activity that aimed to exploit the diversity of cultures, and thus mitigate the global image of a tourist offer solely based on sun, sea, and sand. The budgets for such initiatives promoted by the local and regional municipalities were modest and passed largely unnoticed (Pack, 2007, 57).

A Sub-secretariat for Tourism was created in 1962 (*Ministerio de Información y Turismo*, 1962), alleging that the public organigram was insufficient to attend to the exponential growth of visitors and companies in the sector. "The universal importance of tourism for political, cultural, hygienic, and social order was declared, as the regime's aspirations to become an industrial nation were cast aside" (Loren-Méndez & Pinzón-Ayala, 2020, 7). Manuel Fraga Iribarne, appointed Minister for Tourism, interpreted the success of tourism as proof of the failure of political campaigns that stood against the dictatorship. Unlike most government members, Fraga saw the continued growth of tourism as positive and inevitable. He was aware of Spain's need to adapt to social, cultural, and economic change, as well as the consequences for the political and institutional order (Pack, 2007, 56). Government resistance to decentralization and political reform still existed in 1969. By then, however, tourism had become a key vector of Spanish politics and identity (Sánchez-Albornoz, 1975).

The 1959 *Plan de Estabilización* (Stabilization Plan) by the Office of the Head of State (Jefatura del Estado, 1959) propelled a radical change in direction (Martín & Martínez, 2007). "The regime allowed, promoted, and legislated for an economic liberalization that contrasted with the absence of freedoms of expression and ideology" (Loren-Méndez & Pinzón-Ayala, 2020, 7). The Stabilization Plan thus initiated a process approving a series of norms and laws whose economic discourse, based on the tourist offer, was projected onto the territory and the city. The *I Plan de Desarrollo Económico y Social* (I Economic and Social Development Plan, I ESDP) was applied between 1964 and 1967, confirming the importance of tourism for the economic dynamism of the country by establishing the basis for its promotion, and setting priorities at both local and state level (Jefatura del Estado, 1963a; 1969; 1972). The I ESDP defined the active role of the state in the expansion of the tourist industry, confirming that the net balance from tourism constituted 95% of the net surplus from the entire service balance (Presidencia del Gobierno, 1964a). The deregulation of package tours and charter flights across Western Europe at the same time created optimum conditions for the rapid expansion of Spanish tourism. The state's active role in the promotion of tourism went beyond the tasks of monitoring and regulating private initiatives: it was also pivotal in deciding on locations and delimiting the new areas to be developed specifically for tourism (Presidencia del Gobierno, 1964b). The I ESDP for the period 1964–1967 confirmed that the "development of tourism in Spain has been, without a doubt, faster than in other countries, managing, in 1962, to surpass the income figure obtained by Italy and France in 1958" (Presidencia del Gobierno, 1964c, 8106). By 1960, Spain had already surpassed France in volume of foreign tourists, overtaking Italy in 1964, thus launching a second phase of acceleration in the European and, in particular, the Spanish tourist economy (Pack, 2007, 55). By 1969, Spain had attained the figure of 19 million tourists (Townson, 2007).

Legislation mainly concentrated on the physical aspects of the phenomenon that created the right conditions for the transformation, occupation, and management of the coastal territory. The Ministry of Information and Tourism focused on ensuring that international standards were met. Between 1963 and 1975, a systematic normative development aimed to identify, promote, and regulate potential tourist zones, including their outfitting and infrastructural modifications. The government devised a plan that would accelerate and facilitate the exploitation of the territory for tourism. The *Centros y Zonas de Interés Turístico Nacional* (CyZITN) (Centers and Zones of National Tourist Interest, C&ZNTI) constituted a primary legal instrument that defined and established regulations for the new spaces for tourism: sites for new tourist centers were regulated independently from existing norms already governing local cities, towns and villages of historical or traditional interest.

"The self-sufficient requisites of the C&ZNTI and their required scale resulted in the creation of new autonomous tourist centers" (Loren-Méndez & Pinzón-Ayala, 2020, 2). The *Consejo de Ministros* (Council of Ministers) took direct responsibility for the declaration of the C&ZNTI, concentrating their territorial development on the coast (Jefatura del Estado, 1963b, 18226). Only the Centers of National Tourist Interest would subsequently be created.

Directly approved by central government, their promotion and development not only received independent official endorsement, but were also given priority status over regional or municipal development master plans. New tourist centers were legally authorized to occupy areas of special natural and traditional heritage interest whose original values, while they should remain in evidence to this day, have, "paradoxically, decreased as a direct consequence of the new implantations" (Loren-Méndez & Pinzón-Ayala, 2020, 2). A minimum capacity of 500 beds and an area covering no less than ten hectares including services was stipulated for the new tourist centers, while the government retained the right to intervene favorably if conditions were not entirely met (Jefatura del Estado, 1963b, 18227). Such dimensions were multiplied on occasion by more than a hundred, clearly reflecting the aim of the C&ZNTI Law to create large self-sufficient nuclei, in some cases, these new touristic centers were similar in number of inhabitants and services to the preexistent towns and cities of their municipalities (Galiana & Barrado, 2006). The ZNTI aimed to incorporate two or more centers and provide accommodation for a minimum capacity of 5000 beds (Jefatura del Estado, 1963b, 18227). The declaration of Zones was once more given consideration, even when required conditions were not met. Such a declaration would attempt to activate "the start of works and infrastructural services that required the coordinated action of the Public Administration in diverse spheres" (Jefatura del Estado, 1963b, 18227).

Alongside the concentration of the C&ZNTI on the coast, there also existed a clear drive to modernize coastal infrastructures. The government justified the investment in road infrastructure in light of its social commitment to improving opportunities and access to the most remote and depressed areas; "the disproportionate investment in coastal roads, however, reveals an imbalance in the distribution of resources, and the prioritization of the service offer to tourists" (Loren-Méndez & Pinzón-Ayala, 2020, 12–13): The government proceeded to spend lavishly on road improvement, disproportionately so in insular and coastal areas, heavily trodden by tourists. "For 1965, 53% of road funds were allocated for Mediterranean provinces, which comprise 30% of the national population, while interior provinces, 48% of the population, received 12% of highway funds" (Pack, 2007, 57).

These new independent tourist centers, created under the general plans for urban and territorial regulation and in accordance with new national housing legislation, relied in an unprecedented way on the autonomous power of private capital in urban development destined for tourism, without a comprehensive approach to tourism and thus betraying an integrated vision for territorial development. Such power was further endorsed by its priority status, and the administrative support from the municipalities that they were legally bound to provide for the purposes of urban development in tourism. This period also pointed to the exceptionality of the case of Spain in the role of the state as promoter as well as in setting the regulatory basis that encouraged private international investment, allowing the purchase of property with foreign capital, thus setting future trends for the Spanish coast, the result of which we see today (Loren-Méndez & Pinzón-Ayala, 2020, 17–18)

13.4 Governance and Tourism under Democracy (1975–)

13.4.1 Decentralization and Transversality in Tourism Management

The management of tourism has shifted under democracy, from a rigid centralized system, whereby each urban tourist development had to meet with the approval of Madrid on an individual basis, to an extremely decentralized government, resulting in a complex organigram for tourism (Cortes Generales, 1978, 29333). Spain has been organized into 19 autonomous governments, 17 of which are regions and two autonomous cities. Apart from central government with its own competencies in tourism, Spain has a further 19 autonomous presidents, parliaments, and ministries or departments of tourism. Ministries of tourism – both national and autonomous – are generally combined with other sectors. On the one hand, some ministries combine tourism with other sectors (Figure 13.1)

TOURISM MANAGEMENT IN SPANISH AUTONOMOUS GOVERNMENTS

Region or City English (Original)	Ministry (English translation)	Ministerios, Departamentos y Agencias
Emphasis of tourism as an industry: with energy, economic development, consumption, employment, commerce, innovation		
Explicit reference to tourism		
Balearic Islands (Illes Balears)	Ministry of Economical Model, Tourism and Labour	Consejería de Modelo Económico, Turismo y Trabajo
Canary Islands (Canarias)	Ministry of Tourism, Industry and Commerce	Consejería de Turismo, Industria y Comercio
Ceuta (Ciudad de Ceuta)	Ministry of Tourism and Development	Consejería de Fomento y Turismo
Basque Country (País Vasco)	Ministry of Tourism, Commerce and Consumption	Departamento de Turismo, Comercio y Consumo

Figure 13.1: Tourism management in Spanish autonomous government, 2020. Elaborated by the authors.

TOURISM MANAGEMENT IN SPANISH AUTONOMOUS GOVERNMENTS		
Region or City English (Original)	**Ministry (English translation)**	**Ministerios, Departamentos y Agencias**
No explicit reference to tourism		
Aragon (Aragón)	Department of Industry, competitiveness and Business Development	Departamento de Industria, Competitividad y Desarrollo Empresarial
Castile-La Mancha (Castilla-La Mancha)	Ministry of Economic Affairs, Business and Employment	Consejería de Economía, Empresas y Empleo
Catalonia (Cataluña)	Department of Business and Knowledge	Departamento de Empresa y Conocimiento
Melilla (Ciudad de Melilla)	Ministry of Economic and Social Affairs	Consejería de Economía y Políticas Sociales
Chartered Community of Navarre (Comunidad Foral de Navarra)	Department of Economic and Business Development	Departamento de Desarrollo Económico y Empresarial
La Rioja	Ministry of Autonomic Development	Consejería de Desarrollo Autonómico
Emphasis of tourism as culture and leisure and combines with education and youth		
Principality of Asturias (Principado de Asturias)	Ministry of Culture, Language Policy and Tourism	Consejería de Cultura, Política Lingüística y Turismo
Cantabria	Ministry of Education, Professional Training and Tourism	Consejería de Educación, Formación Profesional y Turismo
Castile and Leon (Castilla y León)	Ministry of Culture and Tourism	Consejería de Cultura y Turismo
Extremadura	Ministry of Culture, Tourism and Sports	Consejería de Cultura, Turismo y Deportes
Galicia	Ministry of Culture and Tourism	Consellería de Cultura y Turismo
Community of Madrid (Comunidad de Madrid)	Ministry of Culture and Tourism	Consejería de Cultura y Turismo
Region of Murcia (Región de Murcia)	Ministry of Tourism, Youth and Sports	Consejería de Turismo, Juventud y Deportes

Figure 13.1 (continued)

TOURISM MANAGEMENT IN SPANISH AUTONOMOUS GOVERNMENTS		
Region or City English (Original)	**Ministry (English translation)**	**Ministerios, Departamentos y Agencias**
Excepcional cases		
Andalusia (Andalucía)	Office of Tourism, Regeneration, Justice and Local Administration	Consejería de Turismo, Regeneración, Justicia y Administración Local
Valencian Community (Comunitat Valenciana)	Agency for Tourism (directly dependent of Presidency)	Agencia del turismo (dependiente directamente de la Presidencia)

Figure 13.1 (continued)

The approach and management of tourism in Spain is extremely diverse, with two predominant tendencies:

- On the one hand, some ministries combine tourism with energy, economic development, employment, commerce, or innovation, with an emphasis on tourism as an industry. This is the case of the central Spanish government *Ministerio de Industria, Comercio y Turismo* (Ministry of Industry, Trade and Tourism), as well as part of the autonomous governments of *Illes Balears* (Balearic Islands); *Islas Canarias* (Canary Islands); the autonomous city of Ceuta and *País Vasco* (Basque Country).

  In some of these autonomous ministries the term 'tourism' does not even appear in the denominations of the department or office, prevailing terms such as Economy, Employment or Development. This is the case of Aragón; *Castilla-La Mancha* (Castile La Mancha); *Cataluña* (Catalonia); the autonomous city of Melilla; Navarra and La Rioja.
- The other tendency makes an emphasis of tourism as leisure, culture and sport, combining it with education, training, and even language. This is the case of the autonomous governments of Asturias; Cantabria; *Castilla y León* (Castile and Leon); Extremadura; Galicia; Comunidad de Madrid and Región de Murcia.

  Comunitat Valenciana (Valencian Community) and *Andalucía* (Andalusia) are two exceptions. In the Valencian government tourism is not integrated in any ministry but under the direct responsibility of Presidency. In Andalusia, where our case study is located, tourism is integrated in the Ministry of Justice, Politic Regeneration and Local Administrations. This can be explained by the fact that in 2019 a new government took power, being the first time that the conservatives (Partido Popular) won an election in this Region. The strange reorganization of the government organigram can be understood from an effort to look for a clear difference with the previous socialist government. On the other hand, the strong link between tourism and urban development has led to corruption. This explicit relationship with Justice and Politic Regeneration may help the new government to present themselves as honest and transparent. Andalusia has a *Secretaría General para el Turismo* (General Secretary for the Tourism) designated "for the quality, innovation, and provision of tourism", and a professional association of members *Consejo Andaluz del Turismo* (the Andalusian Council for Tourism) acting as a consultancy and advisory agency in matters related to tourism.

In terms of economic management, a formula has also emerged. Since a Law approved in 2005, allowing public institutions to be structured and run as publicly owned businesses, some autonomous governments, such as Madrid, Catalonia and Valencia, have created stand-alone agencies. They are run as public firms, so they are more competitive in achieving goals and more focused on the profitability of tourism. In Catalonia, the government has created the Catalan Tourist Board by Law 15/2007 of 5 December 2007. The Valencian government has created the Valencian Agency for Tourism.

Decentralization has been formalized thus through the creation of structures both at provincial level – the *diputaciones* (provincial councils)– and at municipal level – the *ayuntamientos* (town halls) present in every town and city. At a provincial level, the management of tourism on the Costa del Sol depends on the *Diputación de Malaga* (Provincial Council of Malaga), which has created a specific department denominated *Turismo y planificación Costa del Sol* (Tourism and Planning on the Costa del Sol) depending directly on the President of the Provincial Council: tourism is thus directly linked to the urban dimension that prevails on the Spanish coast. The town halls would become the greatest decision-making body in urban development: despite the existence of a national central government ministry and the autonomous regional ministry for tourism as well as a department in the Provincial Council of Malaga, it is in fact, paradoxically, in the town halls on the Costa del Sol, in their departments of urban development, where the power to govern tourism has largely been largely concentrated.

Despite this decentralization process, the state has continued to play an outstanding role in all matters related to tourism, almost as a mediator or coordinator between the different administrations and actors in the industry. Its main function has been to promote Spain's image to the exterior world, both from a marketing angle and by providing economic support to both the private and the public sector. The Spanish Institute of Tourism *(TURESPAÑA)*[2] was created in 1984 to oversee this process as an autonomous body responsible for the international promotion of Spain as a travel destination.

2 Created by Law 50/1984, of 30th December, figuring in the General State Budgets for 1985, as the National Institute for the Promotion of Tourism, subsequently denominated as from 1989 as the Institute for the Promotion of Tourism in Spain, and, finally, in 1991, as the Spanish Institute for Tourism.

A transversal approach with issues related to tourism cutting across a number of administrations and official bodies has been historically present: since the last period of Francoism, tourism has become a key issue on Spain's economic agenda, and this accounts for its relevance in strategies developed by a range of ministries. The national *Comisión Interministerial del Turismo* (Inter-ministerial Commission for Tourism) was created in 1994, having a historical precedent in the first transversal commission created in 1954 under Franco. Its goal is to promote interaction with national ministries in order to coordinate plans, policies, and actions that can have a direct and favorable impact on tourism. The *Conferencia Sectorial del Turismo* (Sectorial Tourism Conference) is another national body that facilitates interaction among the national and autonomous governments. Although this national body attempts to coordinate a common strategy and brand for Spain, the autonomous governments and even certain municipalities develop their own agenda, competing with the others.

Such transversality has also been historically present with the involvement of the private sector. As seen in the previous chapter, since Franco's regime, and in concordance with Healy's approach to the subject (Healy, 1994), the public sector has controlled and regulated processes, but has relied on the private sector for development, creating policies to facilitate its active participation and the commitment of public bodies in offering a friendly framework. It can be confirmed that the power of private capital in urban tourist development that was legally promoted during the dictatorship is still very present to this day. Although the state structure has undergone profound change since the Franco era, territorial development still runs in parallel with official urban planning in the 21st century; private developers still lead the logics of coastal occupation through specific agreements with official municipal and regional bodies and thus prevent the development of overall strategies for tourism in this part of the Mediterranean. Albeit more nuanced, such a way of operating has never been eradicated by any of the democratic governments. The private sector is also invited to participate in governing bodies, this includes the National Tourist Board as an advisory body which includes the participation of the private sector in setting the agenda for tourism in Spain at a national level.

Such cooperation between the public and private sector has also been incorporated as the fourth axis of the National Integrated Plan for Tourism (Ministerio de Industria, Energía y Turismo, 2012), responsible for establishing the concept of *Marca España* (Spain Brand) as a state strategy to improve the image of the country abroad in different fields, including tourism. Also, currently operational is the *Plan Horizonte 2020* (Horizon Plan 2020) with both public and private sectors involved in its elaboration, further consolidating the influence of the private sector in state decisions.

These current plans, like those approved since 1992 – FUTURES I (1992–95), FUTURES II (1996–99), PICTE (2000–06) – insist on transversality with the private sector, supporting and favoring the modernization of the tourist sector and its continuous adaptation to changing international contexts. Problems detected in the succession of plans, including those operational under Franco, appear to be based in four areas: the administration of tourism, the entrepreneur in tourism, the tourist, and the destination (Velasco, 2008, 23). Its varied objectives encompass matters related to competitiveness, quality, innovation, sustainability, with the creation in 2002 of the *Sociedad Mercantil Estatal para la Gestión de la Innovación y las Tecnologías Turísticas* (SEGITTUR) (State Trade Organisation for the Management and Innovation of Technologies in Tourism), assuming the responsibility for promoting R+D+I in tourism, both in the public and private sectors.

At regional level, the involvement of the private sector has even greater impact: in tourist agencies such as the Catalonia Tourist Board CTB, the private sector serves not only as a consulting body but also shares responsibility with the public sector.

> *When the CTB took over from the consortium as the body leading initiatives to raise international awareness of the attractions that Catalonia can offer, it marked a turning point in Catalan promotional strategy. The main innovation, which is now gathering strength, was the involvement of the private sector, which now shares in the tasks and responsibilities for promoting and selling tourism in Catalonia around the world.* (Agència Catalana de Turisme, 2012)

13.4.2 Established and Emerging Products in Tourism

– Sun, Sea, and Sand Mature Destination

As presented above, an interpretation of the area as a potential model for international mass tourism led to a process of accelerated urban development driven by aggressive speculation throughout the second half of the 20th century. The private sector has played a leading role in the transformation of the coastal landscape, with the state participating from the beginning as a main actor in tourism. The state-run network of Paradores was created to initiate the hotel offer in the country, and later, under Franco, the aim was to reach areas with little tourist development, while now they are considered a quality reference in the hotel industry.

At the beginning of the 21st century, more than 100km of coastline constitute a fully consolidated conurbation. In Spain, urban development has accelerated at an exponential rate, with the speed of transformation doubling in the first years of 21st century (Observatorio de la Sostenibilidad en España, 2010). In terms of tourism, great sections of the Spanish Mediterranean coast are now considered a 'destino maduro' (mature destination): an established and recognized location linked to a specific tourist product. Although the intensive occupation of the coast was almost suspended due to the international crisis of 2007, construction is again recovering, and the development in urban tourism is once again the main economic resource for coastal Town Councils.

– Emerging tourist products. Secure tourism and Urban cultural tourism

Over the last few years, the number of tourists has increased exponentially in Spain, from 60 million in 2013 to 82 million in 2017, even surpassing the United States for the first time (El Independiente, 2018 January 10). If we look at the international context, political instability in the Mediterranean accounts for how the Spanish coast has been able to acquire a significant number of tourists expelled by the Arab Spring, armed conflict in the Middle East, the Greek crisis or terrorist attacks in France, and now drawn to more secure destinations. The Spanish government has decided to use security as an emerging asset in tourism. It is noteworthy that the Internal Affairs Ministry has developed a plan to promote Spain as a country that offers 'turismo seguro' (secure tourism), taking advantage of the international geopolitical situation, and developing specific customer services through SATE Offices *(Servicio de Atención al Turista Extranjero* for foreign tourists (Ministerio del Interior, 2019).

On the Costa del Sol, during the international crisis of 2007, Malaga reverted to culture as an emergent tourist product. Concentrated in the capital, the city of Malaga was even coined the New Barcelona in 2015 (Lambert, 2015, April 4). Its urban revitalization with a proliferation of new museums formalizes its latest strategy for tourism, whose core theme is the international figure of Picasso. Picasso was born in Malaga and the new tourist formula relies on him as a recognizable cultural product: detonator of an urban cultural park, The Picasso Museum, was the first of a series of new national and international museums to set up in Malaga, followed by the inauguration in the port area shortly after of the first branch of the Pompidou Center to be established outside Paris, France. The exceptionality of a Picasso museum in a sun, sea, and sand destination has become surprisingly generalized. The theming of the old city center around culture and heritage is framed within the city's globalized tourist market. This process has been fast and successful, taking into account that just twenty years ago the city of Malaga was largely bypassed by tourists, it now receives record numbers of tourists and hugely benefits from the economic impact (Rivera, 2017, August 27). Malaga now supports intense traffic from tourist cruise ships, adding increased pressure to the city. The proliferation of restaurants, bars, ice cream parlors, and souvenir shops to meet the immediate demands of this type of tourism is leading to gradual transformations in the urban landscape. The data confirms a golden moment for Spanish tourism, specifically on the Costa del Sol. However, such success also constitutes one of the main threats for the sector in Spain: an assertion shared by the private sector, which also now considers the continuous growth of the number of tourists completely unsustainable (Page, 2018, April 29).

Historical cities run the risk of theming and gentrification as a result of the success of the emergent cultural offer. In Malaga, the local population is progressively displaced and forced to live outside the old city center, no longer able to afford property prices, while housing is being transformed into tourist apartments. In February 2018, 29% of properties were destined for residential use while 65.4% were for tourist, commercial, or the hotel and catering industries (García, 2018, February 23). Such a process of gentrification is destroying the urban landscape of the city and limiting the right to the city in the 21st century. This has become a problem for historical cities throughout Europe. Vive la Ville Europe is a European platform that defends the residential character of town and city centers against, for instance, hotel industry proposals requesting that the City Council of Malaga declare the old city center non-residential (Vázquez, 2017, March 27). The City Council established a plenary agreement on February 23rd, 2018, which indicates in part two of point U.9:

> *The Malaga City Council will apply the necessary regulations to incorporate zoning as proposed by the Alter Eco report that determines the current density of tourist accommodation, limits, and qualifications for each zone contemplated in the report, and the possible moratoriums that could be granted.* (Ayuntamiento de Málaga & Observatorio de Medio Ambiente Urbano, 2018, 2)

Along with gentrification and the destruction of urban heritage, the dominance of tourism in the city has led to tourism phobia, representing a serious threat in Spain, especially in city areas. Tourism is becoming a national problem as a result of its intensive use of public space, including cultural and leisure spaces.

> *The masses have acquired the shape of tourism. This concerns a sixth continent composed of 1,000 million univocal beings in perpetual movement having become a global pest devouring cities, monuments, temples, palaces, and gardens along the way. The sole purpose of these masses is to consume, dress up, eat, drink, dance, see, hear, and repeat.* (Vicent, 2017, July 7)

Barcelona has approved a daily tax on any tourist who visits the city without spending the night. This means tourism has become a matter of concern to other governing bodies, in this case that of the Commission for Tax and the Economy and the Municipal Government of Barcelona.

13.5 Conclusion

The study has traced the origins and scope of the management of tourism in Spain, signaling the projection of a modern image of the country as its principle objective throughout the 20th century. Successive governments would focus on international tourists and development, linking tourism with the construction of infrastructures and even the role of government as hotel developer. While tourism, with culture and urban heritage as main tourist products, was a marginal activity until the second half of the 20th century, sun, sea, and sand mass tourism would come to rule the Spanish economy, through the creation of governmental structures and the legal and administrative apparatus to facilitate the transformation of the Mediterranean coast and accommodate the international vacation demands of post-World War II societies.

A detailed and critical analysis of the legal apparatus demonstrates how the approaches to tourist management relating to the economy and urban planning initiated under Franco in the last phase of his rule, paradoxically transcending the state of dictatorship, continued under democratic rule, and are still manifest today. In fact, the first plan for tourism created under democracy, the *Plan Marco de Competitivad del Turismo Español* (Plan for a Framework for Competitiveness in Spanish Tourism) (FUTURES 1992/1995) was approved twenty years after the preceding regulatory document, produced in 1974 under Franco (Velasco, 2008, 18). By the end of the 20th century, the state had thus firmly taken hold of the reins of the tourist sector, enshrining it as the main economic engine of the country. Tourism became a state matter, transcending ideologies, with no political party wanting to position against its unsustainable consequences.

Decentralization in tourism management has been a direct result of the new federal vision of Spain, resulting in a decentralized form of tourism management, each autonomous region developing its own scope from within their own ministries of tourism, always within other sectors. Despite the deep shift that has taken place in the state structure, from a centralized system to an extremely decentralized form of government, the study demonstrates that a degree of inertia existed during the previous dictatorial period in its approach to coastal development. Together with the confirmation of tourism as a key economic resource, the proliferation of official bodies related to tourism at national, regional, provincial, and municipal level has not prevented the exponential urban development of the coast, which has culminated in a tourist conurbation that lacks any comprehensive approach or overall vision. The prioritization of tourism over heritage, industry, and even assets of public property still prevails. It is of note that declarations of Centers of National Tourist Interest continued to take place under democracy – even after central government competencies were transferred to the new autonomous governments in the different regions (Galiana & Barrado, 2006). The close collaboration between the public sector and private sector to whom the government had facilitated the exploitation of the coast, is also present today, with new official bodies engaged in the active collaboration of the private sector in tourism.

After a few years adapting to the new norms imposed by the decentralization of competencies, the state reverted to assuming a leadership role in the formulation of the country's image abroad, just as under Franco, promoting Spain as a unified and recognizable brand in the tourist sector, notwithstanding a fresh set of parameters, including quality, innovation and sustainability. In this way, publicity for the national offer is guaranteed, making up for any lack of positioning of regions that historically have not figured as tourist destinations. Autonomous communities, however, while also developing significant marketing strategies to improve their visibility, have ended up taking over the physical formalization of tourist developments, completing and filling any gaps in the urban layouts first designed specifically for tourism in the mid-20th century.

While sun, sea, and sand tourism has become an established tourist product, the tourist offer has changed and is now complemented by urban cultural tourism in mature coastal destinations, and boosted by political instability in other areas of the Mediterranean as well as the 2008 international crisis. The Costa del Sol and its capital are once more paradigmatic of an emergent tourist product with the resurgence of cultural tourism as a resource. The Costa del Sol, especially affected by the real estate crisis, found in the city of Malaga a new tourist challenge. With the success of its urban cultural offer, new global threats have emerged. Even the private sector considers the exponential growth in the number of tourists and the territorial overload unsustainable, given the urban heritage loss, gentrification, and developing phobias against tourism. Official bodies in Spanish tourism are dealing with this context, promoting studies and developing zoning regulations to limit tourist lodgings or establishing daily taxes for tourists. What Jane Jacobs stated in the 1960s, that the success of a place constitutes the main threat in the destruction of its values, is applicable to tourism in Spain, and is framing the challenges for the future of the management of tourism in this country.

References

Agència Catalana de Turisme (2012). About the Catalan Tourist Board. http://act.gencat.cat/act-about-us/act-about-the-catalan-tourist-board/?lang=enm (December 4, 2020).

Ayuntamiento de Málaga & Observatorio de Medio Ambiente Urbano (2018). Aproximación a las intensidades del uso turístico en Málaga. Málaga: Ayuntamiento de Málaga. http://static.omau-malaga.com/omau/subidas/archivos/2/7/arc_7972.pdf (December 4, 2020).

Barciela, C., López, M.I., Melgarejo, J., & Miranda, J.A. (2001). *La España de Franco (1939–1975). Economía*. Madrid: Síntesis.

Chuliá, E. (2007). Cultural diversity and the development of a pre-democratic civil society in Spain. In N. Townson (Ed.), *Spain Transformed. The Late Franco Dictatorship, 1959–1975* (pp. 163–181). New York: Palgrave MacMillan.

Cortes Generales (1978). Constitución Española. *BOE*, *311*, 29313–29424.

Dirección General de Turismo (1956). *Movimiento Turístico en España*. Madrid: Ministerio de Información y Turismo.

El Independiente (2018, January 10). España cierra 2017 con 82 millones de turistas y supera por primera vez a Estados Unidos. *El Independiente*. https://www.elindependiente.com/economia/2018/01/10/espana-cierra-2017-82-millones-turistas-superando-estados-unidos/ (December 4, 2020).

Fernández, L. (1991). *Historia general del turismo de masas*. Madrid: Alianza.

Galiana, L. & Barrado, D. (2006). Centers of National Tourist Interest and the take-off of mass tourism in Spain. *Investigaciones Geográficas*, *39*, 73–93.

García Recio, J. (2018, February 23). La batalla por un modelo para el centro. *La Opinión de Málaga*. https://www.laopiniondemalaga.es/malaga/2018/02/23/batalla-modelo-centro/988864.html (December 4, 2020).

Gobierno del Estado (1938). Ley de 30 de enero, organizando la Administración Central del Estado. *BOE*, *467*, 5514–5515.

Healy, R.G. (1994). The "Common Pool" Problem in Tourism Landscape. *Annals of Tourism Research*, *21*(3), 596–611.

Jefatura del Estado (1939). Ley de 8 de agosto, por la que se modifica la organización de la Administración Central del Estado establecida por las de 30 de enero y 29 de diciembre de 1938. *BOE*, *221*, 4326–4327.

Jefatura del Estado (1963a). Ley 194/1963, de 28 de diciembre, por la que se aprueba el Plan de Desarrollo Económico y Social para el periodo 1964/1967 y se dictan normas relativas a su ejecución. *BOE*, *312*, 18190–18198.

Jefatura del Estado (1963b). Ley 197/1963, de 28 de diciembre, sobre "Centros y Zonas de Interés Turístico Nacional." *BOE*, *313*, 18226–18230.

Jefatura del Estado (1969). Ley 1/1969, de 11 de febrero, por la que se aprueba el II plan de Desarrollo Económico y Social. *BOE*, *37*, 2137–2142.

Jefatura del Estado (1972). Ley 22/1972, de 10 de mayo, de aprobación del III Plan de Desarrollo Económico y Social. *BOE*, *113*, 8239–8276.

Lambert, C. (2015, April 4). Welcome to Malaga, the new Barcelona! Arty regeneration makes unappreciated Spanish city one of Europe's hip locations. DalyMail. https://www.dailymail.co.uk/travel/article-3025411/Welcome-Malaga-new-Barcelona-City-s-multi-million-pound-regeneration-project-makes-one-Europe-s-hippest-destinations.html (December 4, 2020).

Loren-Méndez, M. & Pinzón-Ayala, D. (2020). Spatial Legal History of Spanish Mediterranean Geography: The Last Phase of Franco's Regime. *Geographical Review*, 111(3), 458–478.

Martín, P. & Martínez, E. (2007). La Edad de Oro del capitalismo español: crecimiento económico sin libertades políticas. In N. Townson (Ed.), *Spain Transformed. The Late Franco Dictatorship, 1959–1975* (pp. 30–46). New York: Palgrave MacMillan.

Ministerio de Industria, Energía y Turismo (2012). *National Integrated Plan for Tourism.* https://turismo.gob.es/es-es/servicios/Documents/Plan-Nacional-Integral-Turismo-2012-2015.pdf (December 4, 2020).

Ministerio del Interior (1938). Decreto de 25 de marzo, autorizando al Ministerio del Interior para organizar un circuito para de viaje denominado "Ruta de Guerra del Norte". *BOE*, *593*, 7738–7739.

Ministerio del Interior (2019). *Plan Turismo Seguro.* (December 4, 2020).

Moreno, A. (2010). The National Tourist Board (1928-1932). An economic Assessment of Tourism Policy. *Investigaciones de Historia Económica*, 6(18),103–132.

Pack, S.D. (2006). *Tourism and Dictatorship. Europe's Peaceful Invasion of Franco's Spain.* New York: Palgrave MacMillan.

Pack, S.D. (2007). Tourism and political change in Franco's Spain. In N. Townson (Ed.), *Spain Transformed. The Late Franco Dictatorship, 1959–1975* (pp. 47–66). New York: Palgrave MacMillan.

Page, D. (2018, April 29). ¿Una España con 120 millones de turistas? *El Independiente.* https://www.elindependiente.com/economia/2018/04/29/una-espana-con-120-millones-de-turistas-record/ (December 4, 2020).

Poutet, H. (1995). *Images touristiques de l'Espagne. De la propaganda politique à la promotion touristique*, Paris: L'Harmattan.

Presidencia del Consejo de Ministros (1911). Real Decreto de 19 de junio, creando en esta Presidencia una Comisaría Regia, encargada de procurar el desarrollo del turismo y la divulgación de la cultura artística popular. *Gaceta de Madrid*, *171*, 805.

Presidencia del Gobierno (1964a). Plan de Desarrollo Económico y Social para el período 1964-1967, (Continuación). *BOE*, *7*, 375–378.

Presidencia del Gobierno (1964b). Plan de Desarrollo Económico y Social para el período 1964-1967, (Continuación). *BOE*, *35*, 1767–1770.

Presidencia del Gobierno (1964c). Plan de Desarrollo Económico y Social para el período 1964-1967, (Continuación). *BOE*, *150*, 8102–8106.

Rivera A. (2017, August 27). Málaga, récord de visitantes y negocio . . . y hace 20 años ni existía para el turista. *El Confidencial.* https://www.elconfidencial.com/espana/andalucia/2017-08-27/turismo-malaga-record-turistas_1434383/ (December 4, 2020).

Townson, N. (Ed.) (2007). *Spain Transformed. The Late Franco Dictatorship, 1959–1975.* New York: Palgrave MacMillan.

United Nations (1948). *Universal Declaration of Human Rights*, Paris: United Nations.

Vázquez, A. (2017, March 27). La plataforma vecinal europea apoya que el Centro de Málaga siga como zona residencial. *La Opinión de Málaga*. https://www.laopiniondemalaga.es/malaga/2017/03/27/plataforma-vecinal-europea-rechaza-centro/919405.html (December 4, 2020)
Velasco, M. (2005). *La política turística. Gobierno y administración turística en España (1952–2003)*. València: Tirant lo Blanch.
Velasco, M. (2008). Evolución de los problemas del turismo español. La administración general del Estado como analista y los Planes Públicos como indicadores (1952–2006), *Papers de Turisme*, *43–44*, 7–31.
Vicent, M. (2017, July 7). Todo lleno. La masa adquiere hoy la forma de turismo. *El País*. https://elpais.com/elpais/2017/07/07/opinion/1499427585_520309.html (December 4, 2020).

Louise A. Mozingo & Wilasinee Darnthamrongkul

14 Thailand

14.1 Introduction

14.1.1 The Kingdom of Thailand

Thailand, formally the Kingdom of Thailand, is located in Southeast Asia at the center of the Indochinese peninsula. Its neighbors include Laos, Burma, Cambodia, and Malaysia. The Andaman Sea and the Gulf of Thailand, known as Ao Thai, bracket the country on the east and west. The county's 76 provinces and the capital city of Bangkok form six physiographic regions including northern, northeastern, eastern, central, western, and southern. With almost 70 million total population, the majority religion practiced by its citizens is Buddhism (95%), with minorities of Muslims, Sihks, Hindus, and Christians.

Officially named Thailand in 1939, the country was historically known as Siam and is occasionally still referred to as such. Thailand has its own language, Thai, which is also used as official language in the country, though regional dialects do exist. The term Thai is also used to refer to people of Thailand – the Thai people – formerly known as the Siamese.

The first Siamese or Thai state – Sukhothai Kingdom – was found in 1238 and existed until 1583. UNESCO designated the ruins of this early kingdom in Sukhotai, Kamphaeng Phet, and Si Satchanalai as World Heritage Sites in 1991. In 1351, a new kingdom – Ayutthaya Kingdom – arose. This new kingdom progressively developed and gradually overshadowed the old one. Sukhothai became Ayutthayan tributary in 1378 and then a part of Ayutthaya Kingdom in 1583. The Kingdom of Ayutthaya maintained its prosperity and power for more than 400 years. Defeated in the Burmese-Siamese War, Ayutthaya was burned and destroyed in 1767. UNESCO also declared the ruins of Ayutthaya a World Heritage Site in 1991.

After the fall of Ayutthaya, Thonburi, an area located on the west bank of Chao Phraya River, became the capital of the next ruling kingdom. Thonburi Kingdom existed for 15 years. In 1782, King Rama I the Great established the new dynasty, Chakri, starting the new kingdom, Rattanakosin, and founded the new capital, Bangkok, on the east side of Chao Phraya River. For more than 200 years, the Chakri Dynasty has ruled the country, as Thailand continued to develop and flourish.

14.1.2 The Thai Monarchy and Government

The absolute monarchy of the Chakri dynasty ruled Thailand until 1932. The bloodless Siamese Revolution of 1932 led to significant political reform in the country, marking

https://doi.org/10.1515/9783110638141-014

the end of the absolute monarchy and the start of the constitutional monarchy. Within the current framework of a constitutional monarchy, Thailand has the Monarch as the head of the state and the Prime Minister as the head of the government.

The institution of the monarchy in Thailand existed since the Sukhothai Kingdom in 1238. For almost 800 years, the Thai monarchy has played a very important role in ruling and developing the country. The King of Thailand, thus, is not only the head of the state, but also revered as the soul of the country. In 2016, King Rama IX the Great or King Bhumibol Adulyadej, the world's longest reigning monarch, passed away, resulting in the deepest sorrow all over the country. The Royal Coronation Ceremony of King Rama X took place during May 4–6, 2019.

The government of Thailand is formally known as the Royal Thai Government (RTG). It is a unitary government which is composed of three branches – the executive, the legislative, and the judiciary. The executive branch administers the section of the government headed by the Prime Minister, who is selected by the House of the Representatives and approved by the King. The Prime Minister is usually the leader of the winning political party of the latest election. The Prime Minister is responsible for nominating persons to be appointed by the King as the ministers and deputy ministers, making up the Council of Ministers or the Cabinet. Apart from the members of the Cabinet, the executive branch also consists of the permanent officials of the ministries. The legislative branch is the law-making section of the government. It is officially called the National Assembly, and also known as the Parliament of Thailand. This branch of the government is composed of two houses – the Senate (the upper house) and the House of Representatives (the lower house). These two houses of the legislative branch are made up of both elected and appointed members. The judiciary branch is responsible for law enforcement. This governmental branch consists of four court systems: the constitutional Court, the Supreme Court, the Appeal Courts, and the Trial Courts.

Apart from the central government, each of the 76 provinces as well as Bangkok also has its own local government. The governor of Bangkok is elected while the governors of the other 76 provinces are appointed by the Ministry of the Interior.

14.1.3 The Golden Axe and the Land of Smiles

As the shape of Thailand appears like an axe in the geographical maps, Thai people usually, and proudly, refer to their country as the "golden axe." The term golden refers to the rice fields that once were everywhere in the country – when the rice turns golden, it also turns the country golden, making the country look like a golden axe. Also, the location of Thailand is believed to be a part of an ancient region called "Suvarnabhumi," a toponym literally meaning "golden land." This is also the reason why King Bhumibol Adulyadej named the new Bangkok International Airport, officially opened in 2006, as Suvarnabhumi Airport.

Thailand, undeniably, lies on one of the most fertile regions in the world, recognized as the land which has "fish in the water and rice in the fields." Agriculture, therefore, has been a significant basis of Thai culture and economy. As Singhapreecha (2014, n.p.) noted, "The agricultural sector of Thailand has long been called the country's 'backbone'. This is because it is the most important sector in the economy, one that has generated food and living incomes for most of Thai people. Before the manufacturing sector began to play an increasing role in Thai economy in the late 1970s, the agricultural sector generated almost 100% of the country's export income. At present, it still constitutes a substantial share of Thai exports of almost 30%. However, its most significant aspect is that it has provided almost constant employment for the majority of Thai laborers."

Thailand is newly industrialized and the economy of the country also relies on industrial and service sectors. Even though Thailand is not a very wealthy country, according to Bloomberg it has been ranked as the world's least miserable economy – the world's happiest economy, in other words – for many years (Jamrisko & Saraiva, 2018; Saraiva & Jamrisko, 2016; Saraiva & Jamrisko, 2020).

Tourist literature in particular, but other media as well, refer to Thailand as the "Land of Smile." The country has quite good weather all year long, spectacular natural features, compelling cultural traditions, tolerant and open religious customs, delicious fruit, and world renowned cuisine. Thai people habitually smile as they carry on their everyday lives in their beloved Thailand. In addition, Thai people are unfailingly polite and gracious to foreigners, part of a closely held cultural and religious practice, thereby creating smiles in most visitors to Thailand.

14.2 Tourism Development in Thailand

14.2.1 The Establishment of Tourism Organizations

Prince Purachatra Jayakara, a son of King Chulalongkorn (King Rama V the Great), first promoted tourism in Thailand. Specifically, when he served as the commander of the railway department during 1920s, he began disseminating tourism information in the United States (TAT, 2019). In 1924, the railway department established an advertising division of the railway department to provide information for foreign travelers and promote the country internationally.

In 1936, promotion of tourism began in earnest, when the Ministry of Commerce elevated tourism promotion to the cabinet level. During 1939–1945, establishment of the tourism promotion department paused because of World War II. In 1949, the advertising division, renamed the Government Public Relations Department (PRD) under the Office of the Prime Minister, became responsible for the promotion of tourism. A year later, a specific office of tourism promotion formed under the PRD.

In 1959, the cabinet passed legislation to establish the Tourist Organization of Thailand (TOT) as an independent entity. The TOT, officially responsible for the promotion of tourism and tourism industry, launched on March 18th, 1960.

In 1979, as tourism significantly increased its role in the economy, the cabinet passed the legislation to replace TOT and establish the Tourism Authority of Thailand (TAT), to oversee a comprehensive portfolio of tourism-related matters. The reorganization of the governmental structure in Thailand in 2002 resulted in the establishment of the new cabinet level Ministry of Tourism and Sports along with the new Department of Tourism (DOT). The Ministry is responsible for directing TAT, DOT and also all the provincial offices of tourism and sports. Previously, the government considered sports an educational activity under the jurisdiction of the Ministry of Education.

In addition to the governmental organizations established to work on developing and promoting tourism, in 2001 the tourism industry operators in the country joined hands to establish the Tourism Council of Thailand (TCT), representing all tourism industry operators. Basically, the TCT has an obligation to "propose important guideline policy, promote quality verification system, standard system, and quality assurance system of the businesses related to the goods or services for tourists, and to encourage the tourism industry operators to carry out the operation with quality, morality and ethics, so as to promote the efficiency and development of this industry" (TCT, 2016).

14.2.2 The Structure of Tourism Organizations

The Tourism Authority of Thailand (TAT) is the main organization responsible for the promotion and marketing of tourism. It is a government enterprise under the Ministry of Tourism and Sports, essentially a government-owned private entity responsible for promoting and marketing the country's tourism. TAT manages Thailand's tourism affairs, publicizing tourist attractions, supplying tourist information, and conducting studies about tourism. TAT is divided into 7 divisions:

A. Administration
B. Policy and planning
C. Tourism products and business
D. Marketing communications
E. Domestic marketing
F. International marketing (Asia and South Pacific)
G. International marketing (Europe, Africa, Middle East, and Americas)

The Department of Tourism (DOT) is also responsible for the development of tourism in the country. Established in 2002, DOT is a government agency (not a government enterprise) under the Ministry of Tourism and Sports in order to work on the development of tourism sites, tourism services, and tour guide businesses, which were once the task of TAT but transferred to DOT. In addition, DOT took over the tasks of developing and supporting film and related matters from the Government Public Relations Department (PRD). DOT is divided into 7 divisions:

A. Administration
B. Development of tourism sites
C. Development of tourism services
D. Registration of tourism and tour guide businesses
E. Production of films and related matters
F. Development of tourism personnel
G. Development of tourism information technology

Apart from TAT and DOT, there are also 76 provincial offices of tourism and sports. For each of the 76 provinces, its office of tourism and sports, which is under the Ministry of Tourism and Sports, is responsible for all tourism tasks in the province. For Bangkok, the capital city, all of its tourism tasks are done by the Tourism Division, which is a division under the Culture, Sports and Tourism Department (CSTD) of Bangkok Metropolitan Administration (BMA). The organization charts of the tourism organizations in Thailand is presented in Figure 14.1 and the organization charts of the tourism organizations in Bangkok is presented in Figure 14.2.

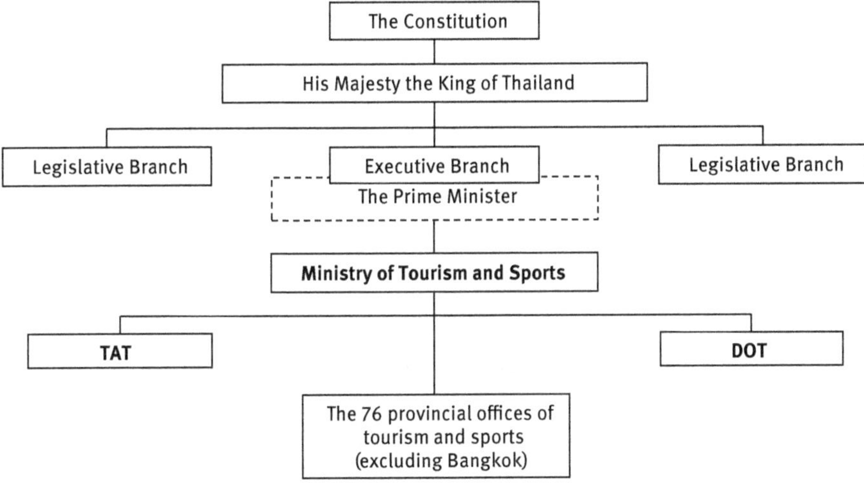

Figure 14.1: The organization chart of the tourism organizations in Thailand.

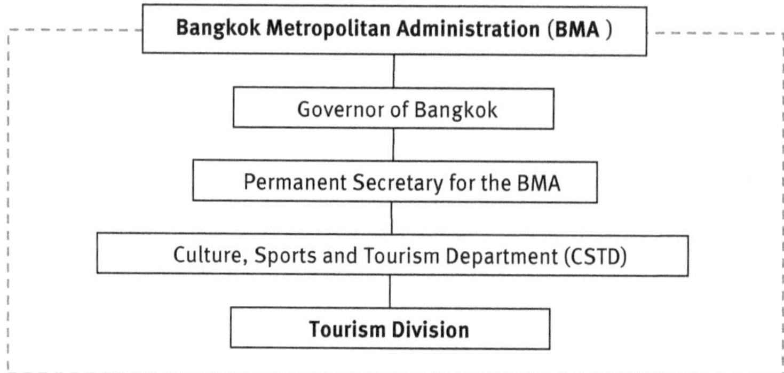

Figure 14.2: The organization chart of the tourism organizations in Bangkok.

The existing structure of tourism governance in Thailand manifests well-organized, hierarchical system that determines tourism initiatives to maintain and grow tourism as a principle economic activity of the country. This extends from the national scale at the cabinet level to provincial organizations concerned with supporting and fostering local programs and initiatives at the behest of the central authorities. The structure builds from the rich history of cultural, religious, culinary, ecological, and geographic conditions of Thailand that have proven to offer a broad variety of attractions to the eclectic populations' global tourism consumers. This rich history emerged from deeply and closely held local practices and their commodification through a highly centralized decision making does not allow for local nuance in decision-making. Nor does it consider the erosion of local practices, the very practices that create the tourist market, under the disruptive forces global tourism. Local people have limited influence on the priorities and policies of tourism governance in Thailand which largely serves Bangkok-based and international investors.

14.2.3 The Tourism Campaigns

TAT launched a series of marketing campaigns to promote tourism (TAT News, 2016), beginning with Visit Thailand Year in 1980. This first campaign resulted in the attraction of 2 million visitor arrivals and the generation of a huge revenues, making the country become one of the top earners from tourism. In 1987, the Visit Thailand Year campaign was re-launched to celebrate the 60th, fifth cycle, birthday of His Majesty King Bhumibhol Adulyadej. This re-launched campaign resulted in the attraction of an even greater number of visitor arrivals, more than 3 million.

During 1998–1999, Thailand launched the new tourism campaign, Amazing Thailand. This campaign marked the 72nd, sixth cycle, birthday of His Majesty King Bhumibhol Adulyadej and also marked the end of the last millennium and the beginning of the new one. Furthermore, this campaign also attempted to use tourism to help the country recover from the 1997 Asian Economic Crisis. During these years, TAT had worked in collaboration with a number of public and private organizations to support the campaign by organizing sale events, announcing new tourism destinations, and promoting unique tourism products including accommodations, entertainment, and souvenirs, for example. This campaign helped increase visitor arrivals to 7.76 million in 1998 and to 8.58 million in 1999.

In early 2003, SARS epidemic significantly affected tourism industry in Thailand. TAT partnered with the Association of Thai Travel Agents, the Thai Hotels Association, the Thai Airways International, and several other tourism-related associations to set up the Crisis Management Committee with the aim of mitigating the impact of the SARS epidemic on the tourism industry. The partnership led to the launch of the "Unseen Thailand" campaign. This campaign featured special promotions offered to attract international visitors and boost domestic travel. Although visitor arrivals dropped 10%, the more than 10 million visitor arrivals in 2003, still above the 1998 levels.

The year 2006 marked one of the most important events in Thailand, the celebration of the 60th anniversary of His Majesty King Bhumibhol Adulyadej's accession to the throne. TAT launched the Thailand Grand Invitation campaign to invite visitors from all over the world to come and join in a number of magnificent year-long, nation-wide, events. In 2006, there were almost 14 million visitors arriving Thailand. The year 2007 marked another important celebration in the country, His Majesty King Bhumibhol Adulyadej's 80th birthday. In this year, TAT launched the "Thailand Talk to the World" campaign to encourage foreigners to join the country's great celebration and more than 14 million visitors arrived Thailand.

In 2010, as TAT, along with Thai Airways International turned 50, Thailand celebrated the Golden Jubilee of its tourism industry. A television commercial entitled Remember was launched to commemorate the vital role that TAT had played in making Thailand become one of the world's most popular tourist destinations. In addition, as TAT aimed to increase the number of repeat visitors, the "Amazing Thailand Always Amazes You" tagline was also introduced to encourage visitors to return to Thailand.

Thailand celebrated the 85th birthday of His Majesty King Bhumibhol Adulyadej in 2012. TAT continued to promote the country's tourism by using the "Amazing Thailand Always Amazes You" campaign. In addition, TAT introduced a series of promotional activities, such as a television commercial entitled "Warm Welcome is all Around You." More than 22 million visitors arrived in 2012.

During 2013–2014, the "Amazing Thailand It Begins with the People" campaign was launched. Under this campaign, TAT continued to join hands with other public and private organizations to promote the country's tourism. The government supported additional pro-tourism policies, most prominently the three-month tourist visa fee waiver for holders of People's Republic of China and Republic of China passports, Thailand Travel Shield insurance coverage policy for visitors, and the

30-day extension of stay for visitors from 48 other countries. In addition, to help improve the visitor experience, the TAT instituted new zoning and regulations at key beaches and the installation of new automatic queue-card taxi kiosks at Suvarnabhumi Airport. The country ended the year 2013 with more than 26 million visitor arrivals and ended the year 2014 with more than 24 million visitor arrivals. Tourist arrivals tripled in 25 years.

The yearly campaigns continue. In 2015, TAT used the "Discover Thainess" campaign to encourage the visitors to experience and explore the glorious cultural assets of the country. This campaign highlighted the unique blend of beliefs and traditions in everyday life of the Thai people that makes the country an attractive tourist destination. For 2016, TAT aimed to promote the overall quality of the visitor experience. The "Quality Leisure Destination through Thainess" campaign emphasized the promotion of the country as a quality leisure destination by offering visitors unique and marvelous "Thainess" in order to make the visitors have a distinct experience during visiting Thailand.

In 2020, in response to the economic effect of the coronavirus pandemic, the government launched a tourism stimulus campaign called "We Travel Together" with the aim of increasing domestic tourism and helping domestic travel and tourism businesses. For this campaign, the government subsidize 40% of normal room rates (with limitation at 3,000 baht per night for up to five nights), 40% of flight tickets (with limitation at 1,000 baht per seat and the quota is limited to 2 million seats), and 40% of other services, including food and beverage (with limitation at 600 baht per room per weekend-night and 900 baht per room per weekday-night).

14.3 Tourism Dynamics and Reality

14.3.1 A World-Class Tourist Destination

With its abundant and unique tourism resources, Thailand has become one of the world's most popular tourist destinations during the past decade. In 2014, the World Tourism Organization (UNWTO) declared Thailand as one of the world's top 10 tourism destinations. In 2013, according to the UNWTO (2014, p. 6) UNWTO (2014, p. 6), "Thailand entered the top 10 arrivals ranking at number 10, climbing an amazing five positions, while it moved up two places to seventh in the ranking by tourism receipts." The political unrest in the country during 2014 resulted in the drop of its rankings, both by tourism arrivals and tourism receipts, in 2015 though it still maintained its position in the top 10 ranking by international tourist receipts, even though it moved down two places to 9th (UNWTO, 2015, p. 6). In 2016, as the political situation in the country stabilized, Thailand climbed to sixth in tourist receipts and to 11th in tourism arrivals (UNWTO, 2016, p. 6). Even though it lies distant from both Europe and North America, the point of origin of most tourist visitors, Thailand's management of its tourism industry over the last decades made it a major player in the global tourist trade. While Australia and New Zealand have always been closer in distance than these other regions, the TAT facilitated the recent explosion of the China market, close by and growing ever wealthier, through visa fee waivers and specific marketing campaigns. In 2018, Thailand also ranked one of the top 10 tourism destinations – the nineth in tourism arrivals and the fourth in tourism receipts (UNWTO, 2019, p. 9). As UNWTO (2019, p. 11) says: "Thailand, the subregion's largest destination, added almost 3 million more arrivals and USD 6 billion more in receipts."

As noted above, the capital, Bangkok, is a leading destination city. Located on the fertile land of the Lower Chao Phraya River Delta with numerous of canals within the city, Bangkok has been known as the "Venice of the East" and has long enjoyed an international reputation. In 2011, MasterCard Worldwide, a global leading financial services corporation, started to develop the Global Destination Cities Index (GDCI) or, in other words, the Global Rankings of Top 20 Destination Cities. For six years since the index was launched, Bangkok has always been ranked in the charts, for both visitor numbers and visitor expenditures. Specifically, in 2011 Bangkok was ranked the third in visitor numbers, estimated at 11.5 million visitors, and the fourth in visitor expenditures, worth an estimated at US$14.4 billion (MasterCard Worldwide, 2011, p. 4–7). Considering the 2012 statistics, Bangkok was still ranked the third in visitor arrivals, yet it climbed up one place to the third in visitor expenditures (Hedrick-Wong, 2012, p. 1–5). Bangkok achieved the top-ranked position in the ranking by visitor numbers in 2013. However, it moved one place to the second in 2014 and 2015. For the ranking by visitor expenditures, Bangkok was positioned the fourth, fifth and seventh in 2013, 2014, and 2015, respectively. Notably, Bangkok grew in visitor spending at the fastest rate, at 12%, between 2014 and 2015 (Hedrick-Wong & Choong, 2015, p. 9). Bangkok beat London again in 2016 as it held the top rank in visitor numbers, estimated at 21 million. Considering the visitor expenditures, Bangkok – following Dubai, London, and New York – was ranked the fourth estimated at US$15 billion. In the 2019 GDCI, Bangkok ranked the first in total international visitors, for the fourth consecutive year, and ranked the third in visitor spending (Mastercard, 2019).

14.3.2 A Variety of Tourism Types

Thailand certainly possesses a large number and a wide variety resources for tourism. Tourism in Thailand can be classified into four types: natural tourism, historical tourism, cultural tourism, and special interest tourism.

14.3.2.1 Natural Tourism

Thailand is rich in natural attractions. With Andaman Sea to the West and the Gulf of Thailand to the East, the country has long been justly famous for beautiful tropical beaches. The provinces in the southern region – such as Phuket, Krabi, and Phang Nga – are among the world's most popular destinations for enjoying sand, sea, and sun. Pattaya is also famous as close-to-Bangkok beaches that particularly gained popularity with Americans during the period of the Vietnam War as a safe tourist destination close to the conflict. In addition to beaches, Thailand also abounds in beautiful tropical islands: prominent among them are Koh Samui, Koh Pha Ngan, Koh Lanta, Koh Phi Phi. Facilitated by favorable exchange rates, the quality of the beaches and associated resorts draws large numbers of winter visitors from wealthy northern European countries.

Besides beaches, Thailand contains an abundance of forests, mountains, rivers, waterfalls, and other nature sites. UNESCO designated two natural World Heritage Sites in Thailand: Thungyai-Huai Kha Khaeng Wildlife Sanctuaries, designated in 1991, and the Dong Phayayen-Khao Yai Forest Complex, designated in 2005. Currenty, there are 127 national parks, including 22 marine national parks, where visitors can experience not only spectacular scenery, but also the distinctive wilderness and wildlife of Thailand. The richness and diversity of natural resources in Thailand undoubtedly offer unique opportunities for ecological tourism activities such as trekking, elephant trekking, bird and animal watching, and wilderness camping. These nature areas also easily accommodate adventure tourism activities such as rock climbing, rafting, and skin diving.

14.3.2.2 Historical Tourism

Thailand's long history bestowed the country with numerous world-class historical and archeological sites. As noted earlier, the ruins of the two ancient capitals of the country were designated as UNESCO World Heritage Sites in 1991; one is the Historic City of Ayutthaya and another is the Historic Town of Sukhothai and Associated Historic Towns. Ban Chiang Archaeological Site, where pottery and other evidence of ancient settlement were uncovered, is the third cultural UNESCO World Heritage Site in Thailand, designated in 1992.

The southern part of the northeastern region of Thailand is the location of series of ancient Khmer cities. The most prominent is Phimai, a large and striking quadrangular ancient Khmer city bordered by walls and moats, and believed to mark of the end of the ancient Khmer route connecting Angkor Wat. In 2004, the Thai government submitted Phimai, (along with the associated temples of Phanomroong and Muangtam), for consideration as a UNESCO World Heritage Site as well as Phuphrabat Historical Park. (UNESCO, 2016a, 2016b) Phuphrabat Historical Park is located in the northern part of northeast Thailand. Phuphrabat means the Hill of Buddha's Footprints and the site's distinguishing features are the two symbolic footprints of Lord Buddha carved on stone. Other significant features in the site include the spectacular overhanging rocks and the prehistoric rock paintings.

14.3.2.3 Cultural Tourism

Thailand possesses a variety of unique and impressive cultural attractions. The most renown is Thai cuisine, regarded as one of the most popular, distinctive, and delicious cuisines in the world. In Thailand, it is not an exaggeration that delicious food can be found on almost every street corner. Therefore, tasting and enjoying Thai food is an unmissable activity for tourists. In 2011, CNN Travel (2011) released the list of the "World's 50 Most Delicious Foods," in which Massaman, a Thai sweet and spicy curry, was ranked number one. Apart from the first rank, there are three more Thai dishes mentioned in the list; they are Tom Yum Gung (ranked the eighth), Nam Tok Moo (ranked the 19th), and Som Tam (ranked the 46th). Pad Thai, even though not on the list, is also another well-known Thai dish.

As Thai culture is deeply associated with Buddhism, Buddhist temples are places of enormous meaning and consequence for Thais. With remarkable architectural styles, splendid figures of Lord Buddha, and distinctive rituals, Buddhist temples in Thailand also welcome and attract large numbers of tourists. In Bangkok, the must-see temples include Wat Phra Kaew, Wat Pho, Wat Arun, Wat Sutat, and Wat Benjamabhopit.

During the past decade, many local villages and communities developed accommodations for homestays, offering tourists an "authentic" travel experience in contrast to the more standardize resort-based tourism. These local villages and communities offer travelers an immersive experience in traditional Thai lifestyles, including lodging in traditional dwellings. In the remote mountainous regions of the north country, hill tribe villages have recently welcomed visitors as well. With their own distinct language and culture the hill tribes include the Akha, Lahu, Karen, Hmong or Miao, Mien or Yoa, Lisu, and Paluang. Local markets – especially the colorful floating markets or marketplaces where goods are sold from boats – attract not only international tourists but also numerous domestic daytrippers. These local markets sell a variety of delicious foods, fresh fruits, and exquisite handcrafts.

As agriculture forms the livelihood of most Thais providing the primary economic and cultural basis of the country, agricultural tourism is an emerging option for tourists. This type of tourism involves visiting agricultural sites, including paddy fields, crop fields, ranches, orchards, plantations, vineyards, and dairy farms. Agricultural tourism also provides a wide variety of activities, such as picking fruits, buying produce, and feeding animals. Because agricultural tourism in becoming more and more popular, farmers now offer opportunities for homestays at the farmsteads.

Among the heritage performing arts, none stands more memorable than the traditional Thai dance of graceful movements, exquisite costumes, and beguiling music. Bangkok offers a number of Thai dance venues and local troupes also offer performances across the country. For centuries, Thais trained and used elephants as beasts of burden. While concerns for animal welfare have curbed egregiously cruel practices and Thai NGOs assiduously protect elephants, trained elephant shows are popular in tourist destinations audiences as appreciate the elegance and intelligence of the giant animals.

Thais enjoy traditional festivals and ceremonies none more so than the Songkarn festival, or Thai New Year festival. Typically, Thais celebrate Songkran by pouring water onto the figures of Lord Buddha and also onto their parents, grandparents, and teachers in order to pay respect to them. In the last decade, international tourists have joined in splashing water on one another on the streets during Songkarn festival.

Because the Thai people consider the King the soul of the nation, traditional ceremonies related to the King and the royal family are of utmost significance. These ceremonies are exceptionally unique, elegant, and beautiful and include royal parades, boat regattas, and elaborate Buddhist ceremonies. As noted above, the TAT launched campaigns to invite visitors to come and join these magnificent royal ceremonies. Additionally, the spectacular Royal Palaces draw millions of both domestic and international visitors every year. In addition, royal projects are also interesting destinations. For the 70th anniversary of his reign, His Majesty King Bhumibol Adulyadej, along with Her Majesty Queen Sirikit and other members of the royal family, initiated thousands of projects to improve the living conditions of people throughout the country. Many of these projects are now open for visitors to learn about development concepts and projects initiated and developed by His Majesty King Bhumibol Adulyadej. Prominent among them are the sufficiency economy concept, alternative economic development, the artificial rain-making projects, the "monkey cheek" projects. Today, these concepts and projects are considered models for solving problems not only in the country, but applicable in other countries, such as Tonga and Jordan (Laylin, 2010; Narayan, 2016).

14.3.2.4 Special Interest Tourism

With a strategic location and rich tourism resources, Thailand has become a popular destination for business tourism and also for an emerging type of tourism known as MICE – meetings, incentives, conferences, and exhibitions or events. As Oxford Business Group (2016) noted, "Thailand's bevy of upscale accommodation, conventions centres and other business facilities, along with its agreeable climate and bountiful leisure opportunities make the country particularly well suited for the MICE sector."

Thailand generates a great deal of shopping tourism. Concentrated in Bangkok, a number of posh shopping centers, including Central World, Siam Paragon, Siam Center, Siam Discovery, and MBK Center, cater to wealthy tourist and Thai customers. Local night markets draw tourists, as well as serving locals, and some tourist resorts, such as Phuket, also provide upscale retail venues. During the past few years, the country has organized several sale and discount events to promote shopping tourism.

In Thailand, excellent facilities also generate sport tourism. Located throughout the country, golf courses not only attract visitors from all over the world, but also support business tourism. Muay Thai or Thai martial art of kickboxing is also very popular internationally, drawing many foreigners to watch and learn about this increasingly popular sport. Thailand also possesses facilities for recreational activities, which include museums, aquariums, zoos, parks, amusement parks, and so on.

Medical and wellness tourism in Thailand has been booming for decades. The campaign called "Visit Thailand, Enhance Your Healthy Life" has been launched by the Ministry of Public Health, in collaboration with the Ministry of Tourism and Sports, the Ministry of Foreign Affairs, and the TAT, to promote medical tourism in the country. A number of leading hospitals and medical centers have joined this campaign and offered a variety of medical treatments for foreign visitors, as well as for Thais. In addition to the medical treatments, Thai massage is also one of the popular health therapies in the world. There are abundance of hotels, resorts, and spas in Thailand that offer superb treatment and massage courses.

Although traditional Thai Buddhist and Muslim customs eschew alcohol consumption, a plethora of pubs, bars, and cabarets forms well-known part of the tourist experience, in both resort areas and Bangkok. Similarly, though prostitution is illegal, sex tourism, for which Thailand is extremely well-known among international travelers, flourishes in the tourist destinations in Thailand. The red-light districts such as Patpong in Bangkok, Pattaya in Chonburi, and resort, and Patong in Phuket are internationally famous. This presents a complicated governance issue for the tourism authorities, who do not officially acknowledge the extent of the sex tourism sector, much less assess its economic or social impact, even though it undoubtedly forms a significant component in the tourism industry. Governance of sex tourism is largely in the hands of local police, who confine red light districts to certain designated areas, and resort owners, who pointedly distinguish their establishments as accepting, or not, of sex workers who accompany tourist clients during their stays.

14.4 Thailand's National Tourism Development Plan

In 2012, the Ministry of Tourism and Sports developed the first National Tourism Development Plan (2012–2016) in order to guide tourism development policies and actions in the country. The aims of this Plan was to raise the quality of tourism offerings, keep the balance of demand and supply, and move up the country's tourism competitiveness. Specifically, the Plan aspired "to place Thailand within ranks one to seven within Asia, to increase revenues from tourism by not less than 5% and to develop eight new tourism segments based on geographic region" (Schuckert & Wattanacharoensil, 2014, p. 15). Delightfully, as stated in the introduction of the Second National Tourism Development Plan (2017–2021), "Upon the conclusion, NTDP 2012–2016 had successfully developed the tourism market, resulting in tourism receipt growth of more than 15%, a beyond-target development of tourism competitiveness" (The Ministry of Tourism and Sports, 2017, p. 4).

The Second National Tourism Development Plan (2017–2021) was launched in March of 2017. This Plan proposed five strategic axes (The Ministry of Tourism and Sports, 2017, p. 2):

- Strategy 1: Development of tourism attractions, products and services including the encouragement of sustainability, environmental friendliness, and the integrity of Thainess in attractions.
- Strategy 2: Development and improvement of supporting infrastructure and amenities without inflicting negative impact to the local communities and environment.
- Strategy 3: Development of tourism human capital's potential and the development of tourism consciousness among Thai citizens.
- Strategy 4: Creation of balance between tourist target groups through targeted marketing that embraces Thainess and creation of confidence among tourists.
- Strategy 5: Organization of collaboration and integration among public sectors, private sectors and general public in tourism development and management including international cooperation.

Apart from these five strategic axes, the Second Plan also set Thailand's tourism vision towards 2036 in order to be a framework for driving the tourism growth of the country. The vision says: "By 2036, Thailand will be a World's leading quality destination, through balanced development while leveraging Thainess to contribute significantly to the country's socio-economic development and wealth distribution inclusively and sustainably" (The Ministry of Tourism and Sports, 2017, p. 12). Accordingly, this vision indicates five key essences which set a clear tourism goal for the next 20 years. These five key essences include (The Ministry of Tourism and Sports, 2017, p. 14):

A. Achieve leading quality destination through the development of tourism products and services; the enhancement of Thailand's tourism competitiveness; and the increase of tourism receipt per head by extending the length of stay and the spending per day

B. Create balanced tourism development through the balance of tourist segments (e.g. between international and domestic, between mass and niche, and among different origins); the balance of tourist areas by improving of second-tier locations and local areas; and the balance among tourism periods and seasons by scattering activities throughout the year
C. Leverage Thainess through the development of tourism offerings organized around Thai culture, heritage, and uniqueness; the raise of understanding, awareness, and appreciation of Thainess among tourists and Thais
D. Contribute to the country's socioeconomic development and wealth distribution inclusively through the development of tourism industry as one main source of income and as key driver of infrastructure improvement and socio-economic opportunity development; and the creation of regional tourism, especially in second-tier cities
E. Promote environmental and cultural sustainability through the preservation of fragile attractions and the local heritages

14.5 Conclusion

Thailand is a geographically small country with a remarkable array of unique tourist features appealing to global travelers – spectacular beaches, historic artifacts, cultural monuments, distinctively delicious cuisine, religious shrines, scenic landscapes, and all manner of resorts. In addition, special interest tourist services for sex, medical care, and shopping add to this mix. The government has devoted resources to tourism promotion since the 1920s but has significantly fostered the aggressive and expansive governance of tourism since the 1970s. This has encompassed planning, education, marketing, and infrastructure development. Because of the array of tourist draws, almost no region of the country has been completely unaffected by tourism, though tourism does concentrate in particular locales. Tourism now forms a significant part of the GDP, over 20% and rising, and Thailand often ranks in the top ten of global tourist arrivals. By sheer economic measures the tourism industry is an astounding success.

This is not accidental but an outgrowth of geographic, cultural, and administrative legacies. Thailand is blessed with a highly scenic landscape whether the white sand beaches, startling blue waters, and rocky upthrusts of the Andaman coast, the emerald rice paddies of Thailand's agricultural heartland, or the lush forests of the highlands. Thai culture practices and artifacts, have flourished for centuries and, in particular, withstood the disruptive occurrences of 20th century history through canny geopolitical positioning, the endurance of the Thai monarchy, and an unflagging devotion to Thai Buddhism.

How these geographic and cultural legacies blossomed into the economic juggernaut of the contemporary status of Thai tourism had everything to do with tourism governance. Tourism has ranked as a cabinet level concern since the 1930s and had a ministerial level governing organization by 1960. Recognizing the need for both a government agency integrated into the bureaucracy and the flexibility of a government-owned enterprise, the Thai government established the separate corresponding Department of Tourism (DOT) focused on development of both sites and businesses and the Tourism Authority of Thailand (TAT) focused on marketing and promotion. Both the DOT and TAT work closely with industry representatives through Tourism Council of Thailand (TCT).

During the past three decades, the marketing of tourism engaged the combined attractions of both sites and culture, spinning Thai geography and practices into an irresistible draw for global tourists. The governance of Thai tourism has ensured that an enormous variety of experiences and accommodations, from homestays to the highest end luxury resorts, meet the demands of global travelers. Over the decades, the government has adjusted its polices to meet whichever shifts in preferences and country of origin might occur. The latest tourism policies reflect the increasing potential of the Asian market, particularly China, the interest in spreading the tourism wealth more evenly across the country, and the intent to focus on higher end tourists concentrating on selective tourists with the potential to spend more per capita. The plans also acknowledge that a number of aging tourist facilities no longer serve the tourist the government sees as optimal, and renovation of "down-market" obsolete facilities forms an important part of the tourism future of Thailand. The plans expect tourism to rise to about 30% of GDP in the next decade.

The strategic planning for tourism in Thailand does not consider several looming issues. It does not coordinate between land use planning for tourism in conjunction with local needs, a woeful oversight. The enormous environmental costs both to local communities and the planet as a whole (especially through the energy consumption) do not figure in the calculations for the future. While Thai local culture and social networks have proven resilient in the face of change, social disruptions are increasingly apparent in some communities. Sea level rise acutely threatens the core beach tourism areas and predictions indicate that Bangkok, as an inland delta, will be significantly inundated. While sea level rise concerns other sectors of the Thai government and NGOs, it has not entered into policies regarding tourism. All of these aspects related to tourism will need adequate national resources to address but, thus far, are not incorporated into the tourism governance policies and goals.

Coordination between tourism planning and land use planning presents an obvious area for improved tourism governance, especially in the face of social transformations and climate change. While, inarguably, tourism provides a fundamental level of economic prosperity for a substantial swath of Thais at every level of the economic spectrum, local impacts and global forces may outpace these benefits in the next generation. Whether the Thai tourism governance "hope-for-the-best" approach, which, to be sure, fundamentally reflects the Thai Buddhist attitude of acceptance, will prevail as wise policy remains to be seen.

References

CNN Travel. (2011). *World's 50 Best Foods.* Retrieved from http://travel.cnn.com/explorations/eat/worlds-50-most-delicious-foods-067535/. Accessed November 1, 2016.

Hedrick-Wong, Y. (2012). *MasterCard Global Destination Cities Index.* Retrieved from http://newsroom.mastercard.com/wp-content/uploads/2012/06/MasterCard_Global_Destination_Cities_Index_2012.pdf. Accessed November 1, 2016.

Hedrick-Wong, Y. & Choong, D. (2013). *MasterCard Global Destination Cities Index.* Retrieved from http://newsroom.mastercard.com/wp-content/uploads/2013/05/Updated-Mastercard_GDCI_Final_V4.pdf. Accessed November 1, 2016.

Hedrick-Wong, Y. & Choong, D. (2014). *MasterCard 2014 Global Destination Cities Index.* Retrieved from https://newsroom.mastercard.com/wp-content/uploads/2014/07/Mastercard_GDCI_2014_Letter_Final_70814.pdf. Accessed November 1, 2016.

Hedrick-Wong, Y. & Choong, D. (2015). *MasterCard 2015 Global Destination Cities Index.* Retrieved from https://newsroom.mastercard.com/wp-content/uploads/2015/06/MasterCard-GDCI-2015-Final-Report1.pdf. Accessed November 1, 2016.

Hedrick-Wong, Y. & Choong, D. (2016). *Global Destination Cities Index by MasterCard (2016).* Retrieved from https://newsroom.mastercard.com/wp-content/uploads/2016/09/FINAL-Global-Destination-Cities-Index-Report.pdf. Accessed November 1, 2016.

Jamrisko, M. & Saraiva, C. (2018). *These Are the World's Most Miserable Economies.* Retrieved from https://www.bloomberg.com/news/articles/2018-02-14/most-miserable-economies-of-2018-stay-haunted-by-inflation-beast. Accessed October 21, 2020.

Laylin, T. (2010). *Thailand to Help Jordan Make Artificial Rain.* Retrieved from http://www.greenprophet.com/2010/09/thailand-artificial-rain-jordan/. Accessed November 1, 2016.

Mastercard. (2019). *Global Destination Cities Index 2019.* Retrieved from https://newsroom.mastercard.com/wp-content/uploads/2019/09/GDCI-Global-Report-FINAL-1.pdf. Accessed November 7, 2020.

MasterCard Worldwide. (2011). *MasterCard Index of Global Destination Cities: Cross Border Travel and expenditures.* Retrieved from http://www.moodiereport.com/pdf/MasterCard_Global_Cities_Report.pdf. Accessed November 1, 2016.

Narayan, R. (2016). *Tonga's Royal Estate to Pilot an Agricultural Project Initiated by the Thailand Royal Family.* Retrieved from http://www.looptonga.com/content/tonga%E2%80%99s-royal-estate-pilot-agricultural-project-initiated-thailand-royal-family. Accessed November 1, 2016.

Oxford Business Group. (2016). *Business tourism on the rise in Thailand.* Retrieved from https://www.oxfordbusinessgroup.com/analysis/dark-horse-full-potential-business-tourism-has-yet-be-realised. Accessed November 1, 2016.

Saraiva, C. & Jamrisko, M. (2016). *These Are the World's Most Miserable Economies.* Retrieved from https://www.bloomberg.com/news/articles/2016-02-04/these-are-the-world-s-most-miserable-economies. Accessed November 1, 2016.

Saraiva, C. & Jamrisko, M. (2020). *U.S. Worse Off Than Russia, Mexico in 2020 Economic Misery Ranking.* Retrieved from https://www.bloomberg.com/news/articles/2020-08-06/misery-ranking-will-show-u-s-getting-worse-versus-rest-of-world. Accessed October 21, 2020.

Schuckert, M. & Wattanacharoensil, W. (2014). *Reviewing Thailand's Master Plans and Policies: Implications for Creative Tourism? Current Issues in Tourism*, 2014. DOI: 10.1080/13683500.2014.882295

Singhapreecha, C. (2014). *Economy and Agriculture in Thailand.* Retrieved from http://ap.fftc.agnet.org/ap_db.php?id=246. Accessed November 1, 2016.

The Ministry of Tourism and Sports, Thailand. (2017). *The Second National Torism Development Plan (2017–2021).* Bangkok: Ministry of Tourism and Sports.

Tourism Authority of Thailand (TAT). (2019). *About TAT*. Retrieved from https://www.tourismthailand.org/About-Thailand/About-TAT. Accessed July 20, 2019.
TAT News. (2016). *Tourism Authority of Thailand: Past and Present*. Retrieved from http://www.tatnews.org/history/. Accessed November 1, 2016.
Tourism Council of Thailand (TCT). (2016). *About TCT: Background – Objective*. Retrieved from http://www.thailandtourismcouncil.org/en/about.php. Accessed November 1, 2016.
UNESCO. (2016a). *Phimai, its Cultural Route and the Associated Temples of Phanomroong and Muangtam*. Retrieved from http://whc.unesco.org/en/tentativelists/1919/. Accessed November 1, 2016.
UNESCO. (2016b). *Phuphrabat Historical Park*. Retrieved from http://whc.unesco.org/en/tentativelists/1920/. Accessed November 1, 2016.
UNWTO. (2014). *UNWTO Tourism Highlights: 2014 Edition*. Retrieved from http://www.e-unwto.org/doi/book/10.18111/9789284416226. Accessed November 1, 2016.
UNWTO. (2015). *UNWTO Tourism Highlights: 2015 Edition*. Retrieved from http://www.e-unwto.org/doi/book/10.18111/9789284416899. Accessed November 1, 2016.
UNWTO. (2016). *UNWTO Tourism Highlights: 2016 Edition*. Retrieved from http://www.e-unwto.org/doi/book/10.18111/9789284418145. Accessed November 1, 2016.
UNWTO. (2019). *International Tourism Highlights: 2019 Edition*. Retrieved from https://www.e-unwto.org/doi/pdf/10.18111/9789284421152. Accessed November 7, 2020.

Ines Mestaoui & Amira Benali

15 Tunisia

15.1 Introduction

Tourism is a fundamental component of many countries' economic systems. In Tunisia, ever since the mid-1950s where independence was obtained, governmental efforts were registered allowing tourism to occupy a prominent place in the country's economy. The Ministry of Tourism was founded three years after the declaration of the Republic of Tunisia in 1956. Bourguiba, the first president of the country, appointed Mohamed Masmoudi Minister of Tourism in 1960. Several ministers have succeeded ever since to reflect the vision of the government in power deploying different strategies rendering Tunisia a competitive tourist destination by the second half of the 20th century. However, recent challenges have emerged in post-revolutionary Tunisia and made the Tunisian economic system a complex one. In this chapter, we aim to unravel this complexity by investigating these challenges and the related impact on the touristic industry. We will start by briefly presenting the Tunisian cultural and economic contexts. We will trace then the evolution of the tourism sector highlighting the key periods that marked Tunisia's history. We will critically analyze how the related institutions, especially the government, are promoting tourism and conclude by proposing different forms of the latter.

Tunisia is an independent, North African country located in the Mediterranean. With an area of 163,610 km2, it is the smallest country in the Maghreb. Despite its small area, its strategic geographical location made it the main gateway for many conquests (Séguin & Guy, 2011). While historians trace the native people of Tunisia to the Amazigh (Cheriguen, 1987), the country is a mosaic of ethnicities that reflect its history (Auzias & Labourdette, 2009). The cosmopolitan character of Tunisia is the result of several civilizations' succession (Dridi, 2006). Punic, Roman, Vandal, Byzantine, Aghlabid, Fatimid, Hafsid, Turkish, Husseinite, and French have conquered the country before its independence in 1956. Its museums, archaeological sites (e.g. Temples, coliseums, forts), palaces, medinas, mosques, along with picturesque and Berber villages in the desert are testimonials of great civilization. This heritage is the great value of Tunisia (Boulaares, 2011). For instance, Carthage, the Punic city, is known worldwide and constitutes a primary motivation to visit the country. While cultural heritage is an essential asset of the country, tourists come for many other reasons. In the following section, we will present how Tunisian tourism has evolved through the years.

Tourism in Tunisia represents 13.8% of the GDP, which outline the importance of this sector, and the will to build strategies that continually boost the tourist activity. The official representative is the Ministry of Tourism, which implements government policy in tourism and recreation. Several organizations are attached to the

https://doi.org/10.1515/9783110638141-015

latter: Tunisian National Tourist Office (a Tunisian public establishment responsible for promoting tourist activity in Tunisia), Hydrotherapy Office (relating to the various spas), Tourist Land Agency (relating to the investment in the tourism sector), Company promogolf Hammamet (Management and operation of a golf club of Hammamet), Company promogolf Monastir (Management and operation of a golf club of Monastir), Company promogolf Carthage (Management and operation of a golf club of Carthage), Tourist leisure company (hotel management company). The diagram of the governance structure of the Ministry of Tourism is presented below in figure 15.1.

15.2 Historical Analysis of the Sector

The first Ministry of Tourism was created in the early 1960s (Belhedi, 1999) in Bourguiba's governance. Tourism was a driver of development after Tunisian independence. In 1962, Tunisia recorded only 52,000 tourist entries (Mzabi, 1977). 20 years later, non-resident entries exceeded 1 million visitors to reach 1.3 million in 1982 (Abdallah and Adair, 2012). During this period, the primary infrastructure was achieved. Noticeably, the resorts and hospitality venues were concentrated in Sahel's region (Jedidi, 1986). The government has focused on the promotion of the country as a low-cost seaside destination. The hotel infrastructure was slowly taking shape. Between 1962 and 1986 the number of tourist hotels was multiplied by six and the accommodation capacity in beds by 24 (from 4000 beds in 1962 to 80,000 beds in 1982), encouraging tourists to consider Tunisia as a Mediterranean destination (Belhedi, 1999). Therefore, the number of tourist nights rose by 21 (from 64,000 to 13 million) (Miossec, 1996). The number of foreign tourists has grown in the same proportions: 52,000 in 1962 to 1.5 million in 1986 (Belhedi, 1999). The tourism industry has thus contributed to the economic and social growth of Tunisia.

15.2.1 Tourism in the Late 20th and Early 21st Centuries

During these years, Tunisia became a popular destination for mass tourism (El Bekri, 2013). For instance, it recorded 4.8 million tourist arrivals in 1998, 5 million tourists in 2002, 6.5 million tourists in 2006, and 7 million tourist admission in 2008) (Othmani and Dhaher, 2018). These were immediate consequences for charter flights, all-inclusive accommodations, and group visits including a quick exploration of Tunisia's key locations. Tourism investments have increased to 300 million dinars (1990–2000s) (Gan & Smith, 1992). The Tunisian State invested in recreational centers and marinas. Several airports were constructed in: Djerba, Tabarka, Enfidha, Gafsa, Tozeur Nafta, Sfax,

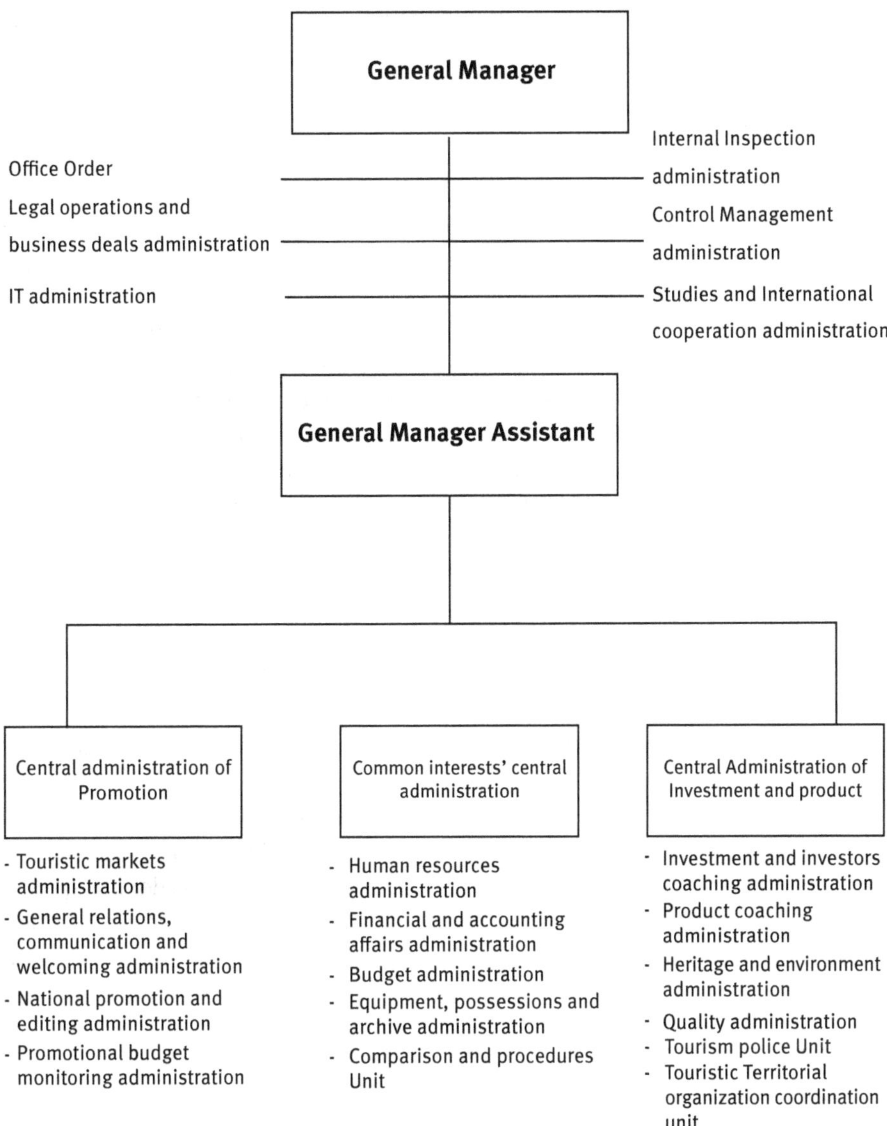

Figure 15.1: Tourism governance structure of the Ministry of Tourism adapted from 2011 annual report of National Office of Tunisian Tourism.

and Monastir to facilitate the reception, transportation, and movement of foreign visitors.

The Tunisian government also encouraged tourist promoters to invest in large hotel complexes, entertainment, and wellbeing spaces. For example, the number of hotels has evolved from 600 establishments in the 1980s to 800 in the 2000s (Khlif, 2006).

The Ministry of Tourism has continued to encourage seaside tourism boasting the beauty of beaches and the affordability of Tunisian holidays. Thus, the village of Sidi Bou Said was considered as a main attraction for visitors. However, in the 2000s, tourism experts began to see cultural tourism (a form of tourism that promotes cultural heritage) as an excellent development opportunity. Thus, it has deployed extensive cultural tourism efforts as a new niche segment. For instance, tours in the archaeological site of Carthage along with ones in the capital are organized by Jamila Binous, a Tunisian historian and urban planner, to discover the medina and its cultural sights (e.g. traditional markets or "souks," palaces, Zitouna mosque, the cathedral of Tunis). Seaside resorts, namely in Hammamet, Sousse and Monastir, have also emerged during recent years and have continued to grow. It is noteworthy that the north-west of Tunisia and many other regions in the center and the South, were less involved in the tourism development plan despite their touristic potential in terms of attractive natural sights, cultural experiences and leisure activities. This difference reflects the very central strategy of both the government of Bourguiba and Ben Ali, former presidents of Tunisia. Moreover, it led the country to enormous socioeconomic regional disparities that were the revolution's premise in 2010.

15.2.2 The Spring Revolution and its Impact on Tunisian Tourism

Tunisia has witnessed during this last decade a prominent political and social transformation. The revolution fueled by aspirations for social justice and democracy by the Tunisian youth has brought massive changes. The country was under a dictatorship that began just before the era of President Ben Ali and reached its peak during the latter. The Tunisians have been emancipated. They won the freedom of speech. They began to vote and to determine the political landscape. The latter has considerably changed. Nevertheless, Tunisia is now experiencing the most significant political, security, and socioeconomic crisis since independence in 1956 (Kasmi, 2014). Notably, its manifestations are traced in the multiplication of terrorist acts even ones including tourists: Bardo Museum attacks (in March 2015 and which left 21 victims) and those of Sousse (which took place on June 2015 and which had targeted 39 tourists) that have dramatically affected the tourism industry (Tamzini, 2013).

Tunisia has become a risky country and thus lost a million foreign visitors, through a drop of 17.7% in tourist arrivals at Tunisian borders (Becheur, 2011). The Tunisian government expressed a local and national determination to arise from this crisis. To "save" the tourist seasons, it has appointed different ministers after the revolution with a new political vision aiming to encourage national tourism and interfering to support the tourism sector. The government also declared emergency measures consisting of help, assistance, and contributions (money donations to hotels and the local flight company) to save the system and preserve jobs.

Awareness-raising campaigns encouraging Tunisians to spend their holidays in Tunisia have succeeded (e.g., the I love Tunisia campaign promoting Tunisia as a secure, warm, welcoming and hospitable destination). Although many hotels closed down, several others have opened and provided a premium service quality for Tunisians and foreign visitors. There was a time when Tunisian hotels were designed only for foreign visitors considering their needs, traditions, and usages as a priority. As the context has already changed, the hotels have become more adapted to their main customers' requirements; the latter recently considered as a combination between Tunisians and domestic tourists.

Today, the global economic situation made Tunisia rethink its strategy, trying to diversify its tourism offer. The latter was long positioned as a mass tourism in Tunisia and demonstrated great success up until the revolution. However, mass and low-budget tourism brought the system to its limits, particularly during the security crises experienced after the revolution. Given its rich cultural heritage, the crisis situation during the revolution can be a good opportunity for reflection and more thoughtful yet less short-term strategies.

15.2.3 Tourism Promotion and Image of Tunisia as a Destination

Although terrorism alongside social and political instability have undermined Tunisia's image, the Ministry of Tourism spared no effort to promote the country as a dreamy touristic destination. Promotional campaigns are broadcast on Tunisian television to make Tunisia known as a unique and singular destination. Besides, many Tunisians share videos promoting cultural, historical, and natural assets on social media. Several ministers have succeeded since the revolution until today: Mehdi Houas (Minister of Trade and Tourism in 2011), Elyes Fakhfakh (in charge of the Ministry of Tourism from 2011 to 2013), Jamel Gamra, Amel Karboul (2014–2015), Salma Elloumi Rkik (from 2015 to 2018), René Trabelsi (from 2018 to February 27th, 2020). All aimed to boost tourism following the difficulties encountered and experienced after the revolution.

To keep its finger on contemporary society's pulse, Amel Karboul and René Trabelsi, took the tour of several events while personally promoting these events on their accounts on Facebook, Linkedin, and Twitter. They share selfies and videos that reflect the ministry's efforts to revive special tourism after tough times. These two ministers took up their posts just after the national security crises (the terrorist attacks in Bardo and Sousse) and their main missions were to save the tourist seasons. These two ministers also oversaw promotional campaigns for Tunisia. The latter enhance the heritage of the various regions of Tunisia and promote the security of the country. They then encourage Tunisians to share these videos on social networks to reach the campaign better and reassure and make foreigners want to visit Tunisia.

Tourism is a transversal sector of primary importance for the Tunisian economy. Its influence on other activity sectors remains considerable. National security problems were undermining tourism. The relevance of tourism in the GDP (13.8%) left no choice for the Tunisian government to spare no effort to revitalize the most important economic sector. After the revolution, all newly appointed ministers took office amidst crises where their primary mission was to save the summer tourist season. While the previous government had invested all efforts on mass tourism, the country lost its tourists who did not feel safe anymore. Thus, tourism experts sought to incite people to discover and value a unique and authentic Tunisia (Esmail, 2016; Mansfeld and Winckler, 2015; Selmi, 2017). The minister of tourism started putting many efforts to communicate what makes Tunisia a unique destination throughout the year. Examples vary. For instance, the ministry committed to participation in international events (in Madrid, Germany, Paris, Tokyo, Moscow, London, Geneva, etc.). The Ministry of Tourism also initiated many festivals such as El Jem Symphonic Music Festival, Tabarka International Jazz Festival, Musical October, Haouaria Sparrowhawk Festival, and International Oasis Festival.

The website of the Ministry of Tourism is the outward-facing platform for everything concerning tourism in Tunisia. Various bilateral and multilateral alliances and co-operations that denote political and economic relations negotiated and implemented from independence to the present day are presented there. We can also access pamphlets detailing the key places to visit and their main features. The brochures are in French and classified by preselected forms of tourism such as the cultural and the Saharan ones. The website of the Ministry of Tourism also includes the various organizations under the supervision of the Ministry of Tourism, such as the National office of the Tunisian Tourism (ONTT), the National Office of Tunisian Crafts, the Tourism Real Estate Agency, and the Tourist Leisure Company. The Ministry of Tourism is in charge of administering and governing the various organizations and stakeholders in the tourism industry. The ONTT also provides to its visitors a website that promotes the different Tunisian regions and gives to its visitors all the activities to do in Tunisia and the various events that might interest them.

15.3 The Tourism Sector in Tunisia

Tourism key performance indicators (KPIs) have been affected by the political instability after 2011. In 2018, the tourism sector represents 13.8% of the GDP of the Tunisian economy. Tourism income in 2018 registered a 45% increase since 2010. According to the Tunisian National Tourism Office (ONTT), Tunisia wrote 8.3 million tourist entries in 2018. About 25% of these tourists are from Algeria, with more than 2 million in 2018 (double that of 2010). About 29% of tourists are from Europe, with a total of 2.7 million in 2018. Despite an increase of 42.3% compared to 2017, this number still far from the 3.82 million European tourists before 2010 (Figure 15.2). French tourist

Entries had increased by 37.4%, Germans by 52.4%, Russians confirming themselves as an emerging market for Tunisia (589,424 tourist entries). Russians admissions increased by 16.2% in 2018 compared to 2017.

On the other hand, Tunisians living abroad are a new category that had not been counted as tourists in 2010 until very recently. More Tunisians are now staying in hotels and guest houses during their holidays. Their massive influx of 1.5 million visitors in 2018 and their participation in Tunisian tourism stimulation as an essential source of foreign currencies has changed the situation.

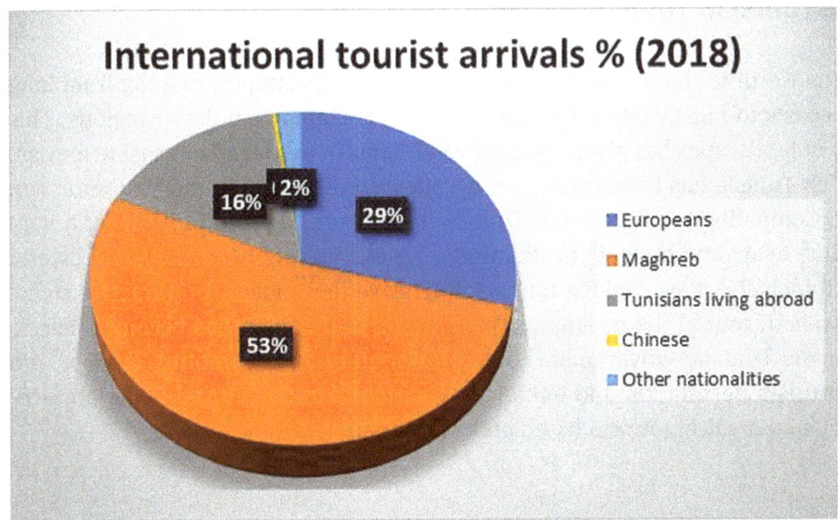

Figure 15.2: Percentage of international tourist arrivals in Tunisia in 2018. (Source: www.ontt.tn/ar).

In the following paragraph, we try to focus on the KPIs of 2018-2017-2016. We then compare the latter with those of 2010 (the year of the Tunisian revolution) and the stability highlighted during 2007–2008–2009. According to the Ministry of Tourism's website: Arrivals at Tunisian borders were 8.3 million tourists in 2018. In 2017, 7 million tourists arrived at our borders, and we recorded 5.7 million tourist arrivals in 2016. A gap of 15% is observed compared to 2017 and of 31% compared to 2016. Revenues recorded in MD were of 4,141.2 MD in 2018. In 2017, revenues were 2.831 MD, and in 2016 we recorded 2,322.9 MD of revenues. The difference recorded was of 31.64% compared to 2017 and of 43.9% compared to 2016.

The gap recorded between 2018 and 2010 (we chose to compare with 2010 because it is the year of the revolution beginning, which recorded 7.8 million) is 5.7%. The latter shows that the efforts made to promote tourism activities were beginning to bear fruit. Tunisia is indeed starting to report the same figures as the stability year: 2010.

216.3 million dinars of tourism investments were spent during the year 2018 for the promotion of tourism. This figure had fallen compared to 2017 when the government devoted 477 billion dinars to boost tourism activity. Likewise, 240 billion dinars were spent in 2016 to revive tourism activity. Large budgets to demonstrate the importance of tourism in the country's economy and the will to build viable strategies to overcome the problems encountered following the terrorist attacks in Bardo and Sousse and the impact that followed on the domain.

15.3.1 Tunisian Tourism Forms

From the North to the South, Tunisia offers various landscapes, ranging from long sandy coasts to northwestern (mountains, valleys, plains) and desert regions. This variety of landscapes has given rise to several forms of tourism. The seaside tourism on which Tunisia has been betting for decades is not unique in the Mediterranean region. Competition is fierce, and Tunisia has chosen to position itself for a very long time as a "low budget" destination with exclusively beautiful beaches and offer all-inclusive stays making tourists only leave their hotels for attractive sights and two-hour tour of the medina of Tunis or Sidi Bou Said. It has become a necessity for the Tunisian government to stop presenting and promoting Tunisia as a mass tourism destination and enhance its various forms: cultural, Saharan, eco-tourism, and medical tourism based on its large variety of landscapes.

15.3.1.1 Seaside Tourism

For many years, Tunisia was promoted as a low-budget Mediterranean destination. Its extraordinary beaches offer a sunny and relaxing experience for tourists. The Ministry of Tourism has invested in several seaside resorts in different cities to provide a welcoming and convenient infrastructure for tourists. Guesthouses, which offer several authentic guest rooms in an excellent location, and provide personalized service, have also emerged in the central coastal towns. We found most of them in the region of the Cap Bon, Tabarka in north-west, and Djerba in the south. Hence, Seaside tourism was and remained the mainstream tourism in Tunisia.

15.3.1.2 Cultural Tourism

As explained in previous sections, the archaeological heritage of Tunisia is rich and varied. For instance, Tunisia is home to seven sites inscribed on the World Heritage List namely El Jem Amphitheater (1979), Kerkouane Punic City and its necropolis

(1985,1986), Dougga / Thugga (1997), Kairouan (1988), Medina of Sousse (1988), Medina of Tunis (1979), and the Archaeological Site of Carthage (1979).

The medina of Tunis is among the best-preserved medinas of the Arab world. The Ministry of Culture and Heritage Conservation oversees various related renovations and make sure of its safeguarding. It is in charge of directing and controlling the different projects because some sites (contemporary inhabited houses included) are considered as an architectural inheritance. For example, the houses nearby are not allowed to do any repair work. They have to submit their requests to the ministry.

Since the 2000s, the ministry has been promoting initiatives to promote cultural tourism, noticeably the medina of Tunis. Charming Hotels (authentic houses with a unique traditional architecture, very representative of the region) such as Dar El Djeled Hotel and Dar El Medina Hotel and guest houses (Dar Hayder, Dar Traki) have been created to offer tourists the opportunity of charming houses and accommodation in the heart of the Medina of Tunis. Tours are organized in the medina of Tunis to discover museums, traditional markets and souks, shops, perfumers, mosques, minarets and some typical prestigious houses such as Dar Ben Abdallah. These tours are organized by tour guides or by historians such as Ms Jamila Binous. In the same line, tours are organized to visit the archaeological site of Carthage, the Museums of Carthage and Bardo.

Furthermore, Cultural tourism is being promoted in other parts of the country. For example, the ancient sites in Utica, Dougga/ Thugga, Bulla Regia, Thuburbo Majus, Oudhna, the Great Mosque of Kairouan, the Aghlabid basins, the amphitheater of El Jem, the temple of Chemtou, the fort of Djerba, Tabarka and Bizerte and The Ribat of Monastir are so many sites that reflect the rich and diverse heritage of Tunisia and the splendor of its history.

15.3.1.3 Saharan Tourism

The Tunisian Sahara is the Southern and Western part of Tunisia it cumulates a total of 90,000 km2 constituted by the arid steppes of the West and the deserts of the South. In addition to sand's dunes, rocky plains and oasis, the region is known for its salty lake Chott jerid. While most of the tourists (both local and international ones) are intrigued by the natural landscapes. The region is also rich in cultural and historical sights and heritage. It is a testimony of the Amazigh traditions, from the Grand Erg oriental to the arid mountains of Dahar (displayed through the Ministry of Tourism website). For instance, the city of Matmata is very famous for its troglodyte habitations. Moreover, the ksours of Chenin in Tataouine are a unique historical and architectural heritage characterizing the region. Two of the most touristic places are the cities of Tozeur and Nefta. Their houses built in brick are considered to be exceptional examples of the old Iraqi cities and serve as a tool of reservation of such ancient cultures. Furthermore, these two cities have the leading hotels and tourist attractions in the region. Accordingly, the Ministry of Culture organizes annual festivals in order to attract visitors both local and international ones. Some examples are a great related testimony: the International Festival of Saharan Ksour (organized annually in March in Tataouine), the Festival of Hamma (organized on the period of March-April in Gabes), the Festival of Oases (in March in Tamerza), the International Sahara Festival of Douz (organized annually in December in Douz), and the International Oasis Festival (organized annually in December in Tozeur).

In addition to cultural events, sports activities are also organized every year in the region and can attract many tourists. We cite the Rallye of Tunisia since 1981 and more recently in 2017, the Ultra Mirage El Djerid which is the first Ultra trail of 100 km taking place and entirely organized by Tunisians. These different events and many others denote the efforts in promoting Saharan tourism in Tunisia both for national and international tourists.

15.3.1.4 Ecotourism

Since 1980, the Ichkeul National Park has been classified as a UNESCO World Heritage Site. Tunisia has many natural parks and natural reserves. Accordingly, this has significantly driven the ecotourism or green tourism, especially after the revolution. Indeed, the national and international tourism is evolving towards a family of green and ecological circuits. Individuals in charge are now making different regions' natural wealth available to everyone: Parks, reserves and forest circuits are available for the discovery. Groups of active members have been formed to discover every place of Tunisia. They go for hiking and discovering authentic places of deep Tunisia. Some guesthouses are even developing this concept targeting foreign tourists: Staying away from the tourist area, in guesthouses that offer a different kind of activities such as bicycle tours, hiking in typical places, eating and buying local food and products, such as the Tunisian honeybee, cheese made by local people, and Tunisian artisans.

Dar El Ain is an example of ecotourism promoters. They invite tourists to stay in the heart of Tabarka (a city in the North-West of Tunisia) and offer them the opportunity to stay nearby the local population and to the benefit of activities in the valleys and green circuits, including hiking and camping.

15.3.1.5 Medical Tourism

Medical tourism is considered as one of the most important sectors of the Tunisian economy (Hallem & Barth, 2011). Tunisia is known worldwide for the quality of its medical care services and its unbeatable prices, especially in plastic surgery (Holliday, Hardy, Bell, Hunter, Jones, Probyn, & Taylor, 2013). Over the years, the country has built a reputation for low-cost plastic surgery. More particularly, medical and aesthetic interventions are twice to three times less expensive than in Europe. The medical tourism agencies (or aesthetic tourism) offer a special kind of holiday for patients: cheaper medical procedures, serious follow-up and post-surgery, medical assistance, transport, and small tours to discover some key places and five-star hotel accommodation.

These medical stays at lower prices have made the success of Tunisia. Although medical tourism is not just about plastic surgery stays. Having medical professionals among the most competent in the world, clinics flourished in Tunisia, offering their services to neighbor countries, particularly Libyan and Algerian, among others. Libyan patients represent the majority of patients in Tunisian clinics (Rouland et al., 2016). They seek medical journeys along with relaxation practices and techniques for their mental health as well.

Besides these forms of tourism, the government promoted other forms, mainly internal ones offering promotions to those living abroad to visit their country during their vacation. As we already mentioned, Tunisians living abroad are a new category that had not been counted as tourists in 2010 until very recently. More Tunisians are now staying in hotels and guest houses during their holidays. Their massive influx, estimated at around 1.5 million visitors in 2018, and their Tunisian tourism stimulation participation as an essential source of foreign currencies have changed the situation.

15.4 Conclusion

In this chapter, we tried to give an account of the tourism industry in Tunisia. Particularly the importance of this sector in the economy as well as the governance aspect. Indeed, while the Ministry of Tourism determines tourism development policy, other ministries are also actively participating in the industry. For instance, through the organization of several cultural events (festivals and concerts), the Ministry of Cultural affairs participates in his turn in foreign tourists' attraction and the promotion of Tunisia and tourism sector.

The government should opt for a better use of the Tunisian territory. As past experiences illustrated, an anarchic recovery after a crisis cannot lead to significant results. It is necessary to coordinate, organize and better manage tourist flows: a focus on customers, a sought offer diversification, and a more balanced distribution throughout the territory. The government may also create and enhance the Tunisian Paratourism (additional activities complementary to tourist activities). Tunisia lacks entertainment and events that can arouse tourists' interest and further immerse them in local traditions and customs. The government should encourage private projects that can enhance its historical and cultural heritage. The previous crisis also taught us that Tunisian tourism should not exclusively depend on the European market. The government can provide more diversified offers for Algerian and Libyan tourists and not limit them with all-inclusive seaside stays in hotels. It can also promote internal tourism and promote guest houses, personalized holidays, and family moments of relaxation in the sun for Tunisians living abroad.

Tunisia must capitalize on the past successful experiences (and subsequently the management of previous crises). There is a real desire for Tunisian tourism players to get out of this crisis and build long-term, persistent and viable strategies. Even if they are challenging to experience, the various national crisis can be a moment of deep reflection and an opportunity to think about Tunisian touristic problems differently and, above all, to pose radical solutions enhancing tourism and the Tunisian economy. Several ministers were successive and most of them are part of a handover government. Since they only hold their posts for a few months, the goal was always a short-term one. They seek to save the tourist season and put the efforts on "numbers" to be reached rather than on the quality of the tourists solicited. The solution in Tunisia lies in transforming these aspirations into actions and long-lasting realities through hard work and persistence.

References

Auzias, D., & Labourdette, J. P. (2009). Petit Futé Tunisie. Petit Futé.

Abdallah, A., & Adair, P. (2012). État et secteur touristique en Tunisie: résultats et limites d'une stratégie de l'offre. L'État, acteur du développement, 35.

Becheur, M. (2011). The Jasmine Revolution and the Tourism Industry in Tunisia.

Belhedi, A. (1999). L'espace touristique en Tunisie. Communication. Les nouveaux espaces touristiques.

Boulares, H. (2011). Histoire de la Tunisie. Cérès éditions, Tunis, 444–455.

Cheriguen, F. (1987). Barbaros ou Amazigh. Ethnonymes et histoire politique en Afrique du Nord. Mots. *Les langages du politique*, 15(1),7–22.

Dridi, H. (2006). Carthage et le monde punique. Belles lettres.

El Bekri, F. (2013). Le tourisme en Tunisie et son impact environnemental. *Maghreb-Machrek*, (2), 73–93.

Esmail, H. A. H. (2016). Impact of Terrorism and instability on the tourism industry in Egypt and Tunisia after Revolution. *The Business & Management Review*, 7(5), 469.

Gan, R., & Smith, J. (1992). Tourism and national development planning in Tunisia. *Tourism Management*, 13(3),331–336.

Hallem, Y., & Barth, I. (2011). Etude netnographique du rôle d'Internet dans le développement du tourisme médical: Cas du tourisme de chirurgie esthétique en Tunisie. In Proceedings of the 27ème Colloque International de l'Association Française du Marketing, Bruxelles.

Holliday, R., Hardy, K., Bell, D., Hunter, E., Jones, M., Probyn, E., & Taylor, J. S. (2013). Beauty and the beach: Mapping cosmetic surgery tourism. In *Medical tourism and transnational health care* (pp. 83–97). Palgrave Macmillan, London.

JEDIDI M 1986: Croissance économique et espace urbain dans le Sahel tunisien depuis l'indépendance. PUT, FLSH, 2 t 377 et 392p (Thèse d'Etat Paris 1983).

Kasmi, MS (2014). Tunisie. Editions l'Harmattan.

Khlif, W. (2006). L'hôtellerie tunisienne: radioscopie d'un secteur en crise. L'Année du Maghreb, (I), 375–394.
Mansfeld, Y., & Winckler, O. (2015). Can this be spring? Assessing the impact of the "Arab Spring" on the Arab tourism industry. Turizam: međunarodni znanstveno-stručni časopis, 63(2), 205–223.
Miossec, J. M. (1996). Le tourisme en Tunisie: un pays en développement dans l'espace touristique international (Doctoral dissertation, Tours).
M'Zabi, H. (1977). Le tourisme en Tunisie. Internationales Jahrbuch für Geschichts-und Geographie-Unterricht, 266–270.
Othmani, W., & Dhaher, N. (2018). Le tourisme en Tunisie: menaces anthropiques majeures versus capacité de résilience. Études caribéennes, (2).
Rouland, B., Jarraya, M. & Fleuret, S. (2016). Revue francophone sur la santé et les territoires.
Séguin, Y., & Guy, M. J. (2011). Tunisie 2e édition. Ulysse.
Selmi, N. (2017). Tunisian Tourism: At the Eye of an Arab Spring Storm.
Tamzini, W. (2013). Tunisie. De Boeck Superieur.

İsmail Kervankıran, Gülsel Çiftci, & Azade Özlem Çalık

16 Turkey

16.1 Introduction

Turkey, with its convenient location, rich natural and cultural tourism attractions and increased investment in tourism has become one of the most popular destination countries in the world. According to the statics publish by the World Tourism Organization (UNWTO, 2020) for 2019 in terms of international tourist arrivals Turkey ranked in sixth in the world and 13th for tourism revenue. Most of the tourism in Turkey caters to mass tourism. However, Turkey offers many alternative tourism opportunities, particularly due to its rich and extensive cultural heritage. Indeed, tourism investments, the number of tourists, and tourism income have increased continuously since 1980 due its diversifying offerings. The country's Ministry of Culture and Tourism, along with tourism associations, non-governmental organizations, and tourism enterprises, continue to develop tourism in the country and will remain a preferred country as long as they abide by values in accordance with "use-protect" principle.

According to the data of TURKSTAT (Turkish Statistical Institute, 2020) the population in Turkey was 83 million in 2020. Most of the population is comprised of Turks and Muslims. The most populated cities are Istanbul, Ankara, Izmir, Bursa, and Antalya. Turkey is a member country of international organizations such as the United Nations, NATO, OECD, Organisation of Islamic Cooperation, Economic Cooperation Organisation, D-8, and G-20.

Today, Turkey has changed very rapidly and profoundly with the Republican era, born from the remnants of the Ottoman Empire (Karpat, 2010; Zürcher, 2003). In this aspect, the Republic has been an era during when modernization has been accelerated (Özkazanç, 2014). Indeed, modernization has been one of the fundamental values (Tekeli, 2005). The Republic of Turkey was established in 1923 under the leadership of Mustafa Kemal Atatürk and has realized fundamental reforms in many areas from politics to culture and from law to economy on its way to becoming a modern state (Dodd, 2012). Even though some interruptions have been experienced from the declaration of the Republic to the present day, we can say that Turkey, with the effect of these reforms, has covered a great distance in terms of integration with the economic, political, and social systems in the world and development. Turkey, which is a regional power due to its geopolitical position, continues to attract the world with the investments it has made in the fields of industry, tourism, energy, transportation, and technology as well as the steps it has taken towards modernization.

Although numerous economic crises have taken place in the world the rate of growth of Turkey has been 0.9% as of 2019 (TURKSTAT, 2020). With these values, the Turkish economy ranks among the top 20 in the world in terms of GDP. Although the economy has developed various problems such as external dependency, technological

https://doi.org/10.1515/9783110638141-016

backwardness, internal turmoil, and lack of innovation have a negative impact on the economy of Turkey. Spite of the economic recession in the world seems to be a significant increase in tourism in Turkey. Despite the political, social, and economic problems in Turkey, the number of domestic and foreign accommodation numbers has increased continuously. Thus, the contribution of tourism sector in Turkey's economy is increasing day by day. According to data from 2019, Turkey's tourism revenues constitute about 4% of the world's total tourism revenue and the share of tourism revenues in the GNP in Turkey has been 4.5% (TURKSTAT, 2020).

16.2 Government System

The form of government in Turkey is Republican, its official language is Turkish, and its capital is Ankara. The Republic of Turkey is a democratic, secular, social state of law. It has a structure based on the principle of separation of forces. In the Republic of Turkey, "legislation" is made by the GNAT (Grand National Assembly of Turkey), "execution" is done by the government whereas "jurisprudence" is the responsibility of independent courts (Constitution of the Republic of Turkey, 1982).

The general principle of the governing structure of Turkey has been determined in the Constitution of 1982. Accordingly, the administration is an embodiment comprised by its institutions and tasks and is regulated by law (Constitution of the Republic of Turkey, 1982). After a referendum that took place on April 16th, 2017, a new government system assumed power. The name of Turkey's new government system is "Presidential System of Governance." Turkey has a uniform structure in terms of administration. The largest administrative units are provinces and there are 81 provinces in Turkey.

During the era of the republic, the first attempt to establish a tourism governance authority was the Travelers' Association in 1923, which changed its name to the Turkey Touring and Automobile Club in 1930 and acted as a governmental body for many years (Roney, 2011). Tourism management in Turkey, which started as a small tourism office in the Ministry of Economy in 1934, became institutionalized with the establishment of the Ministry of Tourism in 1963 and in 2002 its name was changed to the Ministry of Culture and Tourism (Duman & Kozak, 2010; Göymen, 2000). Currently the authorities, duties, and responsibilities regarding tourism management are the responsibility of the Ministry of Culture and Tourism.

16.3 Tourism Profile

According to data from UNWTO (2020) Turkey ranked 20th in the world in terms of international tourist arrivals in 2000 and 14th in terms of revenue from international tourism, while in 2019. According to data from TURKSTAT (2020), 51 million foreign visitors came to Turkey in 2019 and income from tourism reached US$34 billion. Most tourists coming to Turkey come from Germany, Russia, and the United Kingdom. The most frequented cities by foreign tourist in Turkey are Istanbul, Antalya, Muğla, and İzmir. The target of Turkey for 2023 is to be one of the five leading countries to attract the most tourists in the world and generate the most revenue from tourism.

According to the Ministry of Culture and Tourism (2020) during 2000 and 2019 there was a steady increase in the number of tourists visiting accommodation facilities with tourism licenses as well as facilities with operation licenses in Turkey and the country witnessed an increase of 205% in bed capacity from 2000–2019. In 1980 Istanbul was the primary tourism destination in Turkey and during the past 35-year process tourism has shifted towards the Mediterranean and Aegean coasts in Turkey and these coasts have become the most preferred venues in Turkey tourism (Figure 16.1).

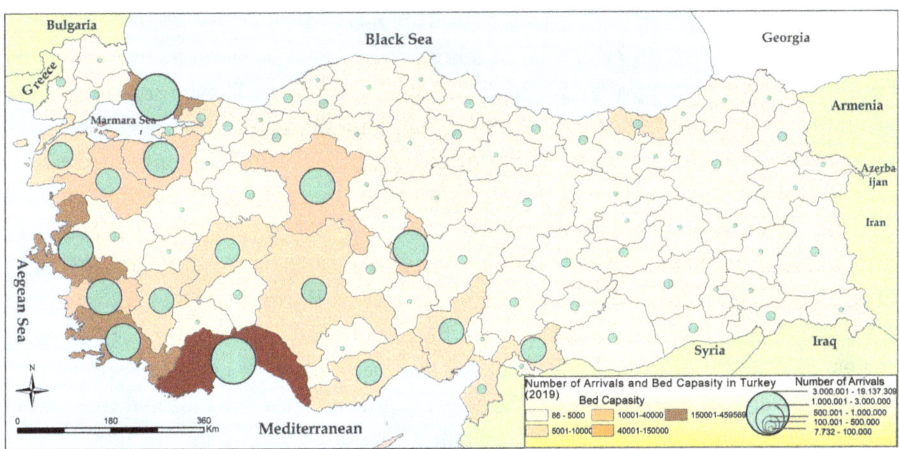

Figure 16.1: The Distribution of the Number of Arrivals and Bed Capacity in Turkey (2019). (Source: Prepared by the authors using the Ministry of Culture and Tourism data (2020)).

In addition to its geographical location, Turkey, is in an advantageous position in terms of tourist mobility as it is close to the tourism market of Europe, Russia, and Middle East (Akkemik, 2012). Turkey has an exceptional wealth of tourism assets which compare favorably with those of competing countries in the Mediterranean region. Its natural attractions include a vast, varied, and unspoiled landscape with forests, rivers, and mountains as well as an extensive selection along its 8,333 km of coastline extending on three sides (Alipour, 1996).

Turkey's largest city in terms of population, Istanbul, has many cultural sites owing to the city's historical status. These include the Blue Mosque, the Hagia Sophia, and the Topkapi Palace. Other cultural and historical attractions elsewhere in the country include the sites of Ephesus, Troy, Pergamon, the House of Virgin Mary, and religious places in Mardin, Konya, Hatay vd. (Abbott et al., 2012). Turkey offers a vast array of cultural resources and activities. However, the number of tourists visiting Turkey for cultural tourism alone is not very high, only about 6% of incoming tourists visiting Turkey choose for solely cultural tourism in 2011 (Okumuş et al., 2012).

The tourism sector, which has an important position in the development of the Turkish economy, also plays an important role for its contribution to the national income as well as closing the deficit in the balance of foreign exchange income and payments. Particularly as a result of the structural changes in 1982, the contribution of tourism to the Turkish economy has continuously increased. According to TURKSTAT (2020), the average expenditure amount of tourists coming to Turkey as of 2019 has been US$666, the share of tourism revenue in GNP has been 4.5%, the ratio of exports and tourism revenues has been 20.1%, the total contribution of the tourism sector in terms of employment has been 8%.

In order to accelerate mass tourism development, in 1982, the government enacted the Tourism Encouragement Law. This has induced many private and public entrepreneurs to undertake large amounts of fixed investment in tourism by building hotels, yacht ports, swimming pools, and so on. These incentives were given to tourism investments that took place in tourism regions, tourism zones, and tourism centers. This ushered in spatial concentration at the expense of regional balanced development (Tosun, 2001). As a result, the number of accommodation facilities (hotels, motels, holiday villages, etc.) are concentrated on the coasts of Antalya, Alanya, Kemer, Belek, Fethiye, Marmaris, Bodrum, Çeşme, and Kuşadası in particular, and central areas such as Ankara, Denizli, Ürgüp-Göreme, Konya, and Afyonkarahisar provinces. Furthermore, the 2023 Turkey Tourism Strategy foresees the establishment of tourism cities based on the attractions in various provinces, including Datça Eco-Tourism City. As a result, tourism has transformed the spatial characteristics of these cities and their cultural and traditional fabrics, replacing it with postmodern styles which are dominated by popular cultural values.

16.4 Shaping Tourism Development

Turkey has great potential of tourism. Although some important attempts were taken before the planned period to benefit from the economic, social, and cultural effects of tourism, it was not until the 1980s that serious and sound attempts were made to enhance the tourism industry (Düzgünoğlu & Karabulut, 1999). It is possible to deal with tourism development in Turkey in three stages:

16.4.1 Pre-Republic Period (before 1923)

Although traveling for tourism purposes in the modern meaning started during the Republican era in Turkey, it can be argued that the start of tourism mobility in Turkey was initiated in the Ottoman era when mostly the sultan and his family, senior statesmen, wealthy families, and non-Muslims traveled to resort areas in and around Istanbul to stay at what was called summer houses (Kervankıran et al., 2018). Also, the development of transportation information increased the demand of travel. The Orient Express was a beginning of luxury and comfort in trains, and people could travel with it from Europe (Paris) to Turkey (Istanbul) in 1883, which was of major significance in the development of tourism in Turkey as a destination (Yolal, 2016).

16.4.2 Republic and Pre-Planned Period (1923–1960)

During this period, tourism was put on the backburner compared to other sectors because priority was given to industrialization in Turkey and consequently the Import Substitution Industrialization model was adopted and the closed economy model understanding dominated (Aykaç, 2009). Furthermore, the fact that the economic policies of the first period of the Republic were statist led to the development of tourism policies and investments connected to public institutions during this period.

In 1939, the Tourism Directorate was transformed into the Tourism Office affiliated with the Ministry of Commerce. In 1943, the Department of Tourism is linked to the General Directorate of Press and Publishing. In 1953, the Tourism Industry Promotion Law entered into force. This law included some incentives and tax reductions for domestic and foreign investments. In 1955, Tourism Bank was established to provide credit support to investments. In 1957, The General Directorate of Press and Tourism became the Ministry. From 1949 to 1957, the General Directorate of Press and Tourism continued tourism activities (Sezgin, 2001). This Republic and pre-planned Period was just a beginning of the tourism development in Turkey.

16.4.3 Planned Period (1960 and after)

During this period, daily practical everyday life had started to change with the impact of urbanization and modernization, and together with the development opportunities in transportation and technology, major changes incurred in the holiday and travel habits of people in Turkey. As a result, not only upper-class people but also the middle- and lower-class people started to participate in tourism activities (Kervankıran et al., 2018). The planned period is divided into two periods: the period between 1960–1980, when tourism development was determined by national plans and statist tourism policies continued, and the period after 1980, when neoliberal tourism policies were effective.

Although Turkey became involved in the tourism sector later than other European countries, it is evident that the tourism sector has developed rapidly. As a result of efforts which took many years, Turkey has started to occupy a significant place in international tourism (Özgüç, 2015). Although planning for tourism in Turkey started before 1963, planning for tangible targets such as the number of tourists, tourism revenue, infrastructure services, diversification of tourism types, increase of promotion, and foreign investments started together with the planned period after 1963 (Kervankıran, 2015). After 1980 Turkey captured a significant speed of development in tourism and within a few years major developments were achieved in touristic demand capacity, acquired foreign exchange as well as the number of visiting foreigners (Kozak, 2012). As a result of the liberal policies in the economy and together with the enactment of Tourism Incentive Law no. 2634 in 1982 for the development of the tourism industry, the private sector was ensured numerous incentives, loans, and facilitation in terms of tourism investments (Erkuş Öztürk & Eraydın, 2010; Yolal, 2016). With this law the Central Administration provided significant incentives for tourism entrepreneurs who wanted to invest in areas defined as tourism zones or centers and allocated areas to attract tourists of the highest level (Tosun et al., 2003). As a result of these incentives tourism investments increased and mass tourism started especially from Western Europe to the west and south coasts (Naumov & Green, 2016). In later years Turkey became one of the countries attracting the most tourists as a Mediterranean tourism destination together with France, Spain, and Italy (Manera, 2016). Increase in the importance given to the tourism sector and the policies implemented by tourism management have been effective in the development of Turkey tourism in recent years. Furthermore, the fact that tourism has been represented on a ministerial level since 1963 has facilitated the application of plans and projects as well as decision making. This has had a positive impact on the success of tourism policies. From 1962 to 2019, 11 Five-Year Development Plans were implemented. In the 11th plan period (2019–2023), the expected number of tourists is 75 million and the estimated tourism income is US$65 million by 2023.

Turkey initiated the process of expanding abroad after 1980 which led to a change and transformation in the tourism policy. While Turkey's tourism management, development, and planning were determined by statist economic policies in the previous period, after 1980 they started to be determined by neoliberal economic policies which are influenced by global capital.

16.5 Tourism in Public and Private Sectors

Small local business owners established the initial base of tourism development in many popular destinations and benefited from this for a short period. However, these small local investors were replaced by large foreign and non-local domestic capital owners at a later stage of tourism development, as is commonly the case in destination communities (Butler, 1980). Turkish tourism is mostly shaped by the private sector. Especially since the beginning of 2000, foreign investors have started to take part in tourism sector in addition to local investors.

Today Turkish tourism development responses to law; however, until the Tourism Encouragement Law of 1982, there were no laws or regulations related to tourism in Turkey. The purpose of this law was to ensure that necessary measures are taken for the regulation and development of the tourism sector and for giving this sector a dynamic structure and mode of operation. Since the 1980s, tourism has also been the focus of successive governments' policies to achieve exported industrialization. That law gave generous incentives to tourism investment, has resulted in exceptionally rapid growth in tourism in terms of volume, value, and physical infrastructure (Şahin, 1990).

The objectives of tourism development in Turkey are determined together with other actors related to tourism, especially the Ministry of Culture and Tourism. Turkey Travel Agents Association (TÜRSAB), Turkey Hotels Association (TÜROB), and Tourist Guides Association (TUREB) are main actors. In addition to them, meetings and workshops are held each year, where public and private sector ideas are exchanged. Although the decision-making mechanism in Turkey is public, private sector investors have the freedom to develop and implement new ideas in the shaping of the sector. Especially private sector investors play a big role in Istanbul and Antalya where tourism is developed intensively.

16.5.1 Tourism in Central Government

The Turkish tourism sector is managed under the auspices of the Ministry of Culture and Tourism. The Ministry has three organizations: central, provincial, and international.

The duties of government in relation to tourism are as follows (Ministry of Culture and Tourism, 2017):

A. To research, develop, protect, live, evaluate, disseminate, promote, embrace national, spiritual, all kind of touristic values,
B. To guide and cooperate with other bodies which are related to culture and tourism issues, to develop communication and cooperation with local governments, non-governmental organizations, and private sector,
C. To protect historical and cultural assets,
D. To direct all kinds of investment, communication and development potential in culture and tourism areas,
E. To acquire immovable related to culture and tourism investments, to expropriate them when necessary, to do their studies, projects, and construction,
F. To carry out promotional services related to culture and tourism by exploiting the touristic assets of Turkey in every field and taking advantage of all kinds of opportunities and means.

The Ministry of Culture and Tourism is the decision maker. The local and foreign organizations are obliged to implement the decisions taken by the Ministry. Local municipalities can implement their plans and decisions in parallel with the Ministry's plans and strategies. Central and local government work in harmony with each other. They do not regard each other as an element of competition but rather act as collaborators.

16.5.2 Tourism and Other Sectors

The tourism sector alone is one of the strongest actors of the Turkish economy. However, there are many different sectors that are in cooperation such as textile, agriculture and stock raising, construction, health, furniture sectors, etc. In other words, Turkish tourism is a sector with a multiplier effect. This system also works very well. It has an important role especially in the employment rate and the rise of the economy.

The authority of tourism sector in Turkey is the Ministry of Culture and Tourism (Law #4848, April 16th, 2003). The aim of the Ministry is to live, develop, spread, promote, evaluate all kind of touristic values, to protect historical and cultural assets, to evaluate all chances as a positive contribution to the country's economy, and to take necessary measures to develop, market, promote and support tourism. While today, the name of the ministry is Ministry of Culture and Tourism, it has changed several times over the years. In 1963 it was Ministry of Tourism and Promotion, in 1971 the Ministry of Culture, in 1982 the Ministry of Culture removed and merged with Ministry of Tourism and Promotion and its name changed to the Ministry of Culture and Tourism. In 1989 it changed to Ministry of Tourism, and most recently, in 2003, the Ministry of Culture and Tourism was established.

16.6 Turkey's Tourism Stakeholders

16.6.1 Ministry of Culture and Tourism

The year 2023 marks the 100th anniversary of Turkish Republic founded by Mustafa Kemal Ataturk and his companions. In 2007, the ministry released an official document about Turkey's tourism detailed plan, named "2023 Turkey Tourism Vision," to become one of the top five visited countries in the world, therefore generating mass tourism revenue. The Turkish Tourism Strategy was extremely detailed because it targets almost every type of travel genre. The strategies include planning, investment, organization, domestic travel, research and development, transportation and infrastructure, marketing, education, service quality, city branding, diversification, host rehabilitation efforts, tourism development zones, and corridors, tourism cities, ecotourism zones (Ministry of Culture and Tourism, 2017).

16.6.2 Blue Flag Program

The Blue Flag works towards sustainable development at beaches/marinas through strict criteria dealing with water quality, environmental education and information, environmental management, safety and other services. The Blue Flag Program includes environmental education and information for the public, decision makers and tourism operators (Mavi Bayrak, 2021). It is an exclusive eco-label given to beaches and marinas having reached to a standard stated in the criteria. The Program is run by the independent non-profit organization Foundation for Environmental Education (FEE). One national FEE member is responsible for the implementation of the program, which is called TURCEV in Turkey. In 2021, there are 519 blue flag beaches, 22 blue flag marinas and 6 blue flag yachts (Mavi Bayrak, 2021).

16.6.3 Green Star Hotels

Green Star is an environmental label given to environment-friendly accommodation enterprises within the scope of sustainable tourism and environment-friendly accommodation enterprises' initiated by Ministry of Culture and Tourism. The enterprises which have a Tourism Establishment License from Ministry of Culture and Tourism can receive a Green Star Certificate when they complete the required criteria (Ministry of Culture and Tourism, 2017). The number of "Environment Friendly" facilities increased to 455 in 2020 in Turkey (Tourism News, 2021). Turkey's hotel number is increasing year by year. In 2016, there were 2,304 3-4-5 star hotels. In 2020 this number increased to 2,738 (TUROB, 2020).

16.6.4 Corporations and Tour Operations

Among the non-governmental organizations operating in the tourism sector in Turkey is the Turkish Travel Agencies Association (TURSAB) established by Law no. 1618 and the Turkish Hotel Association (TUROB) established by the hospitality management associations. Apart from these, the Association of Tourist Guides (TUREB), the Association of Tourism Investors (TYD), and the Tourism Development Foundation (TUGEV) undertake the representation of tourism branches or business lines. There are many national and regional organizations, such as Turkish Promotion Foundation (TUTAV) and several workers unions.

The numbers of Turkish Airlines in 2021; 5643 pilots, 12,113 cabin crew, and in 2018, 75.1 million passengers flying to 127 country, 326 city, and 331 airport (Turkish Airlines, 2021).

Tour operators are the operators which carries international tourists to Turkey. There are 4,077 travel agencies in 2000, and in 2019 there are 11,410 travel agencies in Turkey. It is possible to say that travel agency number increased with great speed every year (Ministry of Culture and Tourism, 2020).

16.7 Tourism Impacts

16.7.1 Ecological Impacts

The conditions of ecological sustainability are the application of "pollutant pays," application of the environmental impact assessment method, protection of biodiversity, consideration of ecological footprint and carrying capacity concepts, and preference of clean energy resources. At this point, carrying capacity in sustainable tourism, ecological footprint calculations, environmental impact assessment, eco-labels can be seen as a tool (Coccossis et al., 2001). The excessive density that can arise from tourism in natural areas can cause degradation or destruction of natural resources after increasing the specific carrying capacity, ecosystems can suffer significant damage. Studies on the negative effects of tourism on the environment in Turkey are being carried out. Demolishing the cafes around the Caretta Caretta (Loggergead Sea Turtle) habitats or Hierapolis Traverses are good examples but Trabzon, Uzungol, as a natural beauty, come into a concrete area. The precautions were not taken for the region.

16.7.2 Social-Economic Impacts

As social impacts, tourism develops an atmosphere of hospitality, accelerates the urbanization of rural areas, allows women to progress in their rights, improves habits of leisure time use, provides the development of cleaning consciousness, ensures the emergence of new social institutions, allows new jobs to emerge, helps the development of environmental awareness, generates consciousness that has the historical and cultural values of local people develops. In Turkey, in recent years, it has been observed that significant changes have occurred in traditional structures of the society in the touristic regions.

Turkey, which started to attract interest in the world with the incentives made in the 1980s, has experienced a certain increase every year in the number of tourism revenues and tourists since these years. Depending on these developments, the number of employees in tourism has increased and administrations aware of the importance of education have opened dozens of tourism schools over the years. Between 1963 and 2017, it was seen that the share of tourism revenues in GNP increased from 0.1% to 6.2%, which means a huge share difference. Also, these revenues' export ratio increased from 2.1 to 21.9 in 54 years. The datas show that, tourism is a very important sector with contributions to the balance of payments and employment and it is both inward and outward has been a major contributor to Turkey. Istanbul Antalya and Mugla are sharing the top five places in terms of arrivals.

16.8 Types of Tourism

16.8.1 Mass Tourism

The most typical characteristic of mass tourism is that people prefer package tours. In mass tourism, the number of people participating is not only superior in terms of number, but also of the continuity of established groups. For this reason, it constitutes the most popular type of tourism in the course of development of tourism. However, countries attach importance to mass tourism after a certain stage need to pay careful attention to natural and cultural balances. The main target mass of tourism in Turkey is the mass tourism groups. However, today, it is observed that in tourism activities, there is a shift towards mass tourism and smaller groups and individual boutique tourism (Kozak, 2012).

16.8.2 Leisure Tourism

It is for the main purposes of recreation and leisure and typically thought of in terms of the residential vacation, but it may also include day trips (UNWTO, 2007). Turkey's tourists are mostly leisure tourists.

16.8.3 Advnture and Cruise Tourism

There are 46 national parks in Turkey in 2020 (Republic of Turkiye Ministry of Agriculture and Forestry, 2021). Bird watching is a popular activity. In Turkey, where over 465 bird species live (Trakus, 2017). There are about 40 thousand caves. The number of caves that all domestic and international cave interest groups have examined and documented is 800, and 31 of them are opened for tourism (Ministry of Culture and Tourism, 2016).

Paragliding is done in several cities, especially in Fethiye. Balloon tours are held in many cities, especially in Cappadocia.There are many hiking locations. The Lycian way is one of the most popular location, where is at the West side of Mediterranean Region. It is 540 km. and takes 29 days to finish it. Clow (2014) wrote a book about this way and it is one of the Sunday Times World's Ten Best Walks. Mugla, Marmaris, Bodrum, Soros,

Gökçeada, Mersin are the scuba diving districts. Rafting is very popular in Turkey. Coruh, Dalaman River, Fırtına Valley, Koprucay are the main rafting rivers of Turkey.

In 2019, 328 cruises and 283.774 cruise passengers visited Turkey (TUROB, 2020). Istanbul, Kusadasi and Izmir are the popular ports.

16.8.4 Religious Tourism

Throughout its long history as a nurturing homeland to Islam, Christianity, and Judaism, Turkey has embraced their diverse beliefs and preserved their holy sites. For more than a millennium, Turkey has been at the crossroads of civilization, a melting pot of eastern and western traditions and a place where faiths converge. There are temples dedicated to ancient gods, churches of many denominations, synagogues, monastaries and, plenty of mosques. As civilizations succeeded each other over a period of 11,500 years, they bestowed their religious legacy and following the monotheistic domination of Anatolia, the religions coexisted in harmony (Go Turkey, 2019). Sultan Ahmet Mosque and Hagia Sophia are the popular historical values of Istanbul. Today, Turkey's Jewish communities are predominantly found in the major cities of Istanbul and Izmir, with a small community remaining in Bursa. Istanbul has more than 16 synagogues (Tourism Turkey, 2016).

16.8.5 Gastronomy Tourism

Anatolia has been a center of many civilizations throughout the history of Hittite, Troy, Urartu, Frig, Lydia, Ionia, Kingdom of Commagene, Persia, Byzantium, Roman Empire, Great Seljuk Empire, Ottoman Empire. The cooking culture shaped by the help of this diversity. Today Turkish cuisine is a very specific cuisine combined from Aegean, Mediterranean, Black Sea, Mesopotamian culture and it is famous worldwide. In 2016, tourists spent US$5.1 billion for dining in Turkey (Tour Mag, 2017). In the UNESCO's Intangible Cultural Heritage List, Traditional Toren Keskegi, Mesir Macunu Festival and Turkish Coffee Tradition are seen (UNESCO, 2016). Gaziantep has also been nominated for "World, Food and Drink Tourism 2019 Year" (Gastro Antep, 2016).

16.8.6 Health and Thermal Tourism

Turkey is one of the most popular destinations of health tourism in recent years. Turkey offers cost advantages, travel opportunities and quality technological infrastructure. Moreover, in some areas treatment in Turkey is cheaper by up to 60% than in many European countries. Turkey ranks ninth in the world with 1,200 plastic surgeons (Turgut & Akbulut, 2016). Health tourism in 2018–2019, increased its capacity by 40%. Turkey hosted 1.5 million health tourists during 2019 and revenues are a total of US$6 billion. Average expenditure per person was US$4,000. 60% of the patients came from the Middle East, 20% from Russia and Ukraine, 10% from Europe (Ministry of Culture and Tourism, 2020).

Turkey is among the richest countries in terms of geothermal resources and has 8% of the world's geothermal potential. Many thermals have been established around thermal water resources which, based on their temperature and on the quality of the spring waters, have been classified as "potable" (mineral springs), "spa," and "çermik" (hot springs). Visitors who used to visit Turkish hamams (baths) as a tradition and subsequently spas and baths for treatment purposes still like these venues but prefer the more developed and modern facilities in recent years (Kervankıran, 2016).

16.8.7 Soap Opera Tourism

Turkish cinema and TV series, exported to more than a hundred countries, have become one of the most effective ways of promoting tourism in Turkey. The increasing number of tourists coming to Turkey parallels with the number of hours of exported Turkish TV drama series. It is clear that Turkish drama presents a unique way of promoting Turkey's cinema-landscape which in the end influences millions. They are visiting the cities where those dramas were filmed (Anaz & Özcan, 2016).

16.8.8 Sustainable and Ecotourism

In sustainable tourism, ecological, economic, and social objectives should be considered. The economic sustainability of tourism means that tourism activities continue. When social sustainability is mentioned, it is understood as the protection of social and cultural identities, the priority of local people, protection of cultural life. Sustainable tourism refers to the necessity of preserving environmental values as a basic element of tourism, carrying out tourism activities in an environmentally responsible manner, and coordinating the efforts of economic development and preservation of environmental values (Çalık, 2010). One of the basic principles of sustainable tourism is the use of clean energy. Among the new and clean energy sources available in Turkey, there are different sources located in almost all parts of the solar energy. Geothermal, small stream, wind, and so on, are also important and resources are relatively local.

Ecotourism can be done in Turkish culture, hospitality that people have strongly represented especially in the wisdom of rural traditions. In consideration of the numbers of comparison between mass production-services with the sustainable ones, Turkey is at the beginning of the development of a good alternative. Turkey has a large ecosystem and four seasons happen all year. Ecotourism also encourages the local economies. In Tourism Strategy of Turkey 2023, the aim of creating ecotourism zones is to develop nature tourism with reference to development plans (Ministry of Culture and Tourism, 2016).

16.8.9 Agro-tourism

Ecologic farm tourism provides the opportunity to do farm tourism in more than 90 farms located in Turkey to those who are looking for alternative tourism options. Bugday Association was launched in 2004 with the support of the United Nations to support farmers engaged in ecological agriculture in a material and spiritual sense (TaTuTa, 2017). The "TaTuTa" project, started with 25 farms in 2004. 60% of farm visitors were international tourists (Çalık, 2010). TaTuTa program includes women of the host families from the beginning in all decision-making and continues in this way. Tourism and other activities on that level can only sustain with a fair and shared participation of the woman in the family (Eco Club, 2016).

16.9 Conclusion and recommendations

The historical process of tourism development in Turkey and policy changes in management progress in parallel with the historical development of the general economic policies in the country. The importance of tourism which has an international dimension in Turkey increased when Turkey joined the structural adjustment process with globalization and adopted neoliberal economic policies to embrace their incorporation into the capitalist economy. At the same time, the fact that the Turkish Lira is convertible, which is an extension of the liberalization process, has put tourism, a major source of foreign currency, into the center of the economy (Aykaç, 2009). Along with the restructuring of capital accumulation on a global level after the 1980s, investments started to increase in Turkey because of factors such as the wealth of natural and cultural resources, favorable climate, proximity to major centers in Europe, affordable living and property prices, securing favorable conditions for foreign capital by law and the safeguarding of international capital in Turkey (Purkis, 2008). Many incentives, loans and facilitations were provided to the private sector in tourism investments with the Tourism Incentive Law number 2634, which entered into force in 1982 to develop the tourism industry (Erkuş Öztürk, & Eraydın, 2010). As a result, investments in tourism increased and a rapid growth was realized in the tourism sector especially after 1990, with the support of the private sector with incentives and loans and renting state lands to the private sector (Aykaç, 2009). All these developments have improved the competitiveness of Turkey's tourism supply in the international tourism market, the number of tourists coming to Turkey from abroad has increased with each day and Turkey has become one of the world's leading tourism destinations.

The tourism administration in Turkey has always supported mass tourism. Although the policies regarding mass tourism have always been positive in the planned term and the years when tourism first started this positive impact did not continue in later years. Unfortunately, the tourism planning made in the recent past makes decisions in favor of increasing mass tourism. The development of tourism in the world, the changes in tourist profiles and preferences should be taken into consideration in the development of alternative tourism policies.

Although regional and global crises such as the Gulf War, Izmit Earthquake, and September 11th, 2001, interrupted the progress of tourism, the tourism sector in Turkey has continued to develop (Aktaş, 2016). Together with this development, regardless of its economic contribution, mass tourism supported by tourism development projects and legislation prepared since 1980 and encouraged by both national and local governments consisting of low-middle-class tourists trapped in the Mediterranean and Aegean coasts, brought about some negative environmental, social, and cultural aspects (Egresi, 2016). As a result, various problems were generated such as unplanned settlement, destruction of natural areas, commodification of cultural values, pollution, overcrowding, exceeding the carrying capacity, and failure to attract quality tourists with high incomes.

In addition, after the 2000s, "neoliberal" policies that began to dominate Turkey's economy also paved the way for the free market in tourism as well as the other sectors and legal regulations that support these policies were enacted and implemented. The integration of the tourism sector into the free market economy in Turkey has accelerated because of the dominant neoliberal governance, private sector investments in tourism businesses and services and consequently Turkey tourism has come under the influence of global capital. The tourism industry is considered a gateway to the global capitalist economy, consequently the tourism industry in Turkey aims to increase mass consumption and consumer satisfaction. Therefore, tourism policies, management and research focus on the revitalization, improvement and development of the tourism market. However, critical policies and studies are needed on issues that the tourism market has ignored or neglected. In order to ensure an effective, correct, and sustainable tourism management in Turkey it is necessary to develop an understanding of an egalitarian, social, just, and inclusive tourism management that supports sub-capital groups in addition to the aid and incentives given to global capital and the private sector, that takes the environmental and social dimensions of tourism into consideration as well as the economic benefits, that uses resources efficiently in line with the joint decisions of broad society segments and different tourism stakeholders, that appreciates local value, that cares about disadvantaged groups (disabled people, senior citizens, children, etc.), that acknowledges the rights of tourism workers and informal workers.

In COVID-19 year 2020, many tourism companies have either gone bankrupt, have decided not to work in 2020 or have had to decide to work with less capacity in terms of customers and employees. As a result of entering a new process in the tourism sector with COVID-19, companies had to make new decisions that ensure the safety of both employees and tourists. Tourists also have to comply with this decision, and naturally, tourists have started to have different expectations from companies. They are more likely to be more rigorous, sifting, critical and anxious compared to the previous periods when they purchased services. Tourism sector employees, who were worried about losing their jobs, had to do twice as much as possible jobs due to lack of personnel. Qualified tourism staff, who saw the uncertainty of the sector, started to look for jobs in other temporary or permanent sectors, and this caused difficulties for companies that are in service provider positions. At the end of these, more working staff may result in more difficult service and more difficult satisfied tourists.

When tourism revenues and the number of tourists reach a recalculable level, it seems likely that complaints will increase as there will be a resumption process for companies. Because the number of less staff serving in more difficult conditions will increase. Tourists will start to prefer a type of travel that is far from crowded places, more isolated, away from public transport, and where they travel with their rented or own vehicles, staying in a boutique hotel. Due to the obligation to work according to social distance and 50% occupancy rate, 20 people are admitted to 40-person bus groups.

16.9.1 Could Tourism Improve More Under Existing Government System?

Governance entails a guidance process that is institutionally and technically structured, that is, based on principles, norms, procedures, and practices to collectively decide about common goals for coexistence and about how to coordinate and cooperate for the achievement of decided objectives (Duran, 2013). According to Piattonni (2009), multilevel governance is a dynamic concept, presenting three dimensions corresponding to the horizontal and vertical delineations of the idea of governance, as well as the frontiers between international and domestic policy. Tourism governance means the process of managing tourist destinations through synergistic and coordinated efforts by governments, at different levels and in different capacities; civil society living in the inbound tourism communities; and the business sector connected with the operation of the tourism system (UNWTO, 2008). The responsibility of tourism governance in Turkey belongs to the organizations and agencies affiliated with the Ministry of Culture and Tourism. The vision, mission, objective, duties, and legislation have been determined with laws in the constitution and the central and local units and organization chart have been established to complete its organization structure. Furthermore, it is necessary to indicate the following.

This new mode of governing is characterized, at a minimum, by cooperation in the formulation and application of public policies (Mayntz, 2001); interaction between governmental and nongovernmental actors (Kooiman, 2003); interdependence and associated action among diverse actors (governmental, private, and social) with the aim of ensuring that problems of interest to them are considered of public importance, exchanging or gathering basic resources for solutions (Rhodes, 2000). The influence and weight of the central administration in Turkey is rather high. Local administrations need to be effective during the decision-making process and other stakeholders in tourism, particularly the local people should take a more active role in the management of tourism.

16.9.2 Are There Non-Tourism Deficits in the Government That Impact Tourism?

When the government is experiencing the difficulties that will affect the tourism sector, it should instantly find scientific solutions and examine the countries that have survived the crisis by experiencing similar problems. By solving the source of the problem from the root, it must focus on long-term solutions rather than on day-to-day solutions, which must be trustworthy, compromising its strong image as a combination with everyone working on this issue. In this regard, 2023 Tourism Vision decisions will help to implement. At the same time, although local initiatives are given much initiative in tourism, it is the end of local formations to wait for everything from the state.

16.9.3 Recommendations

Firstly, it is necessary for the government to develop new strategies in the field of tourism promotion applications in line with developing technology and new generation consumer demand. Especially the use of social media channels should be actively managed by experts. In fact, the Ministry should set up a new department and compete with other developed countries. Government should carry out marketing strategies with maximum success in reaching new markets. With a market-oriented approach, success must be achieved in markets that have not been reached previously or have failed.

For tourism to develop, problems must be solved through cooperation between private and public sector. In addition, a visionary strategy should be established to solve these problems by government. In addition to these, pricing policies need to be determined correctly by the government. In addition, investments towards potential destinations should be made. Misalignment of resources causes both time and money loss.

The crises that have been taking place in Turkey in recent years, especially political crises, are damaging the atmosphere of trust and the touristic demand. Reconstruction of Turkey's image of a safe country and its announcement to potential markets is of great importance. Another important point is to increase archaeological site excavation support to create new tourist products within the scope of cultural tourism. In addition to that, the reorganization of archaeological sites excavated and uncovered will contribute to the increase in demand.

Turkey is a country that is rich in cultural, natural and historical values. A sustainable tourism policy needs to be developed to carry these values from the past to the future. Both tourism enterprises, both government and tourists, should cooperate in the protection of these values.

In terms of cooperative culture and the thorough explanations of the benefits of scale economies, the Turkish government should educate the staff to provide professional advice for new jobs. It should be told very simply that the important thing is to enlarge the cake and that a big share of the cake will be taken (Türkmen, 2017). Active and passive tourism stakeholders should seek local partners' interests in Anatolia seeking tourism development options in their region and local councils registered as "national organization models" in tourism. Despite the complexity structure of the tourism industry, the destinations are "easy to grow together with the local economy" by being easily managed at the scale and with the civil society dynamics (Cengiz, 2016).

As well as creating positive results in the societies realized by tourism structure, it also brings some negative results, especially in ecological and social environments. To minimize these situations, some measures should be taken, and suggestions should be developed. Some were given below. Socio-economic, physical, and ecological carrying capacities should not be exceeded. Local culture should be protected, and slow tourism should be improved. Ahlat, Akyaka, Eğirdir, Gerze, Gokceada, Göynük, Gudul, Koycegiz, Halfeti, Mudurnu, Persembe, Savsat, Seferihisar, Tarakli, Vize, Yenipazar, Yalvac and Uzundere towns are the 18 members from Turkey in the Cittaslow movement. Geographic Indication activity should be supported.

16.9.4 Best Practice to Maximize the Tourism Potential

The new tourist has become a universal human rights observer under the concept of "responsible tourist." In the future, they will be able to present the countries with preserving their existing values, as well as those who realize the new tourist's preferences and tourism. Below, many of the proposals to be made in Turkey can be applied to many countries as well.

A. The tourism sector's banking debt; three years grace period, a total of 10 years must be spread.
B. The debts of the enterprises in the sector to the government should be deleted.
C. The unemployed people who became unemployed while working in the sector shall be paid 2/3 of their wages for three years from the unemployment fund.
D. The establishments that are in a closed or idle statement should be considered by the government as student dormitories.
E. The Unused airplanes of Turkish Airlines should be rented to domestic and foreign tour operators for lower prices and tours should be supported with very reasonable prices from various countries (Türkmen, 2017).
F. A coordination and coordination scheme should be widely established by the public and private sectors, to manage tourism.

Today, managing the developments in Turkish Tourism and taking the necessary precautions and taking them into practice is possible with "putting the National Strategy which is still 7 years life into practice." Mega Projects like the Bridge of Civilizations "that reconcile the East and the West" which will stand out with the "wide Mesopotamian environment" and the Silk Road Project which is rising in the world tourism should come to a fast pace. A tourism strategy and local councils should come together to manage the performance (Cengiz, 2016).

Tourism in Turkey is important in terms of economic, political, cultural, and social development. The ability of the tourism sector, which is of such importance for Turkey, to get a larger share of the world's tourism market and compete depends on its correct, rational, efficient and effective management. Tourism management has its own special features. The tourism sector is a fragile sector; therefore, tourism administrators need to pay close attention to these features. The tourism sector, which changes very much according to the time and conditions, should constantly renew itself and develop a strategy in accordance with the changing conditions, evaluate tourism management, plans and strategies in a multifaceted way, tourism should not only be planned quantitatively but in a qualitative way as well and applicable and realistic tourism policies should be developed with achievable targets.

References

Abbott, A., Vita, G.D. & Altınay, L. (2012). Revisiting the Convergence Hypothesis for Tourism Markets: Evidence from Turkey Using the Pairwise Approach, *Tourism Management*. 33, 537–544.

Akkemik, K. A. (2012). Assessing the İmportance of İnternational Tourism for the Turkish Economy: A Social Accounting Matrix Analysis. *Tourism Management*, 33, 790–801.

Aktaş, G. (2016). Turkey. In J. Jafari and H. Xiao (Eds.), *Encyclopedia of Tourism*, (pp: 974–976). Switzerland: Springer International Publishing.

Alipour, H. (1996). *Tourism Development Within Planning Paradigms: The Case of Turkey*, Tourism Management, 17, 367–377.

Anaz, N. & Özcan, C. (2016). Alternative Tourism in Turkey. In I. Egresi (Ed.), *Alternative Tourism in Turkey Role, Potential Development and Sustainability*, (p: 247–258). Switzerland: Springer International Publishing.

Aykaç, A. (2009) *Yeni işler, yeni işçiler: turizm sektöründe emek (New jobs, new workers: labor in tourism industry)*, İstanbul: İletişim Yayınları.

Butler, R. (1980). *The Concept of a Tourism Cycle of Evolution*. Canadian Geographer, 5–12.

Çalık, A. Ö. (2010). *Özel Ilgi Turizmi*. Ankara: Ankara Üniversitesi Uzaktan Eğitim Yayınları. No: 51.

Cengiz, Z. (2016). Tourism Newspaper. Retrieved from. http://turizmnews.com/yazar/turizmde-cikmazlar-ve-cozumler311.html. Date accessed: 01.02.2017.

Clow, K. (2014). *The Lycian Way*. Turkey: Upcountry Ltd.

Coccossis, H., Mexa, A., Parpairis, A., & Konstandoglou M. (2001). Defining, Measuring and Evaluating Carrying Capacity In European Tourism Destination Final Report. Athens: Environmental Planning Laboratory of the University of Aegean, Greece.

Constitution of the Republic of Turkey, (1982). Article:1, 2, 3, 5, 7, 8, 9, 123.

Dodd, C.H. (2012). The Turkish Republic, In M. Heper and S. Sayarı (Ed.) *The Routledge Handbook of Modern Turkey*, (pp:53–64). New York: Routledge.
Duman, T. & Kozak, M. (2010). The Turkish Tourism Product: Differentiation and Competitiveness, *Anatolia: An International Journal of Tourism and Hospitality Research* 21(1), 89–106.
Duran, C. (2013), Governance for the Tourism Sector and its Measurement, Retrieved from UNWTO Statistics and TSA Issue Paper Series STSA/IP/2013/01. Date accessed: http://statistics.unwto.org/en/content/papers.
Düzgünoğlu, E. & Karabulut, E. (1999). *Development of Turkish Tourism*. İstanbul: Türkiye Seyahat Acentaları Birliği.
Ecoclub (2017). Retrieved from: http://ecoclub.com/news/102/ananias.html, Date accessed: 01.02.2017.
Egresi, I. (2016). Tourism and Sustainability in Turkey: Negative Impact of Mass Tourism Development, In I. Egresi (Eds.), *Alternative Tourism in Turkey Role, Potential Development and Sustainability*, (p: 35–56). Switzerland: Springer International Publishing.
Erkuş Özturk, H., & Eraydin, A. (2010). Environmental Governance for Sustainable Tourism Development: Collaborative Networks and Organization Building in the Antalya Tourism Region. *Tourism Management* 31, 113–124.
Gastro Antep (2017). Retrieved from: http://www.gastroantep.com.tr. Date accessed: 01.17.2017.
Göymen, K. (2000). Tourism and Governance in Turkey. *Annals of Tourism Research*, 27(4), 1025–1048.
ICCA World (2019). Statistics Report. Retrieved from https://www.iccaworld.org/newsarchives/archivedetails.cfm?id=1100291 Date accessed: 01.17.2019
Karpat, K. H. (2010). *Türk Demokrasi Tarihi*, İstanbul: Timaş Yayınları, p: 15.
Kervankıran, İ., Sert Eteman, F. & Çuhadar, M. (2018). Türkiye'de iç turizm hareketlerinin sosyal ağ analizi ile incelenmesi (Examining Domestic Tourism Mobility in Turkey with Social Network Analysis). *Turizm Akademik Dergisi*, 5(1),29–50.
Kervankıran, İ. (2016). Between Traditional and Modern: Thermal Tourism in Turkey. Egresi (Ed.), *Alternative Tourism in Turkey Role, Potential Development and Sustainability*, (p: 23–33). Switzerland: Springer International Publishing.
Kervankıran, İ. (2015). Contribution of The Five-Year Development Plans to Tourism in Turkey, *Turkish Studies- International Periodical for the Languages, Literature and History of Turkish or Turkic*, 10(2), 587–610.
Kooiman, J. (2003). *Governing as Governance*, London: SAGE Publications Ltd.
Kozak N. (2012). Genel Turizm Bilgisi. (Edt. Kozak, M.) *Anadolu Üniversitesi Yayını No: 2472 Açıköğretim Fakültesi Yayını No: 1443.*
Manera, C. (2016). Mediterranean, In J. Jafari and H. Xiao (Eds.), *Encyclopedia of Tourism*, (p: 604). Switzerland: Springer International Publishing.
Mavi Bayrak (2021). Retrieved from: http://www.mavibayrak.org.tr/en/icerikDetay.aspx?icerik_refno=10. Date accessed: 16.09.2021.
Mavi Bayrak (2021). Retrieved from: http://www.mavibayrak.org.tr/tr/Default.aspx. Date accessed: 16.09.2021.
Mayntz, R. (2001), El Estado y la Sociedad Civil en la Gobernanza Moderna, en Reforma y Democracia, Caracas, CLAD. Retrieved from: http://www.clad.org/portal/publicaciones-delclad/revista-clad-reforma-democracia/articulos/021-octubre-2001/0041004. Date accessed: 07- 02-2017.
Milli Parklar (2017). Retrieved from: http://www.milliparklar.gov.tr/korunanalanlar/korunanalan1.htm. Date accessed: 01.11.2017.
Ministry of Culture and Tourism (2016). Retrieved from: https://yigm.ktb.gov.tr/TR-10335/magara-turizmi.html Date accessed: 11.01.2017.

Ministry of Culture and Tourism (2016). Retrieved from: http://www.goturkeytourism.com. Date accessed: 11.11.2020.
Ministry of Culture and Tourism (2016). Retrieved from: https://www.ktb.gov.tr/TR-133349/stratejik-plan.html Date accessed: 11.01.2017.
Ministry of Culture and Tourism (2017). Retrieved from: https://www.kultur.gov.tr/TR,96133/merkez-teskilati.html. Date accessed: 11.11.2020.
Ministry of Culture and Tourism (2020). Retrieved from: Tourism Statistics, https://www.ktb.gov.tr/TR-96695/istatistikler.html. Date accessed: 4.11.2020.
Ministry of Culture and Tourism (2020). Retrieved from: Tourism Statistics, https://yigm.ktb.gov.tr/TR-201116/turizm-gelirleri-ve-giderleri.html Date accessed: 4.11.2020.
Ministry of Culture and Tourism (2020). Retrieved from: Tourism Statistics, https://yigm.ktb.gov.tr/TR-243988/yillara-gore-seyahat-acentasi-sayilari.html Date accessed: 14.09.2021.
Naumov, N. & Green, D. (2016). Mass Tourism. In: *Encyclopedia of Tourism* (Ed: Jafar Jafari and Honggen Xiao), Switzerland: Springer International Publishing, p: 594.
Okumus, F., Avci, U., Kilic, I. & Walls, A. (2012). Cultural Tourism in Turkey: A Missed Opportunity. *Journal of Hospitality Marketing and Management*, 21 (6), 638–658.
Özgüç, N. (2015). *Turizm Coğrafyası*. 8. Baskı, İstanbul: Çantay Kitabevi.
Özkazanç, A. (2014). Cumhuriyet Döneminde Siyasal Gelişmeler: Tarihsel-Sosyolojik Bir Değerlendirme, In F. Alpkaya and B. Duru (Eds.), *1920'den Günümüze Türkiye'de Toplumsal Yapı ve Değişim*, (p:74). Ankara: Phonix Yayınevi.
Piattoni, S. (2009), Multi-Level Governance: A Historical and Conceptual Analysis, *En Journal Of European Integration*, 31, 163–180.
Purkis, S. (2008). *Turizmle kalkınma ve kentlerin kimlik arayışları: Marmaris örneği*, Muğla: Marmaris Ticaret Odası Ekonomi Yayınları-III.
Republic of Turkiye Ministry of Agriculture and Forestry Official Web Site, 2021. Retrieved from: https://www.tarimorman.gov.tr/DKMP/Menu/27/Milli-Parklar, Date accessed: 10.11.2021.
Rhodes, R. A. W. (2000). Governance and Public Administration, (Edt.) Jon Pierre, Debating Governance. *Authority, Steering, and Democracy*, New York: Oxford University Press.
Roney, S. A. (2011). *Turizm: Bir Sistemin Analizi (Tourism: analysis of a system)*. Ankara: Detay Yayıncılık.
Şahin, A. (1990). İktisadi Kalkınmadaki Önemi Bakımından Türkiye'de Turizm Sektöründeki Gelişmelerin Değerlendirilmesi. TOBB.
Sezgin, O. (2001). *Genel Turizm ve Turizm Mevzuatı*, Ankara: Ankara: Detay Yayınları.
TaTuTa (2017). Retrieved from: http://www.tatuta.org/?p=1&tc_aratext=&sayfa=1&sayi=81&lang=en. Date accessed: 01.02.2017.
Tekeli, İ. (2005). Cumhuriyetin Çağdaşlaşma Projesinin Mekansal Boyutu Üzerine Bir Değerlendirme, V. Türk Kültürü Kongresi: Cumhuriyetten Günümüze Türk Kültürünün Dünü, Bugünü ve Geleceği, 17–21 Aralık 2002; *Mimari ve Çevre Kültürü*, Cilt: VIII, Ankara: Atatürk Kültür Merkezi Yayınları.
Tosun, C. (2001). Challenges of Sustainable Tourism Development in the Developing World: The Case of Turkey. *Tourism Management*, 22, 289–303.
Tosun, C., Timothy, D. J. & Öztürk, Y. (2003). Tourism Growth, National Development and Regional Inequality in Turkey. *Journal of Sustainable Tourism*, 11(2–3), 133–161.
Tour Mag (2017). Retrieved from: http://www.tourmag.com.tr/turistler-yemege-51-milyar-dolar-harcadi/. Date accessed: 11.11.2020.
Tourism News (2021). Retrieved from: https://www.turizmnews.com/turkiye-de-cevreye-duyarli-tesis-sayisi-5-yilda-214-adet-artti/20909/ Date accessed: 16.09.2021.
Turkish Airlines (2021). Retrieved from: https://www.turkishairlines.com/de-int/press-room/about-us/turkish-airlines-in-numbers/ Date accessed: 16.09.2021.

TUROB – Turkish Hotel Association of Turkey (2020). Retrieved from: http://www.turob.com/tr/istatistikler. Date accessed: 11.11.2020.

Tourism Turkey (2016). Retrieved from: http://www.tourismturkey.org/experience/religious-travel. Date accessed: 11.11.2020.

TRAKUS (2017). Official web site. Retrieved from: http://www.trakus.org:80/kods_bird/uye/?fsx=. Date accessed: 11.11.2020.

Turgut, M. & Akbulut, Y. (2016). Türkiye'de Medikal Turizme Yönelik Politikaların Analizi. 1st International Scientific Researches Congress Humanity and Social Sciences, Madrid.

Türkmen, N. (2017). Sözcü Gazetesi. Retrieved from: http://www.sozcu.com.tr/2017/yazarlar/nedim-turkmen/httpwww-sozcu-com-tr2017yazarlarnedim-turkmenmalumun-ilani-turizm-bitti-1657684-%E2%80%8E-1657684/. Date accessed: 01.02.2017.

TURKSTAT (Turkish Statistical Institute, 2020). Retrieved from: https://w.tuik.gov.tr/. Date accessed: 02.11.2020

UNESCO (2016). Retrieved from: http://www.unesco.org/culture/ich/en/lists. Date accessed: 01.11.2017.

UNWTO (World Tourism Organization, 2020). World Tourism Barometer and Statistical Annex. Retrieved from: https://www.eunwto.org/loi/wtobarometereng. Date accessed: 10.11.2020.

UNWTO (World Tourism Organization, 2008), Seminario Internacional sobre la Gobernanza en turismo en Las Américas. Informe Final. Villahermosa, Tabasco, Mexico.

UNWTO (World Tourism Organization, 2007). A Practical Guide to Tourism Destination Management.

Yolal, M. (2016). History of Tourism Development in Turkey. In I. Egresi (Eds), *Alternative Tourism in Turkey: Role, Potential Development and Sustainability* (p. 23–33). Switzerland: Springer International Publishing AG.

Zürcher, E.J. (2003). *Modernleşen Türkiye'nin Tarihi*, (p.15). İstanbul: İletişim Yayınları.

Dean MacCannell

17 Unites States of America

17.1 Introduction

Tourism governance in the United States is bound to the nation's embodiment of a combination of laissez faire capitalism and enlightenment inspired multi-cultural, secular democracy – the oldest and perhaps the boldest such experiment so far devised. None of the ideals of a free market, or of secular democracy, have been fully realized, but the arc of US history bends toward these ideals and especially toward those points where conflicts between them need to be addressed. The story of tourism in the United States provides a high-definition mirror of the co-evolution of capitalism and democracy and the tensions between them as they try to share the same space.

The first thing to be noted is the policies that shape US tourism are radically decentralized and diffuse defying any effort at paradigmatic containment. The United States has no federal department or national policy dedicated specifically to tourism. Federal level policy and/or administration that might affect tourism is folded into larger administrative units that have broader mandates: National Park Service of the United States Department of Interior, US Department of Customs and Border Protection, among others Unlike agriculture, manufacture, health, education, and housing, here are no federal agencies or initiatives that define tourism as a distinct sector, policy area, or worthy of its own development strategy. There is no Secretary of Tourism, Department of Tourism, or even sub-division of "Tourism" anywhere in the US federal system. The departments that are tasked with the administration and maintenance of tourist sites like national parks and monuments focus on the preservation of nature and heritage, and on educational programs, not on ways to boost tourist numbers or revenue from tourism. The diagram in Figure 17.1 outlines the organizational chart of the US central government. While no department is specifically tasked with tourism development, the highlighted boxes in green are some of the departments that intersects with tourism as an industry.

Tourism governance falls to literally thousands of uncoordinated and discrete state and local departments and offices. Most, not all, states have tourism Agencies or Bureaus like the New Jersey Division of Travel and Tourism or Visit California. These are never in the top tier of state departments with personnel or funding levels that come close to higher priority areas of governance. While leading the other states in revenue from tourism, California consigns it to a minor role in governance dealing with it in the State Legislative committee on Arts, Entertainment, Sports, Tourism and Internet Media.

https://doi.org/10.1515/9783110638141-017

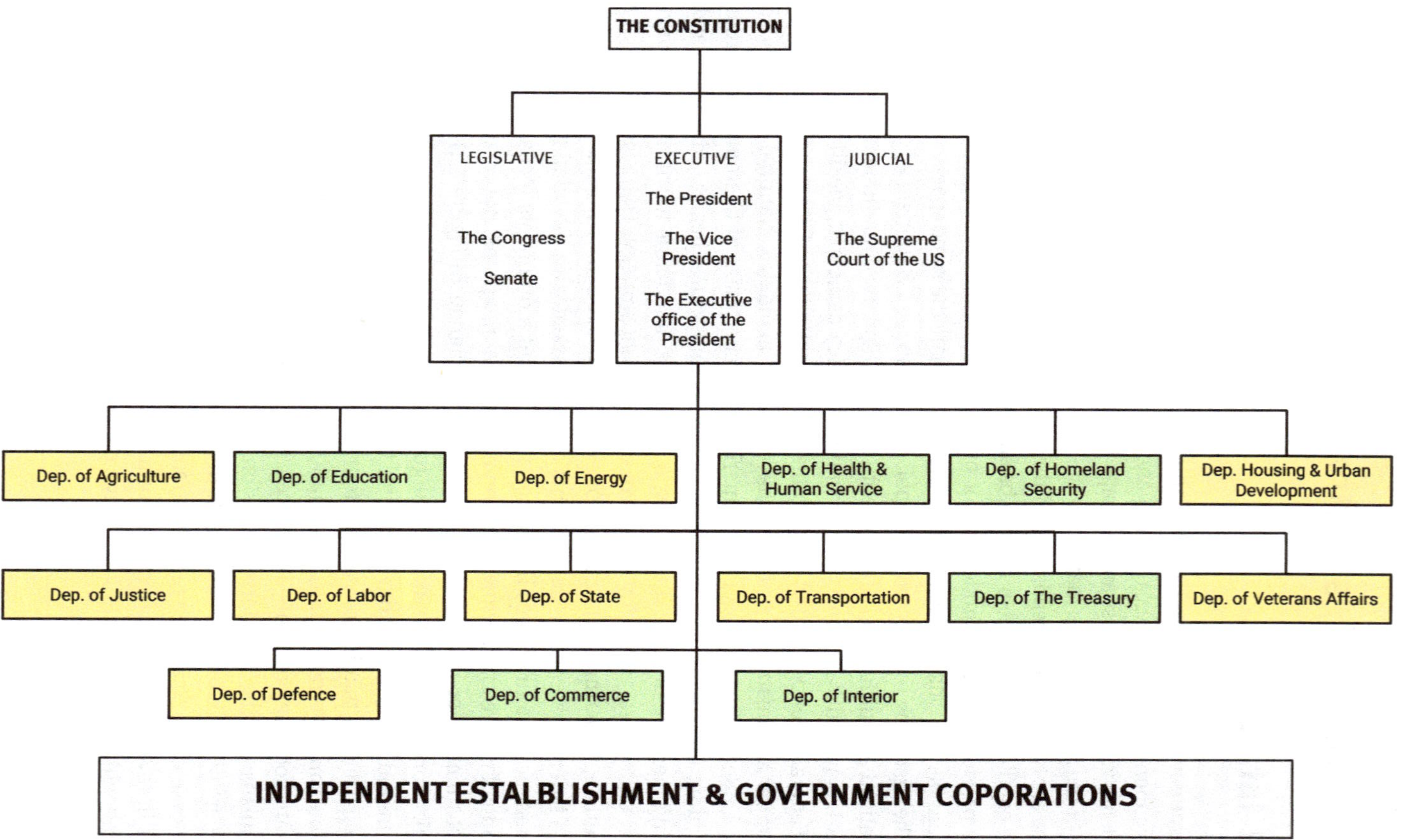

Figure 17.1: US Gopverment Organi Chart – Tourism intersect with all departmens especially the ones in green. (Source: Office of the Federal Register, 2013).

17.1.1 Visit California as Example

According to the Governor of California (2020), the Governor's Office of Business and Economic Development serves as the State of California's leader for job growth, economic development and business assistance efforts. Because tourism is not in the core of the government sectors, the California Office of Tourism is a department within the Governor's Office of Business and Economic Development that supports the promotion of California as a global tourism destination. Accordng to Visit Claifornia (2021) and Anderson (2011), Visit California was founded in 1998 and named: The California Travel and Tourism Commission. In 2011 it was renamed Visit California. Visit California is a nonprofit 501(c) corporation with a mission to develop and maintain marketing programs that keep California top-of-mind as a premier travel destination. Visit California operates a global marketing program in 14 markets on behalf of the tens of the thousands of California businesses who benefit from travel. The Office of Tourism collects the fees that fund Visit California's campaigns and initiatives.

While the most visited State, tourism has a very small footprint in California, as it does in every other US system of governance. In 2019, the direct contribution of recreational travel and tourism (both foreign and domestic) to the US GDP topped all other nations at US$581 billion (www.statista.com/statistics/292461). This is substantially greater than the next leading countries' including China US$403 billion, Germany US$143 billion, and Japan US$126 billion.

17.1.2 New Jersey as Example

Similar to California, another highly visited State, New Jersey, has no stand alone department that advises on or manages tourism. There is a Visit New Jersey program that deals with promoting tourism attractions within the state. According to the Official Travel Guide (2021), Visit New Jersey is an initative to promote all types of attractions in the state and encourage visitors to exeperience a wide range of amentities.

17.2 Private and Public Sectors

Private sector tourism development in the United States ranges from family-owned cafes and bed and breakfasts to very large-scale attractions like Disneyland, sports teams, concert venues, and the casinos in Las Vegas. Very large projects are often granted variances from local codes. Some sports complexes and convention centers receive public contributions, direct for their construction, and in the form of tax breaks. This type of public/private initiative is usually ad hoc or arranged on the spot for specific projects and not a part of the on-going machinery of governments.

Large-scale public/private tourism schemes are democratically supported by the promise of economic spill-over effects of trade shows and professional meetings at a convention center, or the large crowds that converge on sports arenas and concert halls. It is widely understood that large crowds attracted to these venues will patronize local hotels, restaurants, and lesser attractions. Some such events attract tens of thousands of attendees, large enough to fill many hotels and restaurants, close streets, and otherwise take over entire urban centers for up to a week.

US Federal tax policy provides some loopholes and tax breaks for hotel developers, but no more than for the fossil fuel industry, pharmaceuticals or industrial agriculture. Beyond the tourist visa, there are no other federal policies that specifically target tourism with expedited travel approval processes, or hospitality industry relief from labor laws, etc. Tourism is treated as an undifferentiated component of the US economy, like any other component. If a tourism development plan runs up against federal law in the United States, the law is more likely to restrict, not support, development. Vast areas of coastline, forests and mountain ranges are protected from development by federal statute. The rationale for protection is these areas should remain wild and untouched and exist for the enjoyment of all the people, not just those who would profit from their development.

The distinctive way that tourism is administered at the federal level in the United States is based on a separation of tourism as "the pursuit of happiness" (an "inalienable right" of US citizens as per our Declaration of Independence) from tourism as the pursuit of profit. Which brings us back to the tension between capitalism and democracy.

Controversially, early in the Trump administration the Secretary of the Interior proposed a dramatic increase in the entry fees to national parks and monuments. Fortunately, none of the Trump administration's fee increase proposals were implemented. It remains possible to visit Yellowstone and the other affected parks by paying US$20 per car load for a five-day permit. This works out to US$4 per person for a family of five travelling together. By comparison, the Department of Interior under Trump proposed a new rate of US$70 per person or US$350 for a family group of five. This proposal prompted an immediate *New York Times* editorial, "National Parks for the One Percent." Timothy Egan, the editorial writer, makes the following observations:

> [T]his is the Trump ethos. There's always a velvet rope – coming soon at the rim of the Grand Canyon – a place for V.I.P.s, deal-makers and insiders, and too bad for everyone else. . . . All national parks should be free like the great museums of Washington. . . . First we already own these parks – Glacier, Olympic, Mount Rainier, Zion, Yellowstone, the names themselves music to the lovers of magic in the natural world. They are a birthright of citizenship. . . . National parks are not theme parks, market-driven to match the latest entertainment blockbusters. . . . (Egan, 2017)

The language of the editorial indicates that the thriving tourist economy in the United States has (until recently) operated under a pact between democracy and capitalism in which capitalism happily yields to and profits from the democratic principles of equality and inclusion at least in the tourist sector. Our most sacred attractions are free and welcoming to all. Indeed, they are funded by the very taxes US citizens pay. This has made it possible to grow a robust and differentiated tourist economy around them, an economy nurtured by a steady flow of people from all backgrounds and walks of life, indeed from all over the world. Crucial to our great natural and historic attractions is the fact that every kind of human being from paupers to kings may wish to be in their presence. Should an attraction be barred for any reason, especially if it is made prohibitively expensive for most people, the flow of tourists will dwindle, and the tourist economy in its specifically US variant will suffer. Entry fee increases are not shared with the entrepreneurs who make up the penumbra of services in the surrounding tourist economy. Increasing entry fees will only serve to depress the tourist economy.

The historical framework for the development of tourism in the United States is inseparable from the historic deployment of a democratic US pluralistic culture across the continent. The most profound expression of the deep rationale for the absence of coordinated federal tourism policy in the United States is found in the lyrics of Woody Guthrie's (1944) "This Land is My Land":

> This land is your land this land is my land
> From California to the New York island
> From the redwood forests to the Gulf Stream waters
> This land was made for you and me

The United States has never had a tourist sector that was designed and built specifically to attract wealthy foreign visitors. All of US tourism was first built for democratic internal use, from the "See America First" movement started in the 1880s (Shaffer, 2013). Foreign visitors today come to see US culture and landscape pretty much as it was first created as a mirror for ourselves.

17.3 Some History

In assessing the overall shape and direction of US tourism, it is important to bear in mind that the United States was founded by Puritans. Gambling was prohibited in Plymouth Colony, as were theatrical productions. All of leisure was viewed with suspicion: "idle hands are the devil's workshop." The Sunday after Abraham Lincoln was assassinated most of the sermons preached in the United States emphasized that he was killed by an actor and he would still be alive if he hadn't given in to sinful temptation and gone to the theater. (Reported to me in conversation with my paternal great grandmother Emily Amelia MacCannell who was eight years old when Lincoln was assassinated and who attended closely to reports and events in the aftermath.)

The idea of travelling for pleasure was complicated. We were not supposed to do anything for pleasure. But if you were going to experience pleasure you would better be somewhere far away from home where you would not be seen.

Proto-typical tourist culture in the United States is based not on tourists' travel, but on bringing the attractions to the tourists. In the 19th century, travelling ("touring") musical and theatrical productions, burlesque, vaudeville, circuses and fairs, brought entertainment to frontier towns. Shakespeare was performed in the mining camps of the 19th century American West. These were all private, entrepreneurial productions and famously operated outside of local of and national regulation. The meme of "running away with the circus" was a stereotype of youth breaking free from small town roots and restrictions. "Patent" medicine containing cocaine and morphine was available from "snake oil salesmen" who travelled by wagon from town to town proffering entertainment from scantily clad female assistants along with their high pressure sales pitches.

Some of these early touring shows became hugely successful financially both nationally and internationally and created the first entertainment super-stars. E.g., P.T. Barnum's ("This way to the EGRESS. Only 25 cents to see the EGRESS," "There's a sucker born every minute") travelling circus. As the American frontier was disappearing, Buffalo Bill Cody's Wild West Show featured Annie Oakley's and Frank Butler's sharp shooting prowess, "Wild Bill" Hickok playing himself, and Calamity Jane telling fanciful stories of the American frontier. Cody staged re-enactments of Custer's Last Stand with "authentic" Indian combatants before thousands of paying visitors in the US and for Europeans including European royalty.

These shows synergistically became the framework for future cousins like rock-n-roll oncerts and not incidentally laid the foundation of desire of Europeans and East Coast Americans to see and experience the "wild west" for themselves. Cody's Wild West Show began touring in 1872, the same year that the Yellowstone Act was signed into law setting up the US system of National Parks and Monuments, the only Federal program that might be argued to be directly linked to tourism.

Diverse ethnic sub-cultures in the United States began to attract and influence thinking and desire beyond their localized expression via these early travelling entertainments. Black American jazz from New Orleans travelled north by Mississippi paddle-wheel steamers, floating theaters actually, bringing the "Blues" first to St. Louis and eventually to Chicago and beyond. Jewish American song writers penned some now universal standards like "Somewhere Over the Rainbow" for movies and for Broadway shows that travelled the entire continent.

The movement of sub-cultures mirrors the movement of tourists in the US. When the mob still owned the casinos in Las Vegas it was also the primary Western venue for New York Broadway shows.

17.4 "Follow the Yellow Brick Road"

So how does a nation that abjures all central planning and federal support for tourism have one of the most robust tourist economies in the world? To answer this question, it is necessary to look at the history of recreational travel and its fit with a polyglot frontier culture in the United States.

As soon as the automobile was invented, and before there were roads for anything except horse carts, people dreamed of driving from coast-to-coast and the first such trips were undertaken and heavily publicized in the early years of the 20th century. There is a substantial "Romance of the Road" type of tourism in the United States that probably reached its zenith with the publication of Jack Kerouac's *On the Road* in the late 1950s.

The tourists began travelling to the attractions rather than vice versa, and the culture began to acknowledge tourism rather than vice versa. Witzel (1996), Kelly and Scott (1990), and Caton and Santos (2007) explore tourists' experiences along Route 66 and confirmed that a romance and eventually a significant culture industry grew up around " Route 66." Opened in 1926 and called "The Main Street of America" and "The Mother Road," Route 66 connected Chicago to the Pacific Ocean beach in Santa Monica. Route 66 inspired songs ("I get my Kicks on Route 66") and a popular television show. Now it was possible for everyone to travel "Somewhere Over the Rainbow." Route 66 was superseded by the interstate highway system in the 1950s. Today it is no longer a way to get to the attractions of Chicago, Los Angeles, or the Pacific Ocean. Its last remaining, noncontiguous segments have become tourist attractions in their own right. The roadside tourist attractions along 66 were called "catch pennies" and "tourist traps." They included snake and alligator farms, opportunities to photograph Indians wearing buckskins and feathers, etc. They were usually created by the owners of petrol stations or motels to give their business some "destination advantage."

A step up from the "tourist traps" are a proliferation of private collections and museums found everywhere in the US. Examples would be the Museum of the Hot Rod in Florida, the Barbie Doll Museum in California and the former private homes of notable citizens that have been turned into small shrines for the curious. Likewise, many of these roadside attractions became popular and famous. Childrens' "playlands" featuring petting zoos and tame and safe simulacra of carnival rides became the template for Disneyland (Lukas, 2008).

A distinctive US type of attraction is the "Hall of Fame." Some of these have actually become world famous like the Baseball Hall of Fame, or the Rock and Roll Hall of Fame. But ironically and bathetically there are scores of lesser-known Halls like the Sports Mascot Hall of Fame or the Adult Video Hall of Fame, and the Cowgirl Rodeo Star Hall of Fame. The famous Halls of Fame are created and supported by national organizations like National League Baseball and have rigorous standards for inclusion. The lesser known Halls are supported by local clubs or even enthusiastic individuals and may have idiosyncratic standards.

In order to compete with the automobile, the train had to be made into something more than a transportation device. From the early to the mid-20th century train travel became integrated with luxury hotel construction in the West and train advertising emphasized its sightseeing potential.

From the 19th century forward to today there has been a patchwork of specialized summering places in the US roughly equivalent to European Spas. The Catskill villages and resorts offered bourgeois New Yorkers a place to cool off for a few weeks. Dude Ranches in the West gave "city slickers" a chance to learn to ride horses and play at being cowboys and girls. Children everywhere went to "summer camp" some of which offered specialized experiences and services, e.g., music, science, exercise, and special diets. Franklin Roosevelt built a spa in Georgia for himself and others who suffered from polio. Even though he was president, his Warm Springs resort was entirely his own private initiative.

The postmodern versions of this would be the hedonistic temporary "Spring Break" communities that pop up on Florida Beaches, or the mastodon Burning Man festival in Nevada. As mentioned above, at the sub-national level tourism development in the United States sometimes receives state and local municipal support. States and cities give land, tax breaks, and maintenance support to their parks, museums and zoos, and public art.

At the Federal Level, there is a system of hundreds of protected national parks and monuments examples of which can be found in every state. These include a vast array of places of scenic, historic, or scientific value; from fossil fields, to mountain ranges, to forests like Muir Woods in California. On the historic side they include battlefields like Gettysburg, the White House, the Statue of Liberty, etc. These are preserved to a uniform high standard, maintained, and staffed by Federal agencies. They vary in popularity from over 12 million visitors a year (Golden Gate, Blue Ridge, Great Smokey Mountain) to 16,000 (Isle Royale in Michigan), but all are tourist attractions (www.nps.gov/about us/visitation-numbers). Entry to most of them is free of charge (e.g., the White House, the Liberty Bell), or very nearly free (US$20 for a five-day vehicle permit in Yellowstone). The rationale for open access is that US residents have already paid for them with their tax dollars and that they belong to the people not to some commercial enterprise. Park guides must pass rigorous tests of their relevant historic or natural science knowledge before they can interact with tourists. The national parks and monuments are frequently referred to in the popular press as the "Crown Jewels" of the United States.

It should be noted that US national parks and monuments are rigorously protected from commercial encroachment. The public initiatives at the Federal level are explicitly anti-commercial. The Walt Disney Corporation proposed to take over management of Gettysburg and other Civil War battlefield national monuments, but was denied by Congress (Synnott, 1995).

Accordingly, none of the expense and energy that is poured into maintaining the National Parks and Monuments system is connected to or justified as "tourism development." It is all for the preservation of "our priceless" nature and heritage and simply for the educational benefit and enjoyment of the people. Even though they may be the ultimate driver of a multi-billion dollar US domestic tourist industry, nothing so crass as profits from tourism is considered in the policies leading to their preservation and protection.

In the United States, capitalism and democracy are arduously trying to grow up together. So far, the potential for deep structural contradictions between democracy and capitalism has been ameliorated by tourism. This may be the reason why the nation with no explicit tourism development policy also happens to have one of the most robust and flexible tourist economies. Up until now, there has been an implicit understanding that when we put a price tag on the pursuit of happiness, capitalism and democracy will have to go their separate ways.

References

Anderson, M. (2011). *Tourism board changes name to Visit California*. Sacramento Business Journal. https://www.bizjournals.com/sacramento/news/2011/09/26/visit-california-name-change-tourism.html

CA Governor. (2020). *Governor's Office of Business and Economic Development*. California Gov. Office. https://business.ca.gov

Caton, K., & Santos, C. A. (2007). Heritage Tourism on Route 66 : Deconstructing Nostalgia. *Journal of Travel Research*, *45*(May), 371–386. https://doi.org/10.1177/0047287507299572

Crespo, S., Mullen, J., Hanrahan, K., Lloyd, J., & Root, R. (2021). *New Jersey Official Travel Guide. Visiting NJ*, 117.

Egan, T. (2017). *National Parks for the 1 Percent*. New York Times.

Guthrie, W. (1944). This Land Is Your Land. *The Asch Recordings*.

Kelly, S. C., & Scott, Q. (1990). *Route 66: The Highway and Its People. University of Oklahoma Press*, 205.

Lukas, S. A. (2008). *Theme Park*. Reaktion Books Ltd.

Office of the Federal Register. (2013). The United States Government Manual. *National Archives and Records Administration*.

Shaffer, M. (2013). *See America First: Tourism and National Identity, 1880–1940*. Smithsonian Institution.

Synnott, M. G. (1995). Disney's America: Whose Patrimony, Whose Profits, Whose Past? *JSTOR*, *17*(4), 43–59.

Visit California. (2021). *Visit California*. Visit California. https://www.visitcalifornia.com

Witzel, M. K. (1996). Route 66 remembered. *Motorbooks International*, 192.

Kanokwalee Suteethorn & Judith Bopp

18 Tourism Without Governance: WWOOF

18.1 Introduction

WWOOF or "World Wide Opportunities on Organic Farms" is an organization started in 1971 in the United Kingdom. First it was a place where urban habitants connected with weekends working in a farm as the way in which people could simply recover their connection with soil and nature (WWOOF FoWO, n.d. a). Today WWOOF is a platform where hosts and helpers find each other without monetary benefits as it is based on voluntary engagement. While WWOOFers use the opportunities of working in organic farms to travel around the world (Madden, 2010), host farms provide accommodation for the WWOOFers with impetus of sharing the knowledge on organic farming, supporting organic life styles and contributing to the worldwide organic food movement. In some regions, WWOOF is also a mechanism local farmers used against capitalism, political economic inequity, and untrustworthy government (Mostafanezhad, 2016; Yamamoto & Engelsted, 2014).

Tourism industry is a major income of many nations. Even with the effect of global recession, travel and tourism were a direct contribution of 10.3% of global GDP and experienced 3.5% growth outpacing the 2.5% of the global economic growth in 2019 (WTTC, 2019). After World War II, the emergence of the middle class increased the demands for tourism (Theng, Qiong, & Tatar, 2015). In the 1960s, mass tourism became popular as it provided convenient package travels which made tourism accessible and affordable for the broader public. However, mass tourists had less contact with locals and unknowingly caused negative social, cultural, environmental, and economic impact to the natives (Butler, 1990). Some attempts to reduce these problems by completely disconnecting tourists from local residents, such as Club Med style resorts, were intended to avoid conflict and confrontation between mass tourists and destination communities which however make people have even less appreciation of the values of the environment and cultures of the places they visited (p. 41).

Around the 1980s, alternative forms of tourism were initiated with purposes to mitigate the negative impacts of mass tourism. It occurred as a status quo ante of mass tourism where travelers neglect local people, culture, and their impacts on the local environment (Triarchi & Karamanis, 2017). Some types of alternative tourism such as community-based tourism, ecotourism as well as volunteer tourism aim to resist capitalist globalization especially in the Global South (Butler, 1990; Cohen, 1987; Higgins-Desbiolles, 2010). Alternative tourism including backpacking or volunteer tourism has grown significantly over the past years in many countries. Ooi & Laing (2010) found out that making cultural experience, interacting with local people, and broadening one's awareness, personal development or "re-evaluating"

https://doi.org/10.1515/9783110638141-018

(p. 200) personal lifestyles are important volunteer motivations, and that "motivational overlap" often exists between backpackers and volunteer tourists (p. 191). Cohen (1987) argued that even though conventional mass tourism is claimed to cause negative impacts in so many aspects, mass tourism can provide economic benefit and can be the source of major national incomes in many countries. On the other hand, alternative tourists that strived for authenticity of local culture may disrespect the natives by "museumization" of the authenticity of local culture (Cohen, 1987). Similarily, Butler (1990) argued that although it may seem like alternative tourism is a form of tourism that solves the problems of the status quo, alternative tourism itself could cause problems which can cost more than the benefits it provided.

Volunteer tourism is one form of tourism that can provide opportunities for smaller groups of travelers to be part of native communities. Volunteer tourism or voluntourism are portrayed as a combination of volunteering and sightseeing for people who want to explore new unseen cultural and natural destinations (Ooi & Laing 2010; Terry, 2014; Wearing et al., 2020). Within the last decades, the number of volunteer tourism projects and participants has expanded globally, and its organisation changed from "primarily nongovernmental" to a range from "full nonprofits to openly for-profit ventures" (McGehee, 2014, p. 847). There are several agencies that provide different travel and volunteer programs for those who are looking for a unique experience. However, the impacts from volunteer travellers and relationships between visitors and local people are still questionable (Guttentag 2009; Jenkin, 2015; Terry 2014; Tomazos & Cooper, 2011). Visitors staying for a short time at the places cannot see their real impacts on the locals. Besides paying money, many volunteer tourists donate their time and labors on working with locals. However, short-term volunteer workers may "negatively impact labour demand" or "disrupt local economies [. . .] by promoting a cycle of dependency" (Guttentag 2009), and can also cause the full time workers to lose their jobs (Dunford, 2010). Despite its good intentions, volunteer tourism is generally criticized because it causes more negative impact on natives than benefits to the host communities. Volunteer tourism such as WWOOF does not provide national economic benefits as much as it is a form of non-monetary exchange where farm hosts provide accommodation and food while WWOOFers work at their farms. With an informal form of volunteer tourism that is untied to the local government, it lacks a monitoring and evaluation system (Steele, Dredge, & Scherrer, 2017). Volunteer tourism may seem more desirable in comparison with mass tourism but there are more risks of damaging natural and cultural resources. While volunteer tourists feel like they contribute to the localities and are involved with them to a much greater degree, they have privileges to go beyond where conventional mass tourists can go. Some volunteer tourists may become inconsiderate and disrespect local communities and invade prohibited areas or interfere with local traditions (Cohen, 1987). Finally, volunteers' travel motivations are often driven by personal achievement in addition to or rather than altruism (Guttentag, 2009).

WWOOF is a type of volunteer tourism or voluntourism, which is similar to eco-tourism where travellers help with local environmental projects or help local people while travelling. It is part of alternative tourism that started in the early 1980s. Not much research literature has addressed host-volunteer relationships within volunteer tourism constellations thus far (Wengel et al., 2018, p. 46). As Wengel et al. (2018) discovered in their study on WWOOF, "critical tensions regarding the interpretation and practice" (p. 53) of the global WWOOF principles may arise between hosts and volunteers, for instance regarding their mutual expectations "around the nature of the work" (p. 46). Cultural factors may play a role in causing such tensions, as well as the broader socio-structural settings which both hosts and volunteers are used to.

This chapter examines the history and organization of WWOOF in its international and national levels by exploring case studies from four countries: Thailand, Japan, New Zealand, and the United States; the chapter examines the process, opportunities, and constraints of WWOOFing in these countries. It investigates the way different cultures, political economies, and social structures in each place shape the way WWOOFers connect with the host farms. It also explores the aspects of WWOOF from host farms' and WWOOFers' points of view and the effect of local governance on WWOOF as a form of alternative tourism. Through four case studies in these different countries, the following questions are explored: How do cultural, and economic situations, including cost of living that differ in each country affect the WWOOFing experience? Is it a mutual relationship between host farms and WWOOFers? Do they learn from each other and both get their benefits and achieve their goals? Is WWOOF a sustainable and responsible volunteer tourism organization? And finally, can non-governance tourism structures cause ethical and legal issues between hosts and volunteers?

18.2 Story of WWOOF

Started in 1971 in the United Kingdom, WWOOF was initially named "Working Weekends on Organics Farms." Sue Coppard, a London secretary, helped organize and provide opportunities for urban dwellers to participate and stay at organic farms in the countryside (WWOOF United Kingdom, n.d.). WWOOF is one type of alternative tourism that does not occur from monetary motivations (Yamamoto & Engelsted, 2014, p. 967) since it is based on voluntary engagement and exchange between hosts and helpers. It is one of the earlier forms of volunteer tourism that focuses on farms and countryside visiting opportunities for urban dwellers. Initially, Coppard's idea was to provide a meaningful countryside activity, but the exchange soon expanded onto the early organic movement and had common intentions with the UK's Soil Association and the alternative magazine Seed. In the following, Coppard initiated a contact base for interested helpers to get in touch with farms for a weekend. The concept was well received among the like-minded. The intention with this exchange from the beginning was to get first-hand experience in organic farming and to exchange with people in the movement. The WWOOF community thereby supported the organic farming movement which is generally labor-intensive compared to mechanized agriculture (Pier, 2011). Due to much positive response, the movement soon grew beyond its frame of a weekend escape, and the administration of member hosts became overwhelming, analogously. The WWOOF idea spread to an international public. With the broadening frame, the name was adapted to "Willing Workers on Organic Farms," eventually becoming "World Wide Opportunities on Organic Farms," to avoid misunderstanding of the term of workers (WWOOF International). Several factors of WWOOF are notable in terms of tourist experience, and influence farmers' decision to host, and travelers to volunteer; mutual cultural exchange, the sharing of knowledge of organic farming, and eco-friendly living, authentic local experiences, and the connection to where food comes from.

18.2.1 Self-Governance of WWOOF

WWOOF is presented in 130 countries, covering over 12,000 hosts around the world with currently about 100,000 WWOOF members (WWOOF FoWO, n.d. a). Many countries run their organizations independently through national WWOOF websites while sharing their mission of hands-on education, cultural exchange and the support of local organic farmers. There are further 83 countries without their own national WWOOF organizations running under the umbrella of WWOOF Independents (WWOOF Independents, n.d.). These are usually smaller countries, those with relatively few hosts, or those that just started to engage in the organization. In 2012, the Federation of WWOOF Organizations was established. It acts as an umbrella for all national and the Independents groups (WWOOF FoWO, n.d. a). WWOOF is membership-based in exchange for a small annual membership fee that provides a host list for both hosts and helpers for direct contacts. A pre-digital list used to be regularly updated and published in a paper booklet before adopting an online version of it. However, WWOOFing mobile apps exist nowadays for smartphone users like for Australia (WWOOF.com.au). Hosts are supposed to be contacted in advance, but many accept WWOOFers on spontaneous request. Most hosts indicate their settings and requirements, for example preferred months and lengths of stay, diet, number of WWOOFers or acceptance of children, in an online preview. There is also a brief introduction of the hosting farm or a personal call. WWOOF is an example of self-governed tourism in which no intermediary actor such as a governmental body, travelling agency, or tourism corporation is involved: travelers arrange their trips privately and contact hosts directly.

18.2.2 WWOOF in the Context of Alternative Tourism

According to Ooi & Laing (2010), generally, "there is an inclination for backpackers to be volunteer tourism participants" (p. 203) and they point the way towards a new and increasingly sustainable market niche: Low-cost opportunities of traveling such as WWOOF "could benefit from increased backpacker participation, which in turn assists local farmers and communities around the world" (p. 203). While originating in the idea of exchange of work and experience in organic farming, WWOOF evolved into a kind of tourism, particularly increasingly attracting alternative tourists who seek it as a means of traveling the world and exploring unknown places, although attracting tourists was not the network's original intention. This opens opportunities for countries to integrate WWOOF in their tourism governance strategies. Since WWOOF is a rather low impact form of traveling, in terms of negative ecological impact, and has potential to increase cultural awareness and encourage travelers to explore unknown places, this opens opportunities for ecological awareness among travelers and could serve as an alternative strategy for national tourism. WWOOF also advocates local sustainability movements, and therefore more or less unintentionally sustainability strategies of countries. From a traveler's perspective, WWOOFing offers a low-cost way of volunteering in ecological projects during a gap year, while organized volunteer projects often require high fees. However, some travelers see WWOOF as an easy opportunity for budget traveling, and are not necessarily aware of or interested in contributing to sustainability goals which are behind the organic farming engagement. For these travelers, WWOOFing realizes an alternative travel mode where the traveler is accommodated in exchange for working hours, and reduces traveling costs to the in-between journeys. Since the length of stay at WWOOF farms ranges up to several months, travelers are thus able to live in one place without major expenses, although many budget travelers prefer short stays to save a few nights of paid accommodation. McIntosh & Bonnemann (2006) reflect on WWOOF beyond the budget motive as a type of farm tourism: Travelers are "expected to become fully immersed in the day-to-day activities of their hosts," receiving a "hands-on' experience" (p. 84) of organic farming. Since WWOOFing also includes farm and gardening activity, variable according to "WWOOF host and type of farm" (McIntosh & Bonnemann, 2006), it can be seen "as an 'accommodation' and an 'activity' based form of farm tourism (Davies & Gilbert, 1992)" (McIntosh & Bonnemann, 2006, p. 84).

18.2.3 Traveler-Host Relationships in WWOOFing

WWOOFing requires that hosts share their private spaces and often family activities with the volunteers they welcome. The standard suggests four to six hours of work per day with one or two days off per week. In return, WWOOFers are provided food and accommodation which can be in the host's private house or in an outside building, campervan, tipi or tent, depending on weather conditions but mostly at the same site. There is particular emphasis on the act of sharing: living, meals, knowledge, ideology, organic lifestyle, and so on, since WWOOFers get the chance to be involved in the host's daily routine and way of living: meals are often shared at the family table, and WWOOFers participate in leisure activities. WWOOF farms are often located in very scenic settings, have access to the sea or other natural sites, and this is how many WWOOFers get to enjoy the nature experience, and relaxing and peaceful atmosphere. Work can thus be perceived as leisure rather than duty. However, individual experiences vary and depend on both the hosts' and the visitors' personal attitudes, and often their interpretation of the WWOOF principles and the volunteers' ambition to work and to integrate also: WWOOFers' "willingness to work long hours may be dependent upon the perceived value of the experiences they gain and their respect for their hosts" (McIntosh & Bonnemann, 2006, p. 91). The WWOOFers' engagement may hence relate to their perception of personal and social reward of their work.

There is no binding framework on the international scale that sets the dealings, but the adhering countries commit to a standard of common WWOOF principles. Ethical guidelines may be set individually by the participating countries (WWOOF.com.au), and WWOOFers have the opportunity to express complaints about hosts. In case of urgent concerns, hosts can be banned from the network. Hosts welcome unknown travelers with different cultural backgrounds and work ethics, and share their private spheres.

Founder Sue Coppard thinks about the success of WWOOF that it offers a "simple cultural exchange based around organic agriculture" with "positive rewards with fresh and inspiring experiences for both Host and volunteer" (Pier, 2017, Section 3, para. 8). A distinct feature of WWOOF, which makes this kind of travel special, is that of enhanced social interaction between host and traveler, and the opportunity for both friendships and insights into different cultures around the world. Most of the other volunteer tourism organizations focus on Global South countries or developed countries. However, WWOOF members come from countries across the continents covering all, developed, developing, and less developed countries. The organization of WWOOF and the individual experiences that WWOOFers make in different countries depend on the local settings. How do working experiences in these different social, cultural, and economic contexts contrast? The following sections discuss WWOOFing experiences in four different countries.

18.3 WWOOFing Around the World

18.3.1 Thailand

WWOOF Thailand has its own national administration since 2012 (from an interview with WWOOF Thailand coordinator, 2017). The network is operated by a non-Thai person who also engages to establish the permaculture scene in Thailand. The concept of WWOOF is quite different from other countries which the design and content of their website already suggests: "WWOOF Thailand – travel longer, cheaper, better" (WWOOF Thailand) alludes to WWOOF as a travel experience for adventure seeking backpackers while the idea of organic farm stay is not highlighted. Thailand being a popular destination for backpackers, one would expect WWOOF as an alternative tourism finding much response. However, the number of WWOOF hosts there is moderate and many of the WWOOF farms are run either by foreigners or Thai-foreign couples. Generally, WWOOFing has not been taken up as much as in other countries. Several reasons can be given for this. First, volunteer work in Thailand is a common practice to many people, but it is orientated towards social projects and temples rather than towards private people. Second, most WWOOFers are international travelers who do not speak the language, and many local people, especially in rural areas, are intimidated by a language barrier, which plays on their ability to open their places for international volunteers. Third, many farms, although they practice a diversity of natural farming methods, do not declare as organic farms, or the concept is unknown to them, and so is the WWOOF network. Fourth, and importantly, with low agricultural wages in Thailand, hosts usually have high expenses for hosting and feeding the volunteers compared to the expenses for hired labor. Some advantages and challenges to WWOOFing in Thailand are highlighted in the following comments on personal experiences made by the WWOOF Thailand coordinator and the author's[1] own farm stays. In Thailand as in other countries, WWOOF offers opportunity of low cost traveling and the sharing of values between hosts and travelers. It strongly relates to the growing and sharing of food: Eating is "the ideal medium for connecting with others" (from an interview with WWOOF Thailand coordinator, 2017).

According to the coordinator, WWOOF Thailand has been running successfully since its establishment in 2012, and this is in particular because of the general friendliness and culture of welcoming travelers common among Thai people. Most local hosts cannot afford traveling abroad, but they get to meet people from different countries, languages and cultures through WWOOFing. From a traveler's perspective, WWOOFing is a low or no cost way of traveling or volunteer holiday abroad, but it is also a lifetime experience with higher personal meaning compared

1 Judith Bopp (author).

to regular backpacking (from an interview with WWOOF Thailand coordinator, 2017). From my own experience[2] and in comparison to the practices of other countries, the one of WWOOFing for monetary contribution in addition to their volunteer work was new. My first WWOOF farm in Thailand on a permaculture farm East of Bangkok illustrates this experience: The site was run by a Thai-foreign couple and had a garden, pigs, goose, ducks, and chicken. About 10 volunteers at that time were helping with planting, weeding, harvesting, animal feeding, earth building, and general maintenance. However, the farm owner brought in an economic argument in valuing the WWOOFers' work: He found the benefit of hosting WWOOFers dissatisfactory, since the expenses of hosting and feeding WWOOFers do not make up for their accomplished work. The WWOOFers normally did not meet his expectations in terms of productivity. At the occasion of another visit a few months later, the owner had introduced daily contributions from the volunteers to cover for their food or asked them to provide their own food at the local nearby market. Since WWOOF is supposed to be volunteer based, this practice is against its principles. Regardless, in the owner's reasoning, hiring Thai workers for garden and house maintenance, also workers from Myanmar or Cambodia, at the minimal daily wage, is lower than the WWOOFers' meals but mostly returns better labor force.

This shows a disbalance between hosts' and WWOOFers' expectations, and although many WWOOFers undeniably use the volunteering for cheap traveling rather than work experience, the economic argument may evoke feelings of being exploited from a WWOOFer's perspective. The WWOOF Thailand coordinator sees positive effects in taking small contributions or donations, as it adds income to the farmer families and their direct improvement, which is especially positive for rural farmers. This volunteer tourism thereby receives a second, economic connotation. Another WWOOF experience gave a very different impression from the first farm: The organic fruit orchard was set in the Eastern region of Thailand. The young family was about to set up an entire permaculture farm inspired by self-sufficient farming, and had a kitchen garden with vegetables, herbs, and spices, and the orchard to maintain at the time of the first visit. The farm stay gave valuable learning experience about the permaculture concept and many Thai local natural farming practices such as sustainable water management, composting, preparation of indigenous microorganisms, and charcoal making from leftover fruit and casks. The couple, urbanites with academic backgrounds, were seeking an alternative living away from Bangkok and getting local farmers involved in sustainable farming. They had hosted WWOOFers from different countries before, and had mixed experiences. While many WWOOFers were very helpful, eager to work, and genuinely interested in exchanging organic farming skills, others were less reliable or reluctant to get involved with farm work, and searching for free accommodation rather than the

---

2 Judith Bopp (author).

working exchange. To some extent, they regretted lacking commitment to arrangements, and troublesome communication with many volunteers in advance of their stays. Some volunteers cancel their stay at very short-notice or even with no prior notice. According to the hosts, a Thai couple, this is partly due to the Thai WWOOF site that evokes imaginations of budget adventure tourism rather than the rural farm and cultural exchange. WWOOFers also tend to prefer short-term stays compared to the common several weeks or months commitments in many other countries. One reason for this different character of WWOOFing in Thailand might be that the national tourism concept encourages short visits. Since the country is convenient to reach and to travel around, and affordable, many travelers go for short periods. Many WWOOF hosts however prefer longer stays of between two weeks and several months over the fast transit, because effort is taken to pick up WWOOFers at nearby stations and to introduce them into duties and local routines, and also, WWOOFers bring different skills and cultural backgrounds. WWOOFing hence requires engagement from both the hosts' and the volunteers' side.

The challenge of reliability as well as the costs of hosting volunteers led some hosts to retire from the WWOOF involvement, such as the owners of an organic coffee plantation in Northern Thailand who changed from WWOOF to develop an eco- and volunteer tourism site that opens to volunteers who pay at daily rate to engage with a local hill tribe village. Another organic farm in Southern Thailand had a similar experience: "WWOOFers are too expensive" (from an interview with farm owner, 2015), and that is why they changed to receive volunteers who pay for their farm experience.

The issue mentioned above regarding the language barrier arose in many conversations with WWOOF hosts as a barrier to the popularity of WWOOF in Thailand: A number of organic farms and farming communities already exist in Thailand which could benefit from committed volunteers and would return authentic rural experiences to foreign travelers. However, on one hand, English is rarely spoken in remote rural communities, and on the other hand, hardly any travelers have knowledge of the Thai language. This might explain why most of the actual WWOOF hosts are foreign-Thai couples, English speaking urbanites or travelers who stayed in Thailand. These conversations also focused on another challenge for WWOOF, discussed above, that it seems as if the concept of organic farming is less embedded in Thailand compared to other countries participating in WWOOF. The nature of existing small-scale farming in rural Thailand might be one explanation: While organic farming is a concept designed in Western countries, many smallholders practice a variety of sustainable and locally adapted farming methods inherently, without adhering to a broader concept (Bopp, 2016). This might impede farmers' identification with the WWOOF concept and prevent some from joining the network. Apart from that, some hosts might feel shy about connecting to and building friendships with urban people and travelers from other cultural spheres.

Regardless of some challenges that WWOOFing meets in Thailand, where large-scale industrial farming is predominant and ecological degradation puts farmers' livelihoods in high risk, the organic movement can be a contribution to rural and peri-urban sustainability and thus can WWOOF contribute to realizing it. Especially in locations where policies are not in favour of sustainability practices, WWOOFing can help promote civil society engagement. WWOOFing in Thailand features unique experiences in many ways, for its "climate, foods, culture, music and mythology" are very specific and different from other countries (from an interview with WWOOF Thailand coordinator, 2017).

18.3.2 Japan

Cultural differences especially involving societies that do not speak up front about everything could cause communication difficulties, limiting mutual understanding. Japan is one of the most popular destinations for WWOOFing among Thais. There are several Thai books writing about their WWOOFing experience in Japan. Japan is the top destination for WWOOFing because of its safety, weather, and culture with a relatively short and affordable flight. With these advantages, the language barrier seems to be diminished. Many people who have experienced WWOOFing in Japan write reviews on Pantip (a very popular webboard in Thailand). There are found over 3,000 Pantip reviews on WWOOF Japan. Some of these writers had memorable experiences and would love to revisit. However, many of these Pantip WWOOFers were not happy about their Japan experience. The reasons of discontent of the WWOOFing experience were mostly the misunderstanding of the main purpose of the WWOOF program. Most of the disappointed WWOOFers anticipated to have free food and accommodation and spend time traveling around (www.reddit.com, 2015), while the main purpose of WWOOF is to work in an organic farm in trade for food and accommodation. Despite the mission of being mutual relations and cultural exchange, there are some who criticized host farms for taking advantage of the WWOOFers. One WWOOFer wrote about her experience WWOOFing in Japan in www.reddit.com:

> "Two of the farms I went to were literally WWOOFer slave camps. We would work 8 hours every day, alone or with other non-Japanese WWOOFers. The host families hardly noticed the existence of the woofers in the house. We were often called WWOOFer-san[3] rather than by our names" (reddit, 2011, WWOOF, comment 34, para. 5). The common working hours for WWOOFers are usually half day; however, each farm is different so it depends on the agreement between the host and WWOOFers.

3 San is one form of Japanese honorific title. It could be translated as Mr., Mrs., or Ms. and used to show respect (cotoacademy.com, 2016).

One of the WWOOFers in Japan said that for the past two weeks they didn't receive a day off because of "the busy season." She also commented that positive reviews for the host are not always true. Volunteers working eight hours a day should not be called a reasonable workload: "there are places that are actively exploiting the naiveness of foreigners for profit" (reddit, 2011, WWOOF, comment 34, para. 22).

Because the monitoring of WWOOFing is based mostly on the previous WWOOFers's review and the WWOOFers do not want to say bad things as other host farms might hesitate to take them in the future, there is an opportunity for host farms with a commercial operation to fully utilize WWOOFers in place of paid labor. If this is one kind of tourism, should people be able to participate in these situations as tourists if they travel with tourist visas? Should WWOOFers work in the program without working visas or a work permit? In what way could we be in equilibrium on these issues?

18.3.3 New Zealand

New Zealand was the second country where WWOOF was set up, in 1974, and among the popular WWOOF destinations (McIntosh & Campbell, 2001, p. 112). New Zealand is a classical destination for farm tourism, and WWOOF, being integrated in the concept of traveling itself, notably contributes to that reality (McIntosh & Campbell, 2001, p. 112) WWOOFers "can save money, learn about organic farming methods and gain experiences that are unique" while staying with the locals and experiencing their lifestyles (McIntosh & Campbell, 2001, p. 111–112). McIntosh and Bonnemann (2006) relate positive experiences to WWOOFing in New Zealand, such as the experience of "rurality", "opportunity to learn", "personal meaningfulness", and the "element of sincerity" (p. 82). In contrast to many other countries, engagement in sustainable farming as well environmental awareness are commonly inscribed in New Zealand: Although knowledge and practice of traditional farming methods diminished along the 20th century when agriculture became mechanized and technologized, and farmers shifted to chemical inputs and breeded seeds for yield and profit, "problems such as consumer health concerns have led to a recent and increasing return to more traditional organic food farming methods" (McIntosh & Campbell, 2001, p. 112).

New Zealand's greenness, scenic landscape, and moderate climate attract many travelers, including backpackers and other alternative travelers. The idea of work-and-travel has been established for a long time, and also volunteer tourism has become a reliable segment of national tourism, notably in combination with nature experience: As the official tourism website newzealand.com presents, "New Zealand is an amazing place to volunteer, especially if you are a nature lover." (newzealand.com, Section 1, para. 1). WWOOF can be considered one part of this segment, and is even mentioned and linked with the website of WWOOF New Zealand on newzealand.com (Section 1, para. 1). New Zealand WWOOF hosts extend over both main islands and are mostly embedded in spectacular landscapes that appeal to WWOOFers around the world who seek natural scenery. Regular WWOOF work routine allows for days off, so that exploring of the nearby sites is possible. Hosts often take their WWOOFers around, or public transportation as well as hitchhiking are available. WWOOF sites range from individual households with backyard gardens, even urban households, to farms with or without livestock. As a specialty to WWOOF New Zealand, the network provides a section for social volunteers apart from the farm stays. The author[4] made a personal farm experience at a seaside property in a bay near Christchurch. The owners, an elderly couple, maintained an apricot orchard, kitchen garden, and about 50 sheep that were taken for grazing every day. Accommodation was in a separate building on the property, and all meals were shared. The stay gave many opportunities to get involved in enriching conversations, and the author was taken to nearby hikes and to dinner at a friend's house. Thanks to the involvement in the farmers' routines, the stay was sustainably inspiring both in terms of skill sharing and cultural experience.

However, from a WWOOFer's perspective, negative feelings can also occur, such as perceived hostility and the feeling of being exploited: On one farm, the author experienced, along with a travel mate, unusual work hours up to eight hours per day, as well as unusually heavy and sometimes dangerous work such as lawn mowing on a steep and slippery slope. Also on this occasion, the daily portions of shared meals were small and there was no option to shop for extra food, given that the area was remote and no transportation was available. Negotiations were not fruitful and the hosts made us stay for two more days before letting us leave the farm. This reflects a possibility that WWOOF volunteers can be exploited in some cases as cheap labor, and in contrast to Thailand, New Zealand is a high-income country with high labor costs, usually benefiting from volunteers. WWOOFing there represents a balanced opportunity for both sides. Many hosts have steady need for helpers, and the relation between expenses for WWOOFers and their "productivity" is mostly balanced. Compared to Thailand, the economic argument is less pertinent. Yet, the feeling of being taken advantage of sometimes prevails in the experiences of both sides. Volunteers' conflicts with the WWOOF concept are hence of a personal nature and depend on whether the volunteer commits to the work for personal benefit or whether they feel their labor is being exploited, which likely happens when the host-volunteer-bonding is shallow.

18.3.4 United States

In 2015, around 25% of Americans or 62 million people volunteered through different organizations (BLS., US Department of Labor, 2016). More than 1 million Americans

4 Judith Bopp (author).

volunteered overseas (Lloyd, 2008). Today, United States, along with New Zealand, Australia, and Canada, convey the largest numbers of WWOOF host farms in the international realm. Several organizations participate in volunteer tourism by sponsoring trips with low-cost travel and accommodations as part of their Corporate Social Responsibility (CSR). Volunteer traveling trips funded by these organizations have been claimed to be more meaningful. Volunteer travelers can experience local culture and involve in helping with the host work.

There are 2,160 organic farms in the WWOOF USA program[5] (WWOOFusa.org, 2017). Most of the WWOOF farms in the United States are small to medium farms which are located where there are "Bohemian" cultural settings, in the Northwest regions of the country (Yamamoto & Engelsted, 2014). The United States is one of the popular countries for WWOOFing (Davis, 2018). Even though it generally does not have the highest numbers of organic farms (Bialik & Walker, 2019), Hawaii seems to have more host farms for WWOOFing than other parts of the country. There are almost 400 host farms in Hawaii (Davis, 2018). The reasons that farms in Hawaii join WWOOF and be farm hosts are not only to have volunteer workers in their farms but also it is the way in which farmers use to fight against neoliberal capitalism (Mostafanezhad, 2016). The intersection of organic agriculture and volunteer tourism organized by WWOOF became parts of a double movement resisting neoliberal capitalism caused by a lack of trust in government (Mostafanezhad, 2016).

Although most of the host farms prefer WWOOFers to stay longer to learn and exchange their cultures and knowledge on organic farming, there are many one-day WWOOF programs especially in the urban areas (WWOOFusa.org). Numbers of urban farms are found asking volunteers to work for one day or half day and do not provide either food or accommodations. These farms emphasize the opportunities for WWOOFers to work in the farms so they can be closer to nature which helps enhance their connectedness to nature as the original concept of WWOOF. On the contrary, many farms only accept long term WWOOFers because they have to teach and train the volunteer workers. I[6] sent a few requests to organic farms and creameries in Northern California to work and stay for one to two weeks but most of them mentioned that they prefer long term volunteers despite the information on their profiles that they welcome one week WWOOFers. I had a conversation with one host who said he just took a male volunteer who planned to stay for three months. The host had limited accommodation so he would prefer another male WWOOFer who can share the room with the current WWOOFer. I understand that the host farms want to make sure WWOOFers have enough time to understand and acclimate to the

5 The host numbers are constantly fluctuating as hosts join, renew, and expire. With the effects of COVID-19 and a new host approval process with more reviewing procedure, the registered host numbers in November 2020 is 1,941 (Pollard, 2020).
6 Kanokwalee Suteethorn (author).

farms' culture and customs and have time to learn to work efficiently in order to work happily and efficiently and leave the farms with a memorable experience.

Visa is another issue for WWOOFers especially in the United States. There are several comments on the topic "Does WWOOFing illegal? How risky is it?" in pantip webboard discussing the visa issues and opportunities to travel through WWOOFing (https://pantip.com/topic/31216265). One board member (member1103428, 2013) mentioned that the WWOOF program was not responsible for the visa after a WWOOFer contacted them and asked about it. WWOOF organization asked her to apply for visas by herself and told her to not mention WWOOF as her purpose of travel. A similar answer can be found on the WWOOF website in the FAQ section: 'Can you arrange a VISA for me?' It says "it would be impossible for WWOOF to assist volunteers with individual visa requirements", and "[m]ost WWOOFers travel using a tourist visa; however, it is your responsibility to determine the correct visa for your visit depending on the country you wish to visit. WWOOF is not responsible for any problems you may experience with immigration" (WWOOF FoWO, n.d. b). Hence, volunteers have to travel with tourist visas which mean illegally working in the US. Besides, there is no guarantee and no governmental process that supports and validates the agreement between the host farms and WWOOFers. This travel form without government acknowledgment reduces the preparation process but can be dubious.

18.4 Discussion and Conclusion

WWOOF is one of the alternative tourism programs that connect tourists with locals where both can get economic, social, and cultural benefits while jointly working for ecological purposes through organic farming on different scales. However, research literature on WWOOFing as well as the four case studies in this chapter illustrate concerns about and potential conflicting interests within the exchange program from communication and cultural complications to legal, financial, and ethical issues. Possible questions are whether it is safe and worthwhile to WWOOF, both for host farms, WWOOFers, and for the environment; despite that WWOOF provides opportunities for travelers to travel without borders, who would be accountable for the program? From a governance perspective, should governments be involved in the WWOOF program and other forms of alternative tourism? And finally, if governments support the program, how much should they be involved?

A. How do cultural, economic situations, or cost of living that differ in each country affect the WWOOFing experience?

Although most travelers and hosts are open for the cultural experience, problems may arise from cultural differences or misunderstanding. Potential conflicts may for instance arise "between host and visitor on the WWOOF farm" around "eating habits, language difficulties, witnessing arguments" within the host family, and a "lack of privacy for hosts and visitors" (McIntosh & Bonnemann, 2006, p. 91). Sharing private space is not familiar to every traveler and can be a source of becoming overwhelmed.

While the main purpose of WWOOF is to learn and share knowledge on organic farming, cultural exchange is also emphasized here. One of WWOOF principles is: "No money changes hands, it's an exchange. In return for your help on the land and with other tasks you receive bed and board, and a lot more besides" (Pier, 2017, Section 1, para. 5).

And still, not all hosts adhere to this principle, and the monetary aspect in WWOOFing seems to be present as well as prone to conflict. As the case study of WWOOF Thailand shows, economic disparities among the participating countries often give an explanation: while in some high-income countries, WWOOF hosts can actually save money by avoiding expensive labor, in low-income countries, the expenses for WWOOFers sometimes exceed those for hired paid workers. In addition, the daily workload of hired workers is often higher than what can be done by the WWOOFers.

B. Is there a mutual relationship between host farms and WWOOFers? Do host farms and WWOOFers learn from each other, both getting their benefits and achieving their goals, and is their relationship mutual?

McIntosh and Bonnemann's (2006) study on WWOOFers' motivations in New Zealand for instance reveals "interest in learning about organic farming", "meet and stay with local New Zealanders and experience their daily life" while traveling the country, or "experience farm life" (p. 90). For some travelers, WWOOFing inspires them to start growing their own food: "that would never have entered my head if I had not gone WWOOFing" (McIntosh & Bonnemann, 2006, p. 92). Others describe their experience and learning of organic farming approaches and sustainable living (p. 92). Hosting the travelers notably brings a cultural exchange, which can have mind-opening benefits for the hosts. It can bring economic advantages, although it does not always. The learning experience seems less mutual between hosts and WWOOFers, since WWOOFers usually bring less farming knowledge and therefore need instruction to work. However, WWOOFers can sometimes bring in their experiences from other organic farms, thereby initiating a knowledge transfer and enriching the hosts' own knowledge. Probably the most mutual contribution of WWOOF is by personal, noncommercial relationships that share cultural values. As the WWOOF Thailand coordinator (from an interview, 2017) knows, these "are priceless and are incomparable to a booked tour or any other form of travel. The 'experiential value' reaches far beyond a five-star spa, cruise or hotel. WWOOFing is a personal mind-expanding experience. It is sharing one's humanity and culture, it is upfront and personal and the experience and relationships last long after the trip is over."

C. Whether WWOOF is a sustainable and responsible volunteer tourism management?
WWOOF organization does not accommodate travel documents and visas for WWOOFers so it could be illegal working in some countries. WWOOFing focuses on working on organic farms where WWOOFers can learn and exchange their experience reciprocally with the hosts. However, some other forms of volunteer tourism may result in negative opinions about the WWOOF system resulting from unfair benefits. Dunford (2010) discussed in the ABC news article, "Good deeds or guilt trips?" that more people are interested in taking a "volunteer vacation." Visiting orphanages as ethical or responsible travelers instead of mass tourists is one of the popular local attractions for volunteer tourists in Cambodia. However, when there are visitors, the orphans were sometimes arranged to dance for them in this "child zoo" where the welfare of these children are not as important as how to make the visitors donate more. However, for the WWOOF organization, both host farms and WWOOFers participate in the program with initial mutual interests on organic farming and sustainable living with no financial exchange. Hence, from an environmental aspect, the WWOOF program is a considerable sustainable and responsible volunteer tourism organization.

D. Can non-governance tourism structures cause ethical and legal issues between hosts and volunteers?
Concerns of being taken advantage of happens on both the host's and the WWOOFer's side: For example, there is concern that WWOOF hosts abuse the WWOOFers as cheap labor, like commercial organic farms. From other perspectives, it might happen that volunteers want to come just for a few days but are not willing to work or learn (McIntosh & Bonnemann, 2006, p. 91). The relatively loose administration organization facilitates the direct exchange between hosts and helpers. However, it also renders it prone to abusive use (many applicants could not be trusted).

Visa, other travel documents, and working permits are one of major issues of WWOOF. Since WWOOFers have to prepare travelling documents by themselves, they travel with tourist visas which do not allow them to legally work. The WWOOF organization states that they cannot provide or help with the documents so these kinds of volunteer tourists are not legitimate. Host farms or the WWOOF organization are not responsible if the WWOOFers have issues with their travel documents.

Since the WWOOF organisation is not monitored by the state government, its monitoring system mostly depends on WWOOFers and host farms to review their experiences. However, with these mutual benefits they tend to provide good reviews that might not reflect the true experience or address the issues they found.

WWOOF as a form of alternative tourism can reduce obstacles and allow people to travel further and gain the unique experience that conventional tourism cannot provide. However, this form of tourism without governance invites for being further explored with more observations from empirical evidence to create a system that supports and makes tourism without governance be more responsible and accountable.

References

Bialik, K. & Walker, K. (2019). Organic Farming Is on the Rise in the US Pew Research Center (blog). Retrieved from: https://www.pewresearch.org/fact-tank/2019/01/10/organic-farming-is-on-the-rise-in-the-u-s/. Date accessed: 2020-11-20.

BLS, US Department of Labor. "Volunteering in the United States- 2015." Bureau of Labor Statistics. US Department of Labor, February 25, 2016. https://www.bls.gov/news.release/pdf/volun.pdf. Date accessed: 2020-11-20.

Bopp, J. (2016): New momentum to Bangkok's organic food movement: interspersed scenes led by mindful pioneers. Dissertation, Cologne. http://kups.ub.uni-koeln.de/6935/. Date accessed: 2020-10-15.

Butler, R.W. (1990). Alternative Tourism: Pious Hope Or Trojan Horse? *Journal of Travel Research*, 28(3),40–45.

Cohen, E. (1987). Alternative Tourism'-A Critique. Tourism Recreation Research 12 (2), 13–18. Date accessed: 2020-11-15.

Cotoacademy.com, 2016, Japanese study. Date accessed: 2020-11-15.

Davis, D. (2018). What to Know About WWOOFing Around the World. Overseas: Volunteer Abroad (blog), Retrieved from: https://www.gooverseas.com/blog/WWOOF-around-the-world. Date accessed: 2020-11-25.

Dunford, G. (2010). Good Deeds or Guilt Trips?" ABC News, September 2010. Retrieved from: https://www.abc.net.au/news/2009-01-23/37886. Date accessed: 2020-11-20.

Guttentag, D.A. (2009). The Possible Negative Impacts of Volunteer Tourism. *International Journal of Tourism Research* 11, 537–551.

Higgins-Desbiolles, F. (2010). *Justifying Tourism: Justice through Tourism. In (Editors names) Tourism and Inequality: Problems and Prospects*. Cambridge, Mass.: MPG Books Group, pp.

Jenkin, M. (2015). Does voluntourism do more harm than good? The Guardian. Retrieved 5 May 2015 from http://www.theguardian.com/voluntary-sector_network/2015/may/21/western-volunteers-more-harm-than-good. Date accessed: 2015-05-05.

Lloyd, C. (2008). Status Quo: VOLUNTOURISM What's behind the Motivation to Work While Allegedly at Leisure? SF Gate, Retrieved from: https://www.sfgate.com/magazine/article/Time-On-3267021.php. Date accessed: 2020-10-20.

Madden, J. (2010) "WWOOF Your Way around the World!" CNN, June 16, 2010. https://edition.cnn.com/2010/TRAVEL/06/15/wwoofing.volunteer.farming/index.html. Date accessed: 2020-11-12.

McGehee, N.G. (2014). Volunteer tourism: evolution, issues and futures. *Journal of Sustainable Tourism* 22 (6), 847–854. DOI: 10.1080/09669582.2014.907299

McIntosh, A. & Campbell, T. (2001). Willing Workers on Organic Farms (WWOOF): A Neglected Aspect of Farm Tourism in New Zealand."*Journal of Sustainable Tourism* 9 (2), 111–127. https://doi.org/10.1080/09669580108667393.

McIntosh, A.J. & Bonnemann, S.M. (2006). Willing Workers on Organic Farms (WWOOF): The Alternative Farm Stay Experience?" *Journal of Sustainable Tourism* 14 (1), 82–99. https://doi.org/10.1080/09669580608668593.

Member 1103428. (2013). Does WWOOFing Illegal? How Risky Is It? Pantip.com (blog). Retrieved from: https://pantip.com/topic/31216265. Date accessed: 2020-11-20.

Mostafanezhad, M. (2016). "Organic Farm Volunteer Tourism as Social Movement Participation: A Polanyian Political Economy Analysis of World Wide Opportunities on Organic Farms (WWOOF) in Hawai'i." *Journal of Sustainable Tourism* 24, no. 1: 114–31.

newzealand.com (n.d.). Things to do. Volunteering. Retrieved from: https://www.newzealand.com/int/feature/volunteering-in-new-zealand/. Date accessed: 2020-11-03.

Ooi, N. & Laing, J.H. (2010). Backpacker Tourism: Sustainable and Purposeful? Investigating the Overlap between Backpacker Tourism and Volunteer Tourism Motivations. *Journal of Sustainable Tourism* 18 (2), 191–206. DOI: 10.1080/09669580903395030.

Pier, M. (2017). "Sue Coppard, Founder of WWOOF (World Wide Opportunities on Organic Farms)". Retrieved from: https://www.peopleareculture.com/blog/sue-coppard-wwoof. Date accessed: 2020-11-11.

Pier, M. (2011). "WWOOF Founder Sue Coppard: Peer-to-Pier Interview". Retrieved from: https://petergreenberg.com/2011/07/19/WWOOF-founder-sue-coppard-pier-to-peer-interview/?fbclid=IwAR2P_DrhuSPCXFANRBLWBGgW-aiKTSsYWr0vNDdgzGaPqQlTIg0vcN6zrz8. Date acessed: 2020-11-11.

Pollard, J. "Numbers of Farm Hosts" Membership Program Manager WWOOF-USA. Email, November 19, 2020.

Steele, J, Dianne Dredge, and Pascal Scherrer. "Monitoring and Evaluation Practices of Volunteer Tourism Organisations." *Journal of Sustainable Tourism* 25, no. 11 (2017): 1674–90. https://doi.org/10.1080/09669582.2017.1306067.

Terry, W. (2014). Solving labor problems and building capacity in sustainable agriculture through volunteer tourism. *Annals of Tourism Research*, 49, 94–107. https://doi.org/10.1016/j.annals.2014.09.001.

Theng, S., Qiong, X. & Tatar, C. (2015). Mass Tourism vs Alternative Tourism? Challenges and New Positionings. Études Caribéennes [En Ligne], Août-Décembre, 31–32.

Tomazos, Kostas, and William Cooper (2011). Volunteer Tourism: At the Crossroads of Commercialisation and Service? *Current Issues in Tourism* 15, 405–23. https://doi.org/10.1080/13683500.2011.605112.

Triarchi, E. & K. Karamanis. *"The Evolution of Alternative Forms of Tourism: A Theoretical Background."* Greece, 2016.

Wearing, S., Beirman, D., Grabowski, S. (2020). Engaging volunteer tourism in post-disaster recovery in Nepal. *Annals of Tourism Research*, 80, 102802.

Wengel, Y., McIntosh, A., & Cockburn-Wootten, C. (2018). Tourism and 'dirt': a case study of WWOOF farms in New Zealand. *Journal of Hospitality and Tourism Management*, 35, 46–55. https://doi.org/10.1016/j.jhtm.2018.03.001

WTTC, World Travels & Tourism Council. "Economic Impact Reports," 2019. https://wttc.org/Research/Economic-Impact. Date accessed: 2020-10-25

WWOOF Australia. (n.d. a): WWOOF Mobile App. Retrieved from: https://wwoof.com.au/mobile-app/. Date accessed: 2020-11-20.

WWOOF FoWO – Federation of WWOOF organizations (n.d. a): About WWOOF. Retrieved from: https://wwoof.net/about-2/. Date accessed: 2020-11-03.

WWOOF FoWO – Federation of WWOOF organizations (n.d. b): Can you arrange a VISA for me? Retrieved from: https://wwoof.net/doc_articles/can-you-arrange-a-visa-for-me/. Date accessed: 2020-11-15.

WWOOF Independents (n.d.): About WWOOF Independents. Retrieved from: https://wwoofindependents.org/about-wwoof-independents. Date accessed: 2020-11-03.

WWOOF International (n.d.): History of WWOOF. Retrieved from: https://wwoofinternational.org/history-of-wwoof/. Date accessed: 2020-11-11.

WWOOF Thailand (n.d.): WWOOF Thailand – travel longer, cheaper, better. Retrieved from: https://wwoofthailand.com/. Date accessed: 2020-11-26.

WWOOF United Kingdom (n.d.): History of WWOOF. Retrieved from: https://wwoof.org.uk/about/history. Date accessed: 2020-11-03.

WWOOFusa.org. (2017). WWOOF USA. Retrieved from: https://WWOOFusa.org/. Date accessed: 2020-11-20.

www.reddit.com,u/razzek. (2015). My (mostly Bad) Experiences WWOOFing in Japan across 3 Farms for 6 Weeks, Retrieved from https://www.reddit.com/r/WWOOF/comments/3f9d7x/my_mostly_bad_experiences_wwoofing_in_japan/. Date accessed: 2020-11-15.

Yamamoto, D. & Engelsted, A.K. (2014). World Wide Opportunities on Organic Farms (WWOOF) in the United States: Locations and Motivations of Volunteer Tourism Host Farms. *Journal of Sustainable Tourism* 22 (6), 964–82. https://doi.org/10.1080/09669582.2014.894519.

Nelson Graburn

19 Tourism and Governance Beyond National Boundaries

19.1 Introduction

This chapter on supernational forms of governance shaping tourism is perhaps the most complicated topic in this volume. Above the national level of governance – laws, regulations – are regional regimes, such as ASEAN (Association of Southeast Asian Nations), the British Commonwealth, CIS (Commonwealth of Independent States [nine Post-Soviet republics]), ECOWAS (Economic Community of West African States), the EU (European Union), and larger organizations which attempt international, even global governance.

When we talk about tourism as an industry or an activity – e.g. 7 billion trips a year, employing 11% of the world's workers – we are falsely creating a unity, which soon breaks down when we enquire into governance, especially supernational forms of legal and regulatory governance. Secondly, in a global enquiry, we may have to rethink what we mean by governance, comprising global and international treaties, NGOs, directed funding, the nature of borders, and the nature of nature and so on. In considering the governance of tourism, we must separate who is governing and how governance is imposed by starting with what is being governed.

19.2 Governing Travel

The movement of human beings is governed mainly at the international level, through the national (or supernational) issue of passports, visas, and other regulations such as currency controls. Domestic (intra-national) movements are much less regulated, though there are always zones forbidden to the general public, such as governmental and military spaces, and of course privately-owned areas. International tourism travel regulations are of prime importance, e.g. the freedom of movement across national boundaries within the EU, the governance of lengths of stay through visas accorded to different nationalities, and the distinctions made between different kinds of visas, e.g. for tourism, diplomacy, work, study, refugee and so on. The regulations of movement are subject to changes of great importance for tourism depending on international relations, such as reciprocal granting of 10-year tourist visas between China and the USA, or on economic relationships, such as the recent elimination of the requirement for tourist visas for US travelers to Brazil, a requirement that had not applied to EU tourists for decades. Restrictions on international movements, through visas or the issuance of passports, may also be regulated in

https://doi.org/10.1515/9783110638141-019

cases of diseases (SARS, MERS, COVID-19, Tuberculosis, Ebola, etc.), other disasters such as earthquakes or tsunami, and, of course, outbreaks of civil disobedience or war. The possibility of changing international regulations is of prime importance especially to those countries which are highly dependent on tourism, as it used to be said, "They have nothing to sell but their beauty," [i.e. landscape or youth] (Graburn, 1983, p. 441).

19.2.1 Borders

Borders are a measure of movement: they may facilitate, impede, or prevent tourism and other forms of travel. Some borders are seen as "natural," e.g. shores, rivers, mountains, and need technological devices to overcome them: bridges, ships, airplanes, etc., but most borders of concern to tourists are "man-made," lines drawn on a map as it were, on land or on water, dividing otherwise contiguous travel spaces. These borders, almost always marked and manned, are usually constructed barriers, like the partial wall at the US-Mexico border or the electronic detectors of the Finnish-Russian forested borderlands. But some are still unmarked except on maps. And it may be legal to cross them as inside the EU or the CIS, or they may be illegal but remote as parts of the US-Canada border, the Brazil-Peru border, or the Russia-China border.

Borders, too, like visa and customs regulations, are subject to changes over time, because of conquest or decolonization, or mutually agreed adjustments, as may happen to Northern Ireland after Brexit. Other borders are not agreed to by the nations on either side of them, very seriously in the cases of India, Pakistan, and Kashmir, and India and China [Tibet]. Other changes and disagreements may concern not the location of the border but its existence or significance, for instance the borders between China and Macau and Hong Kong. For the purpose of counting tourist flows, some would call them domestic and others international. In a recent table of annual tourist intakes, Macau was placed after the fourth nation, but unnumbered and Hong Kong similarly after the sixth. For China, they are all domestic tourists, whereas the UNWTO lists them as international. And there are similar ambiguities about tourists between China and Taiwan.

19.2.2 Transportation

Transportation is subject to governance for a vast number of reasons: staffing, equipment, weather, safety, pollution, numbers, costs, and so on. Air transport is governed for instance by regulators from International Air Transport Association (IATA), International Civil Aviation Organization (ICAO) (working with the national counterpart Federal Aviation Administration (FAA) in the USA), concerned with cargo, landing rights, prices, taxes, air space, GPS, military spaces, weather, pollution, and safety, etc.

Shipping has similarly complicated governance. Cruise ships, for example, are subject to many levels of governance: national, international, commercial, civil, and technical (Figure 19.1).

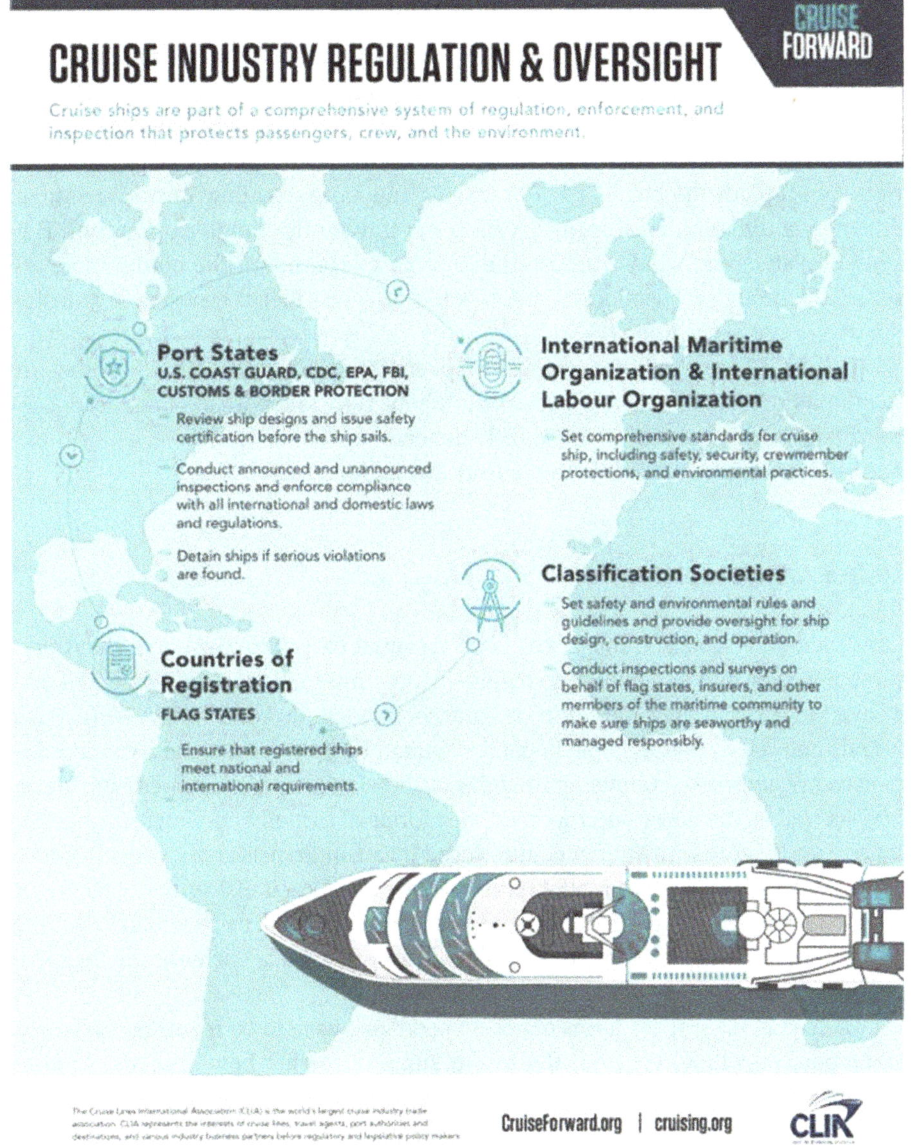

Figure 19.1: The many levels of Governance of Cruise Ships. (Source: Cruise Lines International Association, Cruise Industry Regulations, p. 1).

Other passenger ships, ocean liners, excursions, ferries and so on are also regulated at the national and international level, but with less complexity.

Rail transportation is mainly governed nationally. International rail transport is common in Europe, even going under the sea as in the case of the Eurostar, but is not so common on other continents (of course Russian trains travel through Europe and Asia within one nation). International trains don't always have an easy time, even today. For instance, travel from Western and Central Europe via Poland to Russia is complicated by the Russian railways gauge of 5 feet vs. Europe's of 4′ 8.5", meaning that no rolling stock from Europe can run directly into Russia, thus thwarting possible military invasions. So, each coach has to be hoisted in the air while workmen replace the bogies (sets of wheels and axles) to allow onward passage, drawn by a different, local engine. This is not true for the simpler connection Helsinki-St. Petersburg which runs on the Russian gauge, but at the border there are severely thorough inspections of the coaches to detect illegal travelers, e.g. inside the roof or under the floors!

In Asia, Chinese rail transport now covers Hong Kong and Tibet but they are treated as domestic, and we know little about rail transport to North Korea. In North and South America, there is little dependence on rail transport, as air transport is dominant for tourists both domestically and internationally.

19.2.3 Consumption

Consumption of goods and services is an essential part of tourism and, like transportation, is regulated nationally, regionally, and internationally. An essential part of most tourism, except for day trips, is accommodation. Apart from VFR (visiting friends and relatives), perhaps the most common form of domestic tourism, accommodation comprises, commercial lodging at hotels, inns, hostels, bed and breakfast, as well as the newer internet-mediated forms of Airbnb and Couchsurfing. The latter, free lodging with strangers introduced by a San Francisco network company, appears to be a post-modern attempt to recreate a pre-capitalist form of hospitality (Graburn, 2013). The latter is regulated by its enabling company as would be Airbnb or any hotel chain, but not as thoroughly by governmental agencies as are the other commercial forms.

Of course, all of these forms of accommodations have to be regulated for safety, sanitation, crowding, etc., but the recent surge of private home rentals, such as Airbnb, VRBO, Flipkey, as well as the hosted forms such Homestay (and Couchsurfing) are increasingly coming under severe new regulations following the accusations that they are driving up the price of residential housing and favoring tourists, often from richer countries, over the local residents. This particularly true in areas of "overtourism," and both cities and nations have begun to impose severe limitations (McKercher et al., 2015). The UNWTO, originally a promoter rather than a regulator, has become more involved with governance of international touri sm in the past decades. With a membership of 156 nations and 500 business/NGOs, it is mainly concerned with sustainability, peace, human rights, and the regulation of development and overtourism. National and international regional organization such as Pacific Area Travel Association (PATA) perform similar functions.

Eating and drinking are increasingly made available to tourism by international chains of restaurants and cafes. Regulation for investment, taxation, conditions of employment, and food safety are overwhelmingly local or domestic (within the city, state, or nation), but the resulting regulatory structures have become built into the functioning of many chains and so have influenced the behavior and regulation of their global locations. Of course, all of these are subject to local variation, but the chains value their reputations that could be damaged by increasingly peripatetic tourists. This in turn leads the locals to expect and adopt certain "international" standards which may globalize the content of governance, in ways that have long been strongly enforced in transportation such as airlines. The same has happened to some extent to other globalized consumption activities such as shopping, souvenirs (safety of materials and forms), valuables, fashion, arts and crafts, heritage property, endangered species (see also Heritage below).

The regulation of the consumption of (commercial) sex has always been problematic both as a legal moral matter of exploitation and a medical biological matter of disease and pregnancy. Though morals and medicine have change, there is still great confusion about permissiveness or prohibitions, with conflicting local, national, religious, and international attempts at governance. More recently a similar set of conflicting attitudes have emerged and are evolving within and between polities for the moral and medical regulation of organ transplants (Scheper-Hughes, 2014) and surrogate motherhood.

The human aspects of tourism are often more strictly regulated, and tourists are sensitive to malfeasance and danger. These services include all kinds of personnel, in entertainment, hotels, waitpersons, tour guides, medical and pseudo-medical facilities. The UNWTO and to some extent the World Trade Organization (WTO) attempt to regulate some international transactions. But it is in the areas of communication that international regulation has developed most strongly. This started with the mail in the 19th century. After the invention of mail by postage stamp (UK, 1837) many countries followed suit, and the International Postal Union (now the Universal Postal Union) was established in 1874 and has governed the global flows of mail ever since, now between 196 countries.

During the same Victorian era, communications and tourism (domestic and international) all developed at a great rate, particularly in concert with the deployment of railways – across the US and Europe – and steam ships across the Atlantic and connecting the far flung–? British Empire. The invention of telegraphy and the Morse code (and soon telegrams, TELEX and facsimile machines, and the telephone) brought a surge of international communications aiding tourism as well as trade and military expansions. These necessitated international governance mostly initiated by the British, French, Germans, and Americans. Tourism and communication further advanced in the 20th century with new competition arising with air transport and television and increasingly complex telecommunications. All of these required international regulation and governance, of voltages, CPS (cycles per second frequencies), codes, electric plugs, and sockets, real and artificial languages, timetables, and so on, usually far out of sight of the touring customers.

Again, at the turn of the 21st century, new forms of communication have required supernational agreements, for the Internet (privately governed by an NGO in the USA), wifi and internet connections, plugs and USBs, VPN and all sorts of security programs. The Internet and social media, which have varied levels of international governance, have transformed tourist experiences, enabling instant communication through text, pictorial, and video throughout most of the world. While this enables people to keep in touch and express their social relations and aesthetic tastes (Lo & McKercher, 2016), it has changed the nature of the travel experience from local, intense, and often "spiritual" to global, instantaneous, and more superficial (Frey, 2017).

19.3 Governing Heritage

The concept of heritage, like tradition, has come to the fore particularly in those societies aware of the speed of sociocultural change. This first occurred with the invention of literacy (Goody, 1977) where physical evidence of past thoughts and events remained to challenge memory. Both population and political expansion, and changes in technical and social lifestyles are measures of change, and, when change becomes rapid, people know that they are living differently from the ways of their childhood or their parents. Though progress is welcomed in many ways – convenient life, easier mobility – changes may also generate nostalgia, regret, and alienation from the present (MacCannell, 1976) and a touristic search for authenticity in nature, history, and in the lives of those supposedly still living in a simpler age. The idea of heritage arose from deliberate efforts to preserve and celebrate things and customs (traditions = things handed down) from the past that are deemed valuable (Graburn, 2000). In Europe this was linked to the inheritance of the means of livelihood and identity (French, patrimoine, presumably from fathers). Tunbridge and Ashworth (1996, p. 7) have cogently shown heritage to be:

> Historic Resources (Traditions) from the past, selected by Agencies, assembled with other Resources (knowledge, contexts, associations) for Interpretation and Packaging, as Heritage Products by the Heritage Industry for the public/audience.

The property, identity markers, and performative rites of the powerful – rulers, priests, officials – have always been protected. These people and their cultural icons "stood for" and represented the social group and were in a sense "sacred." As political groups, at first Traditional (Sacred), and later Rational-Legal (Weber, 1947) expanded and formed empires and nations. The individual possessions, e.g. the Crown Jewels, buildings, museums, libraries, monuments, graveyards, and archaeological sites were subjected to regulations and were usually "priceless," that is, not for sale (Appadurai, 1986). And with conquest and empire, the capture of the heritage of the Other, e.g. The Elgin Marbles, head-hunters' skulls, Cleopatra's Needle, and so on, became national or imperial heritage, signifying the glorious history and power of the new owner. The institutionalization of museums first for royal prestige, then national glory, and later for scientific advancements and democratic education was a forerunner of the heritage movement (Graburn 2012).

During the Renaissance and the Industrial Revolution, links to the past (e.g. in Europe the archaeology of Greece and Rome) gained importance and historical change became more obvious, engendering the nostalgia, pride and preservation movements mentioned above. By the 19th century nations generated social movements and policies of labelling, classification, and preservation of their heritage, envisaging much of the world as a vast museum. Even things that were "modern" went out of date and infrastructures, including roads, railways, bridges, dams, and waterworks, rivers/lake/waterfronts/ports became "heritage." As actions and technology pass from being functional to "old-fashioned" they gain aesthetic and emotional values, think about steam trains, B&W or Polaroid photos, Victorian houses, antiques and so on, following Marshall McLuhan's theory of obsolescence (McLuhan and Powers 1989) and Michael Thompson's Rubbish Theory (1979).

National and imperial organizations of preservation and exhibition spread around the world with the expansion of colonialism and Western education, e.g. the Archaeological Survey of India, founded 1861, now part of the Indian Ministry of Culture. This pattern of valuing the past of both the conquerors and the colonized or defeated was also followed in North America and in Africa. As Tunbridge and Ashworth (1996) emphasize, what becomes heritage is selected, by the powers that be, patriarchs and matriarchs in family-level societies, and by the ruling classes and eventually trained or designated specialists in "rational-legal" nations (Nora, 1989). We can see by now that the creation (and suppression) of heritage is thoroughly political, it is an exercise in control over the symbols and identity of the social body (Lowenthal, 2015), which Smith (2006) has called "authorized heritage discourse." In complex post-colonial multi-ethnic societies, different groups may struggle over the control of what is accepted as heritage, a situation Tunbridge and Ashworth (1996) called "dissonant heritage."

But it was World War II and the global politicization that followed, that brought the international regulation of Heritage that we see today. The United Nations was founded (San Francisco 1945) to bring peace and security to a severely war-damaged world. It contained many agencies including International Civil Aviation Association (ICAO), International Maritime Organization (IMO), International Telecommunications

Association (ITU), Universal Postal Union (UPU), as well as UNWTO and the specialized whose functions we have discussed above, United Nations Educational, Scientific and Cultural Organization (UNESCO), founded 1946, which developed the major institution for the international governance of heritage. In addition to education, intellectual freedom, and social justice, UNESCO soon came to foster and impose ideas of heritage preservation to all 193 members. It works closely with an international NGO professional association International Council on Monuments and Sites (ICOMOS), founded 1965.

Of prime importance to the political, commercial and cultural aspects of world tourism was UNESCO's attention to cultural sites and the development of the World Heritage system (founded 1973–76), which divided sites into those of cultural and natural importance. At first these consisted of famous buildings, and complexes, including archaeological sites and urban parks, deemed to be important to all humanity. Although UNESCO grants some minor direct financial support for heritage sites (World Heritage Committee/Fund), it fostered the foundation of International Centre for the Study of the Preservation and Restoration of Cultural Property (ICCROM) in 1956, an intergovernmental organization (IGO) for the preservation of cultural heritage through training, information, research, cooperation, and advocacy programs. This support and the fame sites gain stimulates other forms of donation and support, and of course, lucrative tourism flows. The UNESCO list also include outstanding landscapes and natural sites, which are of course, culturally defined, and this ties in with the protection of nature and endangered species through Man and Biosphere (1968), Global Geoparks Network (1995), International Union for the Conservation of Nature (IUCN), the World Conversation Union, the World Wildlife Fund (WWF), the Worldwide Fund for Nature, and the World Conservation Monitoring Centre. All this happened coincident with the florescence of international tourism due to the spread of affluence and the operation of jet aircraft.

After 10 years or so of identification and certification, the lists appeared heavily weighted to historical Europe and the centers of civilization in Asia. Countries of the "Third World" (now called "the Global South" or LDCs) including Africa, Latin America, and much of Asia had fewer such landmarks. This was deemed to be unfair, as it might indicate that such cultural regions did not have many important sites and because UNESCO listing became a prime guarantee of authenticity and importance in attracting international tourists and their incomes, so important for economic development in such areas. UNESCO and concerned others sought more forms of threatened heritage which might be important for all humanity; these might well be found in the cultures and practices of other nations which were not well endowed with sites and monuments.

Japan, a country which had modernized fast since the 19th century, and had suffered in the Depression and by its own actions in World War II provided the best example. Conscious of the threats to its cultural properties, Japan had enacted regulations preserving its sites, including "national treasures" and scenic landscapes since 1880s but the catastrophe of World War II and the subsequent American Occupation led to both Japanese and American efforts to resurrect a failing nation. The Law for Cultural Properties was strengthened in 1950 and in the same year the National Ministry of Education (Monbusho) created "Art Encouragements" and "Encouragements for New Artists Prizes" in ten fields the modern arts. In the same year a lacquer-ware master craftsman, Gonroku Matsuda, assisted by General Douglas McArthur's office, created the honorific title "Living National Treasures" in the quest for preserving unique practices of traditional arts. Thus, Japan's passage of the Law for the Protection of Cultural Properties to "designate men and women who are judged as outstanding in their field of arts and crafts or in their artistic performances."

This was most immediate ancestor of UNESCO's next global proposal, the designation of Intangible Cultural Heritage, in 2001 and the passing of the Convention for its protection in 2003 (Graburn, 2005a, 2005b). Thus, "nonphysical intellectual wealth, such as folklore, customs, beliefs, traditions, knowledge and language" came to be recognized and protected (Convention for the Safeguarding of Intangible Cultural Heritage, Official site UNESCO. June 20th, 2007.) And nations and NGOs eagerly sought out and proposed examples from all over the world. Particularly prominent are oral and dance performances, ritual practices and recently "food heritages," such as Mexican cuisine or Japanese Washoku, which enhance a country's pride and add to its tourist attractiveness.

19.4 Conclusion

Our interconnected world has historical precedents, such as the Roman Empire and the spread of Islam, both of which imposed "law" on many subjugated peoples. But it was the British empire of the 19th century which most closely presaged the modern global arrangements. The Roman, Islamic, and British Empires all sought to impose their regulations, on morality, economic relations and movement on subject peoples. The post-World War II world of the United Nations is ideally based on a less monocentric structure. Ideally these governmental and nongovernmental organization consist of associations of independent national members and NGOs, in which all have some input into the contents of governance.

However, the ideal must be modified when some nations have dominant control of e.g. air travel and airplane technology, financial capital for investment in hotels and resorts or heritage preservation, or expertise over computers, the internet, and IT. Thus, following the colonial model, the regulations are often generated and fostered by the powerful nations, resistance is possible, and individual nation have the right to e.g. with withdraw from UNESCO or to withhold or propose sites for UNESCO listing. But these facts themselves greatly affect governance. For most international regulation, it is the nations which propose or refuse to submit to e.g.

listing or international financial support, whereas local polities and membership groups can propose to their nation but have no final say; they can, however, resist, e.g. UNESCO listing by breaking the imposed required rules. International regulation does work, when for instance a member of the World Wildlife Fund living in the UNESCO listed old city of Lijiang in Yunnan, became aware of consistent broaching of the rules of preservation of the historic architecture and buildings, she informed the national government that she would bring this to the attention of UNESCO which might withdraw its approval. This prompted the Beijing government to force the local authorities to reinstate the rules. At the same time, the UNESCO rules rein in national governments, many of which are eager to get as many listings as possible in order to attract fame and lucrative tourist flows. At one point the Chinese government had put forward so many sites of UNESCO listing that, as they were reminded, at the rate of one acceptance per year, the maximum allowed for any one nation, most of the proposers would be dead before their pet project came up. UNESCO then made a rule that no country may propose more than 10 at one time and must list them in order of preference.

Much international regulation is less overtly political in nature. These are those discussed above concerning transportation, safety, the spread of diseases. These professional international organizations are much less prone to bargaining and maneuvering. If standards are not upheld, an airline may be barred from flying outside the country, or international airlines will not fly to an unsafe airport or to a country with poor radar or air traffic control. Similarly, more exacting international regulations are often imposed in cases of epidemics or inadequate disease or drug controls. Exacting regulations may be applied to regions with civil war or pervasive crime. However, in these cases, the regulations may be subverted for political reasons by those who take sides or see a chance to make illegal profits.

Finally, one might ask what these supernational organizations can do, why and how they are distinct? Also, how do they have power in the tourism industry? Many of them cannot "enforce" laws/regulations, so why do destinations, DMOs [destination management organizations], businesses, etc. heed their advice?

The major international organizations (UNWTO, UNESCO, and so on. see Appendix at end of chapter) can influence policies on national tourism level because all UN member states aspire to achieve the UN Sustainable Development Goals. So although it is not compulsory on the ministries, they aim to do so, and this applicable on other sectors too not only to tourism.

There are specific programs that exist between central governments and international agencies that implement tourism strategies on national level, such as USAID, UNWTO, UNESCO, WB, UNEP, and the SDG goals. And there are many other less global organizations, such as the Aga Khan Foundation, which invest money to upgrade tourism cities and change tourism policies such as building densities and hotel densities in heritage sites and national parks.

The tourism private sector quality companies or labelling companies are international corporations that influence resorts, hotels tour operators to go green, or preserve water, or wisely use beaches, etc., and these influence tourism destinations (see Pender & Sharpley, 2005). These are organizations such as Green Globe, Blue Flag, SGS, Iso14000, The Ecotourism Society, all of which provide assessment and certification to organizations meeting legal, structural, and environmental standards.

International governance of tourism, and its associated travel, accommodation, food, and consumer industries are here to stay. Tourist flows are voluntary, compared with trade or migration, and they are easily affected by fears, rumors, and threats. Threats of disease, violence, unreliable travel, destruction of heritage – as well as high prices and poor service – quickly bring the masses to a halt. They will not return until stability and safety appear to be re-established.

Appendix

International Organizations

Aga Khan Foundation [Founded 1967, health, education, rural development, environment and the strengthening of civil society include human resource development, community participation, and gender and development.

Blue Flag, Certification of beaches, marinas, boat tour operations

Green Globe, environmental building certification

IATA (International Air Transport Association, a trade association) (1945), 270 airlines of 117 nations, Global Aviation Security Program. [Air traffic control, Airspace, Environment, Airport slots, Codes-tokens-baggage tags, airport modernization (shops, hotels?), data, fraud, terrorism, customer loyalty]

ICAO (International Civil Aviation Organization) (1947) agency of the United Nations. Codifies the principles and techniques of international communication and air navigation and fosters the planning and development of international air transport to ensure safe and orderly growth

ICCROM (International Centre for the Study of the Preservation and Restoration of Cultural Property, IGO) (1956) 135 nations (Inc US, UK, Canada), [training, information, research, cooperation & advocacy]

Iso14000, certifications of environmental management of companies.

IUCN (International Union for the Conservation of Nature); World Conversation Union, World Wildlife Fund, World Wide Fund for Nature, World Conservation Monitoring Centre, IUCN Red List of Threatened Species.

Passport, Visa, Embassy, Consulate, Border entry system (UNWTO).

Religious Organizations: e.g. Catholic Church, Vatican NGOs: e.g. Medecins sans Frontiers,

SGS, Certification of industries, processes for compliance with standards.

The Ecotourism Society. Certification of environmental management of tourism businesses

UNEP (UN Environmental Program) Sustainable Development, Biodiversity.

UNESCO (UN Educational, Scientific and Cultural Organization) (1945), 195 nations (US withdrew 1984–2002; now does not pay in, but was elected to the Executive Board), Man and Biosphere (1968), World Heritage List (1973/8), World Heritage Committee/Fund, ICH (2003) Global Geoparks Network (1995); 322 NGOs inc. ICOMOS (International Council on Monuments and Sites)

UNDP (UN Development Program) UN, 1965 177 countries, [2012 the Biodiversity Finance Initiative (BIOFIN), UNICEF (United Nations Children's Fund), UNFPA (United Nations Population Fund), WFP (World Food Program); JICA (Japan International Cooperation Agency), ODA (Official Development Assistance).

UNWTO (UN World Tourism Organization) 1975, 2003 (from ICOTT) IUOTO, WTO, 156 nations, 500 business/NGOs, Sustainable Tourism, Peace, Human Rights, Development, World Committee on Tourism Ethics, Visa Openness Report: PATA (Pacific Area Travel Association)

USAID (US Agency for International Development, founded 1961) Administer foreign aid and provide development assistance.

World Bank (founded 1944) [Make loans and investments to reduce poverty and increase standards of living.]

WTO (World Trade Organization; successor to GATT [General Agreement on Tariffs and Trade]); Trade, Tariffs, Protectionism.

References

Appadurai, A. (Ed). (1986). *The Social Life of Things: Commodities in Cultural Perspective*. Cambridge University Press.

Bennett, T. (1995). *The Birth of the Museum: History, Theory, Politics*. London: Routledge.

Frey, N. (2017). Text: Pilgrimage in the Internet Age. Retrieved from: https://www.walkingtopresence.com/home/research/text-pilgrimage-in-the-internet-age. Date accessed: 10/ 25/2020.

Goody, J. R. (1977). *The Domestication of the Savage Mind*, Cambridge: Cambridge University. Press.

Graburn, N. (1983). Tourism and Prostitution: a Review Article. *Annals of Tourism Research* 10: 437–456.

Graburn, N. (2000). Learning to Consume: What is Heritage and When is it Traditional? In Nezar AlSayyad (Ed.). *Consuming Tradition, Manufacturing Heritage* (pp. 68–89) London: Routledge.
Graburn, N. (2005a). *Qual autenticidad? Un concepto flexible en la busceda de autoridad Numero special: "Diversidad Cultural y Turismo."* Cultura y Desarrollo Havana: UNESCO, 4: 17–27
Graburn, N. (2005b). *Whose authenticity: a flexible concept in search of authority [Intangible heritage]. Special issue: "Cultural Diversity and Tourism." Cultura y Desarrollo*, Havana: UNESCO, 4: 17–26.
Graburn, N. (2012) 2012 "Ancient and Modern: The Alaska Collections at the Hearst Museum of Anthropology." *Museum Anthropology* 35 (1): 58–70
Graburn, N. (2013). Anthropology and Couchsurfing – Variations on a Theme. [Afterword], In Picard, D. & Buchberger, S. (Eds.) *Couchsurfing Cosmopolitanisms: Can Tourism make a Better World.* (pp. 105–11) Bielefeld: Transcript Verlag.
MacCannell, D. (1976). *The Tourist: a New Theory of the Leisure Class.* New York: Schocken.
McKercher, B., Wang, D., & Park, E. (2015). Social impacts as a function of place change Annals of Tourism Research. 50: 52–66.
McLuhan, M. and B. R. Powers (1989) *The Global Village: Transformations in World Life and Media in the 21st Century.* Oxford University Press.
Lo, I. & B. McKercher. (2016). Beyond Imaginary of Place: Performing, Imagining, and Deceiving Self through Online Tourist Photography. In Gravari-Barbas, M. & Graburn, N. (Eds.) *Tourism Imaginaries at the Disciplinary Crossroads: Places, Practices, Media.* (pp. 233–247) London: Routledge.
Lowenthal, D. (2015). *The Past is a Foreign Country-Revisited.* Cambridge, Cambridge U. Press.
Nora, P. (1989). Between Memory and History. Les Lieux de Mémoire. Representations 26, 7–24.
Pender L. and R. Sharpley (2005) *The Management of Tourism.* Sage.
Scheper-Hughes, N. (2014). Human Traffic: Exposing the Brutal Organ Trade, The New Internationalist. Retrieved from: https://newint.org/features/2014/05/01/organ-trafficking-keynote. Date accessed: 10/ 21/2020
Smith, L. (2006). *Uses of Heritage.* London: Routledge.
Thompson, M. (1979) *Rubbish Theory: the Creation and Destruction of Value.* Oxford University Press.
Tunbridge J. E. & G.J. Ashworth (1996). *Dissonant Heritage: The Management of the Past as a Resource in Conflict.* John Wiley.
Weber, M. (1947). *The theory of social and economic organization.* (Translated by A. M. Henderson & Parsons, T.). New York: Free Press.

Amir Gohar

20 Conclusion

The conclusion of such a volume with a wide range of geographies requires complex unpacking and packing of complex ideas. This chapter aims to synthesis the presented arguments in the chapters and establish patterns, commonalities, differences, and best practices. The research presented in this book presents an overview of tourism governance across different locales without selecting countries for particular predetermined traits. Because of this a basic understanding of government systems, demographic information, and economic background will allow for a more nuanced understanding of the ways in which tourism fits into the fabric of each country. The following analysis of tourism governance will not constitute ranking or hierarchy. Instead, the varying manifestations of tourism governance across diverse local contexts will be utilized in the drawing of thematic conclusions and correlations within the context of tourism governance globally.

20.1 Tourism Governance in the Government (Stand alone/Mixed/Non-existent)

The management of tourism and tourism related activity is often a government enterprise. In many countries, despite varying cultural and geopolitical differences, there are government agencies with similar or identical titles of "Ministry of Tourism." However, in other countries, even with similar levels of tourism sector activity, there appears to be no trace of tourism policy within any particular government agency. Figure 20.1 utilizes the examples of tourism governance in France, Tunisia, Finland, Italy, and the United States, to aid in an understanding of the diversity of tourism's forms in various countries' systems of government.

For the countries of France and Tunisia, there are explicit and discreet ministries of tourism. In France, the Ministry of Tourism controls all aspects of tourism, and while there are regional and departmental divisions within this ministry, it is not within or combined with any other government entity. This stand alone department in France developed alongside the county's long history of successful tourism. Meanwhile in Tunisia, just three years after the founding of the Republic of Tunisia in 1956, the newly formed government created an independent Ministry of Tourism. The relationship between France and Tunisia could have been a factor in Tunisia's formation of a Ministry of Tourism. France has long been viewed as the destination for tourists worldwide, and that is certainly due in part to the efficient and effective management of tourism by the French Ministry of Tourism. Tunisia's initial founding of a stand alone

https://doi.org/10.1515/9783110638141-020

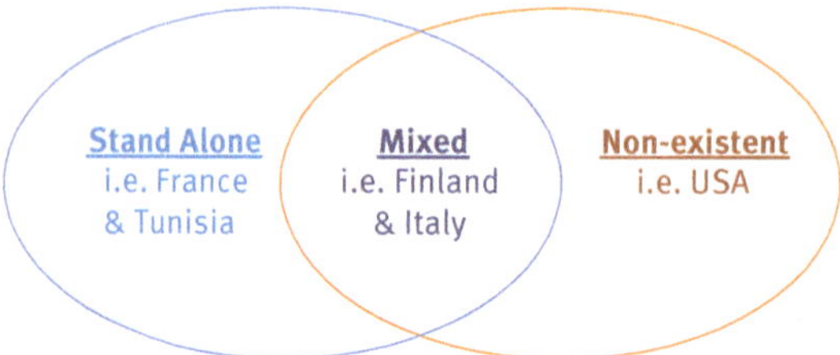

Figure 20.1: Tourism Governance: Standalone versus non-existent.

tourism ministry could point toward an increased prioritization in the importance of the industry for the young country.

In some countries, the control of tourism within the government has fallen under different agencies which have in turn evolved and changed. The cases of Finland and Italy both illustrate these kinds of changes and along with it, the discretion, appearance, and individuality of tourism governance. In Finland, the control of tourism is under the jurisdiction of the Ministry of Economic Affairs and Employment whereas in Italy, this responsibility has fallen to the Ministry of Agriculture. These two cases highlight how for many countries, stand alone tourism ministries are not necessarily the solution to fixing problems in the tourism sector. Rather, the solution lies in combining other government entities in order to secure intergovernmental cooperation for tourism development and activities.

On the opposite end of the spectrum, the United States has no official federal level government agency devoted to tourism. While there are tourism adjacent initiatives at the federal and local levels, the vast majority of tourism regulation comes in the form of the preservation and conservation of cultural heritage as well as natural resources. However, the United States generates a large revenue specifically from tourism and has annual visitors numbering in the millions. The success of the United States tourism sector, considering the absence of a Ministry of Tourism, stand alone or otherwise, may at first be baffling, but in actuality this example serves to show the flexible and contextual nature of successful tourism national tourism governance systems.

20.2 Tourism Governance (Centralized or Localized)

Tourism governance, as explored throughout these diverse contexts, may often appear in one of two forms: either as a centralized agency or multiple agencies which operate in different divisional hierarchies. In some cases, there is a mixture of both centralized tourism governance and localized – or decentralized – tourism governance. In Figure 20.2, five countries have been selected to elucidate the difference in these types of tourism governance.

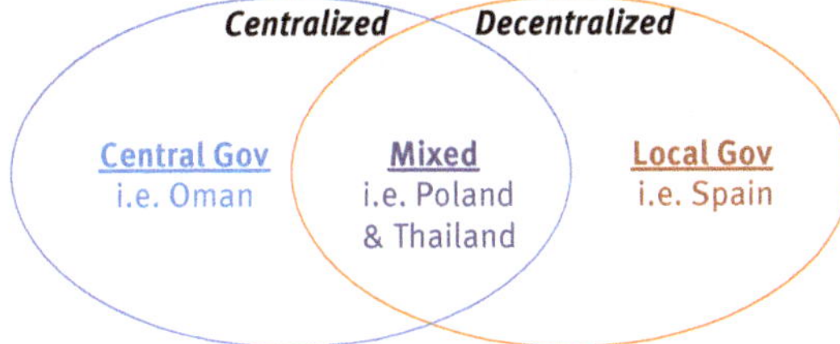

Figure 20.2: Tourism Governance: Centralized versus decentralized.

Oman has been selected as the states which showcase centralized governance of tourism. Oman is an absolute monarchy, ruled by the sultan who is both head of state as well as head of ministries and councils. Such countries share a centralized system of tourism governance, as well as a prioritization of preserving and bolstering their respective cultural heritage. For such local contexts, centralized tourism governance allows a codification of the cultural, religious, and social values that the respective countries want to preserve and upkeep. Centralization, especially in Omen which has relatively small areas and populations, prevents ineffective and bureaucratic systems.

On the other hand, the countries Poland and Thailand illustrate how tourism governance may have aspects of both centralized and localized tourism governance. In Poland, multiple different ministries have held responsibility for tourism, just over the last two decades. The evolution of tourism agencies could be a factor in the decision to decentralize tourism governance. Whereas in Thailand, the mixed decentralization of tourism governance is more a factor of local departments wanting control over their specific band of tourism governance to better account for their own specific tourism initiatives. Thailand also must contend with historic uneven development and the current ramifications of those economic impacts. Municipalities like Bangkok and Phuket are tourist magnets whereas more rural areas barely see any foreign, or even domestic, visitors. Decentralization has benefitted these wealthy, well-visited cities, which can abide by centralized policy that was specifically crafted for them, but this is at the cost of other areas of the country having to deal with inapplicable and outdated regulation federal level, centralized legislation that is unevenly enforced upon them. This form of tourism governance has, in local contexts, may work to benefit certain parts of specific countries while actively harming different parts of those same nations.

Spain, in terms of tourism governance, highlights a strong, but recent, move toward the decentralizing and localization of tourism governance. The localization of tourism governance in Spain is a product of an effort to modernize across very different regions, while accounting for the differences across those regions and the move away from all legislation being centralized under past rule. Highly decentralized tourism governance allows for certain areas of Spain to focus on coastal development toward their goal of increasing mass tourism in the form of sun and sand tourism, while not preventing culturally fragile areas from investing in cultural preservation as a form of tourism.

20.3 Potential Conflicts

There are myriad ways to pursue the overarching goal of promoting tourism within a country. Due to the fact that any given agent within a country has a diversity of priorities to consider when managing tourism, there is a lot of room for potential conflict between tourism governance and national, as well as local government, in addition to other sectors, industries, and organizations.

Tourism governance must contend with other industries where land use is the primary form of income generation. Conflict with these other income generative land uses, especially in cases where the resource extraction visually and severely contributes to landscape degradation, is a major source of concern in the promotion of tourism. For example, in Oman the oil sector has been a long-standing portion of the economy, but in the past decade, tourism has become a new investment and a growing part of the economy. However, there is direct conflict between the differences in coastal usage: for a long time, oil extraction and manufacturing equipment has made coastal areas less attractive to tourists and precluded the opportunity for the explosion of tourism infrastructure. These contrasting motives, which require the same resource to achieve their goals, set the stage for uncomfortable conflict between resource extraction industries and the tourism industry which both depend on resources like coastal areas.

In India, where ecotourism and wildlife tourism are huge phenomena, it becomes evident that the tourism industry itself can act as an environmentally degrading force. In the state of Kerala, a tourist resort was developed without consultation with local and environmental groups, and threatened to damage the fragile ecosystems which the tourists coming to this resort were meant to enjoy. This example set a precedent for more local autonomy over certain aspects of tourism infrastructure development, highlighting another stage of the conflict at the local level. In both cases, for the oil and tourism industries, the goal was coastal development, but these entities must also contend with forces of government which aim to protect and preserve these coasts creating an uncompromisable situation: should there be development or should there be protection?

In addition to resource-based conflict, there is also conflict between tourism and other non-economic values, especially concerning cultural and social values. Although in many cases tourism does benefit the economy, it may also serve as a force used to displace or marginalize locals. Tourism governance must take into account the various motivations of itself as a stakeholder as well as other local and national level stakeholders. Especially when considering natural resources, it is vital that tourism promotion does not come at the cost of total environmental degradation. In that same vein, tourism may be an economic tool to bolster businesses, but there is a potential for tourism – while developing any given area – to price out local residents, destroying not only local social networks but also potential cultural resources and or a potential labor force.

20.4 Concluding Remarks

Despite the diversity of local contexts examined in this text, there are distinct characters in both tourism governance and tourism realities which may be distilled into general themes about systems of tourism management. Many of these thematic similarities concern the interactions between agents who are specifically working in the context of tourism and stakeholders outside of the industry. Competition for resources – including natural resources and labor forces – is common in all industries and especially prevalent in the tourism industry. Examples of this type of competition were highlighted in the above section concerning potential conflicts, but may also occur in any situation where tourism is dependent upon the natural landscape, e.g. agricultural tourism, ecotourism, adventure tourism, as well as sun and sand tourism.

Beyond competition between the tourism industry and other industries which depend on the same resources to function, many themes of tourism governance deal with the forms of tourism governance as they appear in the official government structure of a given country. The graphics above utilize a small sample of countries to comparatively showcase the varying degrees of centralization of tourism governance as well as the varying levels of independence of tourism agencies within the government. These particular samplings are meant to provide both domineering examples, i.e countries with highly centralized tourism governance systems highly independent agencies, and contrast them with middling and exceptional examples, i.e. countries with highly decentralized tourism governance systems and highly integrated tourism agencies. Beyond looking at these specific characteristics, the ideas of domineering and exceptional examples also highlighted specific correlations between other themes.

In countries with strong, centralized tourism governance, and especially in developing and post-crisis countries, it appears that tourism is used as a pathway to modernization. In technologically focused, highly modernized countries like Finland, tourism flourishes because of its inextricable link to well-funded, publicly financed infrastructure. In Italy and Lebanon, their respective tourism industries suffered because of war and the subsequent devastation of their national infrastructure. From these examples it appears that there is a strong link between infrastructure and the success of the tourism industry. Many nations such Oman, Turkey, Spain, and Thailand have seen this correlation and took it upon themselves to invest in their own tourism industries through the modernization of their infrastructure.

Whether it is in a negative perception such as Colombia or a highly regarded destination like France, political will seem to be crucial to the success of countries as tourism destinations. Such political determination leads to high investments in marketing and branding in multiple government sectors (tourism and others) to promote the country as a tourism destination. Moreover, it is apparent that the overall governing system plays a significant role in the success of many economic sectors in which tourism is one. Countries that do not have tourism strongly presented in central government or not as main source of income such as Scandinavian countries, can still succeed as tourism destination more than countries where tourism is the main central sector of the economy but badly managed such as Colombia and Egypt. Throughout the chapters of this book, a directly proportional relationship between a state's economic system, political status quo, and tourism development and sustainability was traced and discussed thoroughly. The strength of a country's political state of affairs can be directly linked to the success of non-political aspects such as the tourism sector, regardless of the tourism governance scheme followed by said country (that is having a stand alone, mixed, or non-existent tourism governmental agencies as discussed earlier). For instance, Scandinavian countries such Finland may have no clear-cut governmental ministry responsible for tourism, and little tourism assets to aid into tourism diversity as opposed to other countries, yet the tourism sector is witnessing a booming increase due to the political and democratic clarity of the nation. In contrast, other countries such as Egypt that may have abundant assets that, if used appropriately, act as strong touristic magnets in addition to an ostensible governmental agency responsible for tourism, may not perform so good as expected due to political ambiguities and lack of governmental success and transparency when managing such assets.

When the government plays the regulator role and leave the implementation of strategies to other governing agencies, the private sector can be highly active and equally keen on developing tourism destinations. This is apparent to varies degrees in different countries; the role played by the private sector into developing tourism destinations could expand to cover different strategies of tourism development such as: being highly involved in establishment of coastal resorts, hotels, and the upgrade of infrastructure such as the case of Egypt. Other private sector agencies can focus on branding, marketing, and repositioning the country's touristic image in the global market such as the case of Colombia and the establishment of ProColumbia to promote a positive, post-war image of the country.

Evidentially, another key factor that aids into the prosperity of a country's tourism sector is international recognition. Institutions such as the UNESCO, UNWTO, Aga Khan Foundation, and multiple other international giants have the ability to highly influence the touristic image of the countries they work within. For instance,

UNTWO's declaration that Columbia's post-war policies and upgrades are a step into the right direction is responsible for promoting the country as a touristic destination. Another illustration of the importance of international recognition is evident in Egypt's case; the UNESCO declared other sites outside of Greater Cairo, such as Saint Catherine Mountain Area (2002) and Wadi Al-Hitan (Whale Valley, 2005), as world heritage sites, which was a key player into the establishment and development of these locations, and nearby sites, as prominent touristic destinations.

While the countries in this book were not chosen with any particular predetermined traits, delving into their tourism governance structures has revealed many global themes. From the bureaucratic organization of tourism management, to the interactions between agents and stakeholders, as well as the impacts of other industries and resources, there are many similarities between tourism industries globally. Importantly, this research does not attempt to rank the differences between these different forms of tourism, but rather to highlight how these differences may help serve these individual countries' varying needs for their specific goals in tourism governance and development. But even given differences in cultural and geographic contexts, there is a cohesive narrative of tourism governance globally and locally.

21 Appendix

1.2.1 Colombia

- Population: 49,084,841 (July 2020 est.)
- Area: 1,138,910 sq km
- Government type: Presidential republic
- GDP: US$711.6 billion (2017 est.)
- GDP Breakdown: Industry: 30.8% (2017 est.)
- Industries: (by ranked ordering) textiles, food processing, oil, clothing and footwear, beverages, chemicals, cement; gold, coal, emeralds
- Tourism analysis: The economy is still in post-war recovery. Tourism remains an underdeveloped industry as it is not a major focus of the current government administration, but private actors are increasingly involved in aiding tourism development.

1.2.3 Egypt

- Population: 104,124,440 (July 2020 est.)
- Area: 1,001,450 sq km
- Government type: Presidential Republic
- GDP (PPP): US$1.204 trillion (2017 est.)
- GDP Breakdown: Industry: 34.3% (2017 est.)
- Industries: (by ranked ordering) textiles, food processing, tourism, chemicals, pharmaceuticals, hydrocarbons, construction, cement, metals, light manufacture . . .
- Tourism analysis: This sector has greatly opened up under recent administrations, but due to stifling centralized regulation, major improvement in the industry would require severe restructuring.

1.2.4 Finland

- Population: 5,571,665 (July 2020 est.)
- Area: 338,145 sq km
- Government type: Parliamentary Republic
- GDP (PPP): US$244.9 billion (2017 est.)
- GDP Breakdown: Industry: 28.2% (2017 est.)

https://doi.org/10.1515/9783110638141-021

- Industries: (by ranked ordering) metals and metal products, electronics, machinery and scientific instruments, shipbuilding, pulp and paper, foodstuffs, chemicals, textiles, clothing . . .
- Tourism analysis: Finland utilizes social media and tech-based strategies to increase tourism in the country. This strategy has thus far generated significant revenue for social welfare and the betterment of the tourism sector.

1.2.5 France

- Population: 62,814,233 (July 2020 est.)
- Area: 551,500 sq km
- Government type: Semi-presidential Republic
- GDP (PPP): US2.856 trillion (2017 est.)
- GDP Breakdown: Industry: 19.5% (2017 est.)
- Industries: (by ranked ordering) machinery, chemicals, automobiles, metallurgy, aircraft, electronics; textiles, food processing; tourism
- Tourism analysis: France has historically had a hugely successful tourism sector. Because it has generated a large revenue for the country it is a priority of the French government, but recent development has been lacking.

1.2.6 India

- Population: 1,326,093,247 (July 2020 est.)
- Area: 3,287,263 sq km
- Government type: Federal parliamentary republic
- GDP (PPP): US$9.474 trillion (2017 est.)
- GDP Breakdown: Industry: 23% (2016 est.)
- Industries: (by ranked ordering) textiles, chemicals, food processing, steel, transportation equipment, cement, mining, petroleum, machinery, software, pharmaceuticals . . .
- Tourism analysis: Given the size and geographical diversity of the country, there are many different kinds of tourism. India's focus on developing tourism oriented toward the outdoors and adventuring has, in many localities, contributed to conflict between ongoing tourism development and forest management.

1.2.7 Italy

- Population: 62,402,659 (July 2020 est.)
- Area: 301,340 sq km
- Government type: Parliamentary republic
- GDP (PPP): US$2.317 trillion (2017 est.)
- GDP Breakdown: 23.9% (2017 est.)
- Industries: (by ranked ordering) tourism, machinery, iron and steel, chemicals, food processing, textiles, motor vehicles, clothing, footwear, ceramics
- Tourism sector: Italy has a strong tourism sector, but has not yet reached its potential due in part to poor cooperation between private and public agents.

1.2.8 Lebanon

- Population: 5,469,612 (July 2020 est.)
- Area: 10,400 sq km
- Government type: Parliamentary republic
- GDP (PPP): US$88.25 billion (2017 est.)
- GDP Breakdown: Industry: 13.1% (2017 est.)
- Industries: (by ranked ordering) banking, tourism, real estate and construction, food processing, wine, jewelry, cement, textiles, mineral and chemical products, wood and furniture products, oil refining, metal fabricating . . .
- Tourism sector: Lebanon experienced a golden age of tourism, interrupted by political unrest. Now, it is again becoming a site for leisure tourism, but at the cost of increased coastal development and neglected cultural preservation.

1.2.9 Mexico

- Population: 128,649,565 (July 2020 est.)
- Area: 1,964,375 sq km
- Government type: Federal presidential republic
- GDP (PPP): US$2.463 trillion (2017 est.)
- GDP Breakdown: Industry: 31.9% (2017 est.)
- Industries: (by ranked ordering) food and beverages, tobacco, chemicals, iron and steel, petroleum, mining, textiles, clothing, motor vehicles, consumer durables, tourism . . .
- Tourism analysis: Tourism is a major part of the Mexican economy. Tourism development has been a strategic tool used to boost revenue, as well as encourage cooperation between private and public entities at the state and national levels.

1.2.10 Oman

- Population: 4,664,844 (December 2019 est.)
- Area: 309,500 sq km
- Government type: Absolute monarchy
- GDP (PPP): US$190.1 billion (2017 est.)
- GDP Breakdown: Industry: 46.4% (2017 est.)
- Industries: (by ranked ordering) crude oil production and refining, natural and liquefied natural gas production; construction, cement, copper, steel, chemicals, optic fiber . . .
- Tourism analysis: The tourism sector in Oman suffers from historic neglect and overly rigid, centralized legislation.

1.2.11 Poland

- Population: 38,282,325 (July 2020 est.)
- Area: total: 312,685 sq km
- Government type: Parliamentary republic
- GDP (PPP): US$1.126 trillion (2017 est.)
- GDP Breakdown: Industry: 40.2% (2017 est.)
- Industries: (by ranked ordering) machine building, iron and steel, coal mining, chemicals, shipbuilding, food processing, glass, beverages, textiles . . .
- Tourism analysis: Investment from the EU benefits Poland's infrastructural development, which in turn aids the tourism sector. However, the country's tourism sector would greatly benefit from better cooperation between tourism stakeholders.

1.2.12 Portugal

- Population: 10,302,674 (July 2020 est.)
- Area: 92,090 sq km
- Government type: Semi-presidential republic
- GDP (PPP): US$314.1 billion (2017 est.)
- GDP Breakdown: Industry: 22.1% (2017 est.)
- Industries: (by ranked ordering) textiles, clothing, footwear, wood and cork, paper and pulp, chemicals, fuels and lubricants, automobiles and auto parts, base metals, minerals, porcelain and ceramics, glassware, technology, telecommunications; dairy products, wine, other foodstuffs; ship construction and refurbishment; tourism, plastics, financial services, optics . . .

- Tourism analysis: Tourism in Portugal is a burgeoning industry, quickly making the country a destination of international note. However, the industry is encumbered by the dichotomy of regulation creators and enforcers at the local and national levels.

1.2.13 Spain

- Population: 50,015,792 (July 2020 est.)
- Area: 505,370 sq km
- Government type: Parliamentary constitutional monarchy
- GDP (PPP): US$1.778 trillion (2017 est.)
- GDP Breakdown: Industry: 23.2% (2017 est.)
- Industries: (by ranked ordering) textiles and apparel (including footwear), food and beverages, metals and metal manufactures, chemicals, shipbuilding, automobiles, machine tools, tourism, clay and refractory products, footwear, pharmaceuticals, medical equipment . . .
- Tourism analysis: Tourism governance in Spain is trending toward decentralization, spurring the growth of autonomous, regionally based tourism organizations. This has helped to produce a modern image of the country as well as to further coastal development for mass tourism based around sun and sand.

1.2.14 Thailand

- Population: 68,977,400 (July 2020 est.)
- Area: 513,120 sq km
- Government type: Constitutional monarchy
- GDP (PPP): US$1.236 trillion (2017 est.)
- GDP Breakdown: Industry: 36.2% (2017 est.)
- Industries: (by ranked ordering) tourism, textiles and garments, agricultural processing, beverages, tobacco, cement, light manufacturing such as jewelry and electric appliances, computers and parts, integrated circuits, furniture, plastics, automobiles and automotive parts, agricultural machinery, air conditioning and refrigeration, ceramics, aluminum, chemical, environmental mangement, glass, granite and marble, leather, machinery and metal work, petrochemical, petroleum refining, pharmaceuticals, printing, pulp and paper, rubber, sugar, rice, fishing, cassava, world's second-largest tungsten producer and third-largest tin producer . . .

- Tourism analysis: Thailand's success as a tourism destination has been combated by visitors' safety concerns. While the country attempts to repair this image and become one of the great tourism destinations, it has occurred at the cost of increasing environmental degradation.

1.2.15 Tunisia

- Population: 11,721,177 (July 2020 est.)
- Area: 163,610 sq km
- Government type: Parliamentary republic
- GDP (PPP): US$137.7 billion (2017 est.)
- GDP Breakdown: Industry: 26.2% (2017 est.)
- Industries: (by ranked ordering) petroleum, mining (particularly phosphate, iron ore), tourism, textiles, footwear, agribusiness, beverages . . .
- Tourism analysis: The tourism industry drove modernization in Tunisia, but uneven development has led to political unrest. Tourism is vital to the economy and has been on a steady incline for the past several decades.

1.2.16 Turkey

- Population: 82,017,514 (July 2020 est.)
- Area: 783,562 sq km
- Government type: Presidential republic
- GDP (PPP): US$2.186 trillion (2017 est.)
- GDP Breakdown: Industry: 32.3% (2017 est.)
- Industries: (by ranked ordering) textiles, food processing, automobiles, electronics, mining (coal, chromate, copper, boron), steel, petroleum, construction, lumber, paper . . .
- Tourism analysis: Turkey has had great and fast success in the tourism sector. However, overdevelopment of infrastructure has had many consequences. In mitigating future issues, Turkey has focused specifically on fostering cooperation between public and private stakeholders.

1.2.18 United States

- Population: 332,639,102 (July 2020 est.)
- Area: 9,833,517 sq km
- Government type: Constitutional federal republic
- GDP (PPP): US$19.49 trillion (2017 est.)

- GDP Breakdown: Industry: 19.1% (2017 est.)
- Industries: (by ranked ordering) high-technology innovator, second-largest industrial output in the world; petroleum, steel, motor vehicles, aerospace, telecommunications, chemicals, electronics, food processing, consumer goods, lumber, mining
- Tourism analysis: The United States' flexible and robust form of tourism governance has been successful, however, tourism initiatives are disjointed and do not prioritize economic factors of tourism development.

List of Figures

https://doi.org/10.1515/9783110638141-022

List of Tables

https://doi.org/10.1515/9783110638141-023

Contributor Biographies

Rashid Al-Hinai holds a Doctor of Engineering Sciences (DSc) degree in spatial planning from Vienna University of Technology (Austria). He is working as a senior planning specialist in Ministry of Tourism in Oman. One of the current key duties of his work is to contribute in the strategic planning of tourism industry including the implementation of Oman Tourism Strategy 2040 and the development and the implementation of the tourism regional strategies and masterplans. He contributes also in the integration between the spatial strategies conducted by Supreme Council of Planning in Oman including Oman National Spatial Strategy 2040 and the strategies related to tourism sector. His research interest is sustainable spatial planning and tourism planning (as a sectorial planning direction) and also the integration between these two fields of planning which is necessary to create more liveable and sustainable touristic communities. His research focus is more on the technical (land-use, environment, activities distribution, etc.) and governance (policies, strategies, regulations, and legal frameworks) dimensions related to these two planning fields.

Joana Almeida is Professor at Instituto Superior Técnico (IST), University of Lisbon, Portugal. She has a PhD in Territorial Engineering, IST (2013): "Tourism versus territory collaborative conflict management: The Troia-Melides Coast, Portugal." She is a Research member of CiTUA – Centre for Innovation in Territory, Urbanism and Architecture and, since 2021, Councilor for Urbanism and for Transparency and Anti-Corruption of Lisbon City Council. Her recent research projects, as principal researcher, are: "*Tur&Bairros*: Tourism and Historical Neighbourhoods Transformation" and "*LxLAB*: Tourism Monitoring for the Historical Centres Sustainable development." The recent papers published in scientific journals and book chapters are focused on conflicts between tourism and territory in coastal areas, on collaborative conflict management and new collaborative leadership approaches, on reconciling the growth of tourism and the sustainable development of historic city neighbourhoods and in different understandings of public interest. Her recent research interests are: active public participation, negotiation and conflict management methods in urban planning, urban governance, quality of life.

Amna Al- Ruheili has a PhD in Landscape Architecture and Environmental Planning, and a Graduate Certificate in Geographic Information Science and Technology from University of California Berkeley in 2017. Currently working Assistant Professor at College of Agricultural and Marine Sciences and serve as a Deputy Director of Remote Sensing and GIS Research Center at Sultan Qaboos University. In my work, I strive to address challenges pertaining to infrastructure and urban planning impacted by natural hazards. My chief focus is directed toward finding how long-term resiliency can be better integrated within Oman's strategic developmental planning. I am interested in design and development of analytical methods embedded in GIS. My previous research included applications remote sensing and GIS to tackle various environmental problems. My current research focuses predominantly on wadi flooding impacts on infrastructure, and agricultural lands with emphasis on managing flood-prone lands ("wise use" of floodplains).

Matilde Córdoba Azcárate is a social anthropologist interested in questions of space, politics, ecology, ethnography, and global capitalism. She works as an Associate Professor in the Communication Department and she co-directs the Faculty and Graduate Research Group Nature, Space and Politics at the University of California, San Diego. Among other publications, she is the author of two recent books: *Stuck with Tourism: Space, Power and Labor in Contemporary Yucatán* (UC Press, 2020), an ethnographic account of tourism as an organizing force in the predatory geographies of late capitalism as well as the moral and ecological entrapments it creates for local communities. And the co-edited book *Tourism Geopolitics* (University of Arizona Press, 2021), an

https://doi.org/10.1515/9783110638141-024

interdisciplinary and international collaboration into the growing centrality of tourism infrastructures, affects and representations in geopolitical affairs.

Magdalena Banaszkiewicz is an Associate Professor in the Institute of Intercultural Studies at the Jagiellonian University in Cracow (Poland). She graduated both in Russian Studies and Cultural Studies. Her research interests focus on dissonant heritage, tourism, memory, and sustainability, particularly in the Central and Eastern Europe region. For the last few years she has been exploring the problem of tourism development in the Chernobyl Exclusion Zone. Recently, she published a monograph *Tourism in Dissonant Heritage Sites* (Jagiellonian University Press, 2018) and co-edited with Sabine Owsianowska a volume *Anthropology of Tourism in Central and Eastern Europe. Bridging Worlds* (Lexington Books, 2018). She is a Chair of the MA program in Heritage Tourism and a member of the Self-Steering Committee in Cultural Heritage Area in the UNA EUROPA network.

Amira Benali is a postdoctoral researcher in the Department of Management, Politics, and Philosophy at Copenhagen Business School in Denmark. She is an interdisciplinary scholar working on issues related to social justice, poverty, gender, and alternative economies in the intersection of management, geography, and socio-anthropology. Drawing on feminist, postcolonial, and post-development theories, her current project focuses on the grassroot movements in Tunisia. In particular, she is studying the challenges and opportunities of the social and solidarity economy in the tourism context.

Judith Bopp is a postdoctoral researcher currently based at University of Vechta, Germany. She received her PhD in Cultural Geography at University of Cologne, Germany, in 2016. Her research interests are in new social movements, alternative urban food systems, smallholder farming and rural livelihoods, as well as environmental and climate justice. She has empirical research experience in Thailand, Myanmar, and India. Her current postdoctoral project explores organic farming practices in the context of household resilience and social-ecological change. During her studies, she discovered WWOOF (World-Wide Opportunities on Organic Farms) as a way of ecotourism that allows practical experience in sustainable agriculture while learning about local culture. She has made her own WWOOF experience in Asia, Australia, South America, and Europe.

Azade Özlem Çalik completed her Tourism Management undergraduate, master's, and PhD degree at Gazi University (Turkey), and she studied in North Carolina -(USA) for postdoctoral studies between 2017–2018. She has been working as a Lecturer at Ankara University since 2004. She also has had a Tourist Guide License since 2003 and trains tourist guides. She received the Cake Decorating Certificate in 2014. She has various publications and projects on tourism in the country and abroad.

Gülsel Çiftci is an Associate Professor working in the Department of Tourism Management at Trakya University in Turkey. She received her PhD in 2015. Her major research interests are crisis management in the hospitality and tourism industry, sustainable tourism, and cross-border shopping tourism. She is also a licensed professional tourist guide. Her area of expertise as a tourist guide is Gallipoli Peninsula Historical National Park in Canakkale-Turkey.

Wilasinee Darnthamrongkul is a faculty member of the Department of Landscape Architecture at Chulalongkorn University, Bangkok, Thailand. She is also a founding member of the Water Resilient City Unit (WRCU) and the Learning Innovation for Thai Society Research Unit (LIfTS). She received a Bachelor of Architecture from Silpakorn University, a Master of Landscape Architecture from Chulalongkorn University, and a PhD in Landscape Architecture and Environmental Planning from the University of California, Berkeley. Her research focuses on the interrelation between people and their environments. She had worked in collaboration with Thailand's Office of Natural Resources and Environmental Policy and Planning (ONEP) to develop the first visual impact assessment manual for building construction projects in the country. She has also worked with several governmental, non-governmental, and local agencies to study landscape and visual resource management and planning for both natural and cultural tourist districts in Thailand.

Priyanka Ghosh is a Senior Assistant Professor of Geography at VIT-AP University, Amaravati, Andhra Pradesh, India. Her areas of interest are political ecology, biodiversity conservation, protected area management, human-animal conflicts, traditional ecological knowledge, tourism, and sustainable development. She had extensively worked in the Indian Sundarbans during her PhD at the University of Kentucky and published her work in peer-reviewed international journals such as *GeoJournal* and *Geographical Review*. She was a post-doctoral scholar (2016–2017) at the Department of Geography, University of Kentucky, where she investigated rural land use of Fayette County, Kentucky. She served as an Educational Advisor (2017–2018) at the IEC Kumamoto International College, Kyushu, Japan, and later worked as an Assistant Professor of Geography & Environmental Science (2018–2021) at Era University, Lucknow, India.

Amir Gohar is a tourism & environmental planner and sustainable development expert with nearly two decades of working experience with municipal governments, research institutes, international development agencies, private sector firms, and local community organizations. He has worked extensively in the MENA region Africa and Europe, with specific focus on tourism planning in the southern region of the Red Sea in Egypt, agrotourism planning for different sites in Saudi Arabia, the Cradle of Humankind Park and the Kalahari in South Africa, and sustainable tourism in Jebel Akhdar (Green Mountains) in Libya. His scholarship focuses on finding the appropriate balance between the trendies of rapid tourism development and maintaining ecological integrity in both dense cities as well as remote nomadic towns. His doctorate research in the Department of Landscape Architecture & Environmental Planning at UC Berkeley focused on understanding tourism development and its direct and relative impact on the environment. Dr. Gohar obtained previous degrees in Urban & Regional planning from Cairo university, a Master's in Urban design from Oxford Brookes University, and a diploma in Land Management from Erasmus University.

Bertram M. Gordon is Professor Emeritus/History at Mills College in Oakland, California. A core member of the Tourism Studies Working Group at the University of California, Berkeley, he is also Associate Editor of the *Journal of Tourism History*, General Secretary of the International Commission for the History of Travel and Tourism, and co-editor of the H-Travel internet discussion network. A specialist in World War II France and the linkages between tourism and war, his books include *War Tourism: Second World War France from Defeat and Occupation to the Creation of Heritage* (2018), *The Historical Dictionary of World War II France: The Occupation, Vichy and the Resistance, 1938-1946* (1998), and *Collaborationism in France during the Second World War* (1980). He co-edited "Food and France: What Food Studies Can Teach Us about History," a special issue of *French Historical Studies* (April 2015) and has published essays on mass tourism and Mediterranean tourism.

Nelson Graburn was educated at Cambridge, McGill and Chicago (PhD 1963). He has taught at University of California Berkeley since 1964, serving as Curator of the Hearst Museum, and Co-chair of Tourism Studies (www.tourismstudies.org). He taught in Canada, France, UK, Germany, Sweden, Portugal, Japan, and Brazil, and 41 Chinese universities. He has researched identity, multiculturalism, museums, art, heritage and tourism, among the Canadian Inuit (1959–), in Japan (1974–) and China (1991–). His works includes: *Ethnic and Tourist Arts* (1976); *Japanese Domestic Tourism* (1983); *Anthropology of Tourism* (1983); *Multiculturalism in the New Japan* (2008); *旅游人类学论文集* [*Anthropology in the Age of Tourism*] (2009); *Tourism and Glocalization in East Asia* (2010); *Tourism Imaginaries: Anthropological Approaches* (2014), *Tourism Imaginaries at the Disciplinary Crossroads* (2016), *Tourism in (Post)Socialist Eastern Europe* (2017), *Indigenous Tourism Movements* (2018), *Contents Tourism: Japanese and International Fandom* (2019) and *Tourism Fictions, Simulacra and Virtualities* (2019).

Kamil Hamati is an independent researcher and international expert in sustainable development. His main research interests are migration and conflict, public policy, human behavioral sciences, management of the commons, sustainable development, indicator-based decision making, governance, rural livelihoods, and food security. Kamil holds two Masters' degrees from the Lebanese University, the first is in Biology (2008) and the second is in Phyto-Ecology (2012). His experience in migration and conflict was refined through a 3-year research stay at the University of Geneva as a Swiss Government Excellence Fellow. His experience in development and governance in the Arab region is derived from his previous work with different agencies of the United Nations' Secretariat, notably ESCWA and UNEP, on monitoring the SDGs, drafting environmental policy, revising food security governance, advising on climate change adaptation and mitigation, and other thematic issues of relevance to the regional context.

İsmail Kervankiran is an Associate Professor working in the Department of Geography at Suleyman Demirel University in Turkey. He received his PhD in 2011. His major research interests are human geography, tourism geographies, mobility, liminality, place and space theory, memory, cultural studies, heritage, degrowth, regional inequality. He has published book chapters and many articles in national and international journals on a wide range of topics including spatial, social, and economic impacts of tourism, spatial analysis of tourism, and regional development of tourism in Turkey. Dr. Kervankıran worked as a short-term scholar at Texas A&M University in 2014.

Mar Loren-Méndes is a Full Professor of the Department of Architectural History, Theory and Composition, School of Architecture at Seville University (Spain). B.Arch., M.Arch, PhD at Seville University; Master Design Studies, Harvard University (1998); Master on Heritage and New Technologies (1996) European Leonardo Da Vinci program. She has taught and researched internationally at Harvard University, Getty Research Institute, BTU, La Villette or UC. Berkeley, among others. Chairholder of the UNESCO Chair on Built Urban Heritage in the Digital Era CREhAR (Creative Research and Education on heritage Assessment and Regeneration). Director of the Research Group HUM-666 *Contemporary City, Architecture and Heritage* (HUM-666). Her research focuses on heritage assessment with emphasis on modern heritage from a methodological perspective and with the conceptual integration of new technologies, developing three research lines: coastal transformation and tourism, transfer processes Europe-USA and cataloguing modern architecture, with competitive projects, national and international publications and papers in Congresses, and awards endorsing her work.

Dean MacCannell is Professor and Chair Emeritus of Environmental Design and Geography in the College of Agriculture and Environmental Sciences at the University of California at Davis. His 1976 book, *The Tourist—A New Theory of the Leisure Class* has never gone out of print and is available in more than ten different languages. In addition to *The Tourist*, Professor MacCannell has published over 100 books and articles on various aspects of tourism and other cultural manifestations of globalization and hyper-modernization—industrial agriculture, the re-arrangement between the sexes, homelessness, reconstructed ethnicity, among others.

Ines Mestaoui is a Doctor in Marketing at the Higher Institute of Management of Tunis (ISG de Tunis) (Tunisia) enrolled in the department of marketing. Her research mainly concerns the tourism experience, mobile technologies, and social media. Drawing on social imaginaries, liminality, and institutional theories, her thesis work deals with travel experience as it is currently experienced by Tunisian travelers under the influence of mobile technologies and in a transitional context namely the post-revolutionary context.

Louise A. Mozingo is Professor & Chair of the Department of Landscape Architecture and Environmental Planning, faculty of the Graduate Group in Urban Design, and faculty and former Director of the interdisciplinary American Studies program at UC Berkeley. Her research concerns two areas: the history of the American designed landscape, sustainable environmental design, and planning. This research has taken a variety of forms including books, book chapters, scholarly articles, criticism, research reports, design and planning documents, and exhibitions. In 2009, Mozingo founded a research center within the College of Environmental Design, the Center for Resource Efficient Communities (CREC) dedicated to interdisciplinary research regarding resource efficient urban design, planning, and policy.

Patrick Naef holds a PhD in Geography from the University of Geneva (Switzerland). After conducting postdoctoral research at the University of California in Berkeley (2014-2016), he is now a Senior Associate Researcher at the University of Geneva (Department of Geography and Environment). As a cultural geographer and social anthropologist, he works on tourism and crime; the critical approaches to resilience; collective memory and violence. He explored these topics principally in Colombia, the United States, Bosnia-Herzegovina and Croatia. He is now leading a research project on urban resilience, collective memory and violence in Medellin and Cali (Colombia), Chicago and New Orleans (USA), and Belfast (Northern Ireland). Patrick Naef is also collaborating with several community organizations active in cultural and urban development in Geneva and Medellin.

Sabina Owsianowska is a Tourism Researcher and an Assistant Professor at the Faculty of Tourism and Leisure, University of Physical Education in Krakow; she cooperates with the Institute of Intercultural Studies, Jagiellonian University (Poland). Her main areas of research, teaching, and training include the theory of tourism and leisure; anthropology of tourism; tourism education and management in a humanistic perspective; multiculturality and the interpretation of heritage, with focus on its dissonance, mainly in Central and Eastern Europe; visuality and the embodiment of travel experiences. Recently, she co-authored monographs *Humanistic Tourism Education. Values, Norms and Dignity* (Routledge, 2021), *Brexit and Tourism. Travel, Borders and Identity* (Channel View Publication, 2020), *Lifelong Learning for Tourism* (Routledge, 2018), co-edited *Anthropology of Tourism in Central and Eastern Europe. Bridging Worlds* (Rowman&Littlefield, 2018, with M. Banaszkiewicz) and several thematic issues of tourism journals (i.e. *Journal of Tourism and Cultural Change*, 15/2017; *Folia Turistica*. 55/2020).

Monica Pascoli, is Research Fellow in Sociology at the University of Udine and lecturer of Sociology and Methodology of Social Research at the Universities of Udine and Trieste, Italy. She is part of the *Euroculture Master Programme* teaching staff (*Erasmus Mundus Programme of Excellence*). She is currently working on an EU founded project on community-based tourism development, focusing on the community involvement in the creation of the tourism image of destinations. The goal is to analyze the social construction of the tourist imaginary, focusing both on the activity of the tourist stakeholders and the role played by the local community in the active creation and re-creation of collective representations. Her research interests comprise tourism and the memory of the First

World War in Friuli Venezia Giulia, with a focus on the role of local tourist guides in the process of memory transmission.

Pedro J. Pinto is a researcher at CiTUA - Center for Innovation in Territory, Urbanism, and Architecture at Instituto Superior Técnico, Lisbon. He has collaborated in research teams studying the relationship of cities and rivers, the evolution of urban waterfronts, and municipal land-use planning. His doctoral research compared land-use patterns and environmental governance structures in the Tagus Estuary (Lisbon) and the San Francisco Bay. His professional experience also includes collaboration in municipal master plans and social facilities planning. Pedro holds a PhD in Landscape Architecture and Environmental Planning, from the University of California, Berkeley. He has a degree in Territorial Engineer (equivalent to a M.Eng) and a MSc in Land-Use Planning, both from Instituto Superior Técnico (Lisbon, Portugal). His current research topics are metropolitan and environmental governance, sea-level rise adaptation in urbanized estuaries, and the study of the human-river interface.

Laura Puolamäki holds a PhD in Landscape Research. In her her doctoral thesis she studied parallel and contested cultural landscape values and local landscape knowledge. Her research focus is on the culturally and socially sustainable landscape evaluation and conservation through shared knowledge, jointly recognised landscape values and landscape stewardship. Laura Puolamäki works as a landscape specialist in NGO ProAgria South Finland. She teaches environmental education and cultural tourism in University of Turku / Degree Programme for Cultural Production and Landscape Research. She has worked in various projects in the fields of environmental education, cultural tourism, cultural landscape, and cultural heritage.

Kanokwalee Suteethorn is a faculty member of the Department of Landscape Architecture at Chulalongkorn University, Bangkok. She is a landscape architect and a researcher at Healthy Landscape and BioPhilia - Research Unit (HEAL-BiP). She received BArch from Silpakorn University, MLA from University of Washington, and PhD in Landscape Architecture and Environmental Planning with a Designated Emphasis in Global Metropolitan Studies (GMS) from University of California, Berkeley. She has worked as a committee for Thai association of Landscape Architects (TALA) and worked with the Bangkok Big Tree in Thailand. She has also worked as a community outreach at Friends of Urban Forests (FUF) in San Francisco. Her research focuses on two areas: Urban forests and Healing landscape. She is interested in how natural environment affects physiological and psychological health and wellbeing. Her current research is on Forest Bathing, exploring how Bangkok's urban forests have impact on public health.

Kristina Svels is a researcher at the Natural Resources Institute, Finland (LUKE), within Rural studies, land use, and natural resource governance. With a background in rural sociology her research interests include World Heritage studies, regional tourism development, environmental sociology, nature resource governance, second homes, and small-scale fisheries. Kristina's research focuses on local perspectives of participation and co-creation of knowledge, e.g. tourism issues involving grass-root participation in decision-making processes as aligned with governance protocol for impact of natural resource use on communities. She draws on her experience from local to transnational laws and their application during tourism schemes to improve economic development involving diverse stakeholders. Kristina supports transnational learning as part of an efficient backdrop to overcome new challenges in local areas affected by international tourism. Svels' doctoral thesis (2017): *World Heritage management and tourism development: A study of public involvement and contested ambitions in the World Heritage Kvarken Archipelago, Finland.*

Index

https://doi.org/10.1515/9783110638141-025

www.ingramcontent.com/pod-product-compliance
Lightning Source LLC
Chambersburg PA
CBHW061746210326
41599CB00034B/6798

* 9 7 8 3 1 1 1 3 5 3 1 2 8 *